SOLID STATE PHYSICS

VOLUME 37

Contributors to This Volume

Philip B. Allen

J. B. Boyce

T. M. Hayes

Božidar Mitrović

J. C. Phillips

SOLID STATE PHYSICS

Advances in Research and Applications

Editors

HENRY EHRENREICH

Division of Applied Sciences
Harvard University, Cambridge, Massachusetts

FREDERICK SEITZ

The Rockefeller University, New York, New York

DAVID TURNBULL

Division of Applied Sciences
Harvard University, Cambridge, Massachusetts

VOLUME 37

1982

ACADEMIC PRESS

A Subsidiary of Harcourt Brace Jovanovich, Publishers

New York London

Paris San Diego San Francisco São Paulo Sydney Tokyo Toronto

ACADEMIC PRESS, INC.
111 Fifth Avenue, New York, New York 10003

United Kingdom Edition published by
ACADEMIC PRESS, INC. (LONDON) LTD.
24/28 Oval Road, London NW1 7DX

LIBRARY OF CONGRESS CATALOG CARD NUMBER: 55-12200

ISBN 0-12-607737-1

PRINTED IN THE UNITED STATES OF AMERICA

82 83 84 85 9 8 7 6 5 4 3 2 1

Contents

Theory of Superconducting T_c

Philip B. Allen and Božidar Mitrović

Spectroscopic and Morphological Structure of Tetrahedral Oxide Glasses

J. C. Phillips

Extended X-Ray Absorption Fine Structure Spectroscopy

T. M. HAYES AND J. B. BOYCE

Contributors to Volume 37

Numbers in parentheses indicate the pages on which the authors' contributions begin.

PHILIP B. ALLEN, *Department of Physics, State University of New York at Stony Brook, Stony Brook, New York 11794* (1)

J. B. BOYCE, *Xerox Palo Alto Research Center, Palo Alto, California 94304* (173)

T. M. HAYES, *Xerox Palo Alto Research Center, Palo Alto, California 94304* (173)

BOŽIDAR MITROVIĆ, *Department of Physics, State University of New York at Stony Brook, Stony Brook, New York 11794* (1)

J. C. PHILLIPS, *Bell Laboratories, Murray Hill, New Jersey 07974* (93)

Preface

This volume contains three articles reviewing a broad spectrum of topics, each of them having considerable importance in the development of contemporary condensed matter science. The first is devoted to a generally accessible account of the present state of the theory of the transition temperature, T_c, of superconductors. Philip B. Allen and Božidar Mitrović, its authors, note that the prediction of T_c remains largely an unsolved problem. Their detailed account of the Migdal–Eliashberg theory is to be welcomed, as is their critical discussion of the more popular approximate T_c equations. In the second article, J. C. Phillips presents a polycluster model, analogous to his earlier model for chalcogenide glasses, for the structure of tetrahedrally coordinated oxide glasses. He critically examines the vibrational spectra of oxide glasses and concludes that these spectra are consistent with his model but inexplicable by the continuous random network models. Though there may be considerable disagreement with his analyses and conclusions, Phillips raises and treats important issues in the spectroscopy of glasses in a stimulating and provocative manner.

Extended X-ray absorption fine structure spectroscopy (EXAFS) has become, since about 1970, a powerful and now widely used technique for characterizing the atomic short-range order of condensed phases. In the last article of this volume T. M. Hayes and J. B. Boyce describe the physical theory and experimental practice of EXAFS and comprehensively review its application to a considerable range of condensed phase structural problems. It is hoped that this review will provide a useful perspective for both experts and nonexperts having an active interest in EXAFS.

HENRY EHRENREICH
FREDERICK SEITZ
DAVID TURNBULL

Supplements

Supplement 1: T. P. Das and E. L. Hahn
Nuclear Quadrupole Resonance Spectroscopy, 1958

Supplement 2: William Low
Paramagnetic Resonance in Solids, 1960

Supplement 3: A. A. Maradudin, E. W. Montroll, G. H. Weiss, and I. P. Ipatova, Theory of Lattice Dynamics in the Harmonic Approximation, 1971 (Second Edition)

Supplement 4: Albert C. Beer
Galvanomagnetic Effects in Semiconductors, 1963

Supplement 5: R. S. Knox
Theory of Excitons, 1963

Supplement 6: S. Amelinckx
The Direct Observation of Dislocations, 1964

Supplement 7: J. W. Corbett
Electron Radiation Damage in Semiconductors and Metals, 1966

Supplement 8: Jordan J. Markham
F-Centers in Alkali Halides, 1966

Supplement 9: Esther M. Conwell
High Field Transport in Semiconductors, 1967

Supplement 10: C. B. Duke
Tunneling in Solids, 1969

Supplement 11: Manuel Cardona
Optical Modulation Spectroscopy of Solids, 1969

Supplement 12: A. A. Abrikosov
An Introduction to the Theory of Normal Metals, 1971

Supplement 13: P. M. Platzman and P. A. Wolff
Waves and Interactions in Solid State Plasmas, 1973

Supplement 14: L. Liebert, Guest Editor
Liquid Crystals, 1978

Supplement 15: Robert M. White and Theodore H. Geballe
Long Range Order in Solids, 1979

Theory of Superconducting T_c

PHILIP B. ALLEN
AND
BOŽIDAR MITROVIĆ

Department of Physics, State University of New York at Stony Brook, Stony Brook, New York

ISBN-0-12-607737-1

I. Introduction

1. Preliminaries

Bardeen–Cooper–Schrieffer (BCS) theory[1] is a remarkably complete theory of the superconducting state, and it raises the possibility that some day we may be able to understand and predict numerical values of the superconducting transition temperature T_c. The Migdal theory[2] of electron–phonon effects in the normal state and the Eliashberg theory,[3] which generalizes BCS theory to incorporate Migdal theory in the normal limit, provide a rigorous basis for understanding T_c; nevertheless, prediction of T_c is still largely an unsolved problem.[4]

This article reviews Migdal–Eliashberg (ME) theory with two aims: First, we hope to make the theory accessible to a wider audience. The last reviews,[5–9] written in 1969 or earlier, remain correct but are somewhat awkward from the present viewpoint of applications to real metals. Second, we critically examine the assumptions of the theory to clarify the degree of rigor in each of its parts and in its more recent additions. This article assumes that the lattice vibrational spectrum and the electronic spectrum of a material are given, the former presumably by neutron scatterers and the latter by band theorists. It is further assumed that the required electron–phonon coupling can be computed. These assumptions are far from trivial and are the topic of a separate article in preparation.[10] Our chief question is this: Assuming that all the normal state properties are calculable, how is the value of T_c then to be extracted?

Table I gives values of T_c for various materials chosen to illustrate a wide range of properties. It is not yet understood why many A15 structure

[1] J. Bardeen, L. N. Cooper, and J. R. Schrieffer, *Phys. Rev.* **108,** 1175 (1957).

[2] A. B. Migdal, *Zh. Eksp. Teor. Fiz.* **34,** 1438 (1958); *Sov. Phys.—JETP (Engl. Transl.)* **7,** 996 (1958).

[3] G. M. Eliashberg, *Zh. Eksp. Teor. Fiz.* **38,** 966 (1960); *Sov. Phys.—JETP (Engl. Transl.)* **11,** 696 (1960).

[4] D. Rainer, *Proc. Int. Conf. Low Temp. Phys. 16th, Physica B + C (Amsterdam)* (in press).

[5] J. R. Schrieffer, "Theory of Superconductivity." Benjamin, New York, 1964.

[6] D. J. Scalapino, *in* "Superconductivity" (R. D. Parks, ed.), Vol. 1, Chapter 10. Dekker, New York, 1969.

[7] W. L. McMillan and J. M. Rowell, *in* "Superconductivity" (R. D. Parks, ed.), Vol. 1, Chapter 11. Dekker, New York, 1969.

[8] G. Gladstone, M. A. Jensen, and J. R. Schrieffer, *in* "Superconductivity" (R. D. Parks, ed.), Vol. 2, Chapter 13. Dekker, New York, 1969.

[9] V. Ambegaokar, *in* "Many-Body Physics" (C. DeWitt and R. Balian, eds.), p. 297. Gordon & Breach, New York, 1968.

[10] B. M. Klein, W. E. Pickett, D. A. Papaconstantopoulos, and L. L. Boyer, *in* "Solid State Physics" (F. Seitz, D. Turnbull, and H. Ehrenreich, eds.). Academic Press, New York (in preparation).

TABLE I. EXAMPLES OF SUPERCONDUCTING TRANSITION TEMPERATURES

Material	T_c	λ	Comments	Reference
Al	1.2	0.4	sp metal, prototype weak coupler	a
Pb	7.2	1.55	sp metal, prototype strong coupler	b
Amorphous Ga	8.6	2.25	Amorphous sp metal (crystalline Ga has $T_c = 1.1°K$)	c
Nb	9.2	0.9–1	d metal, highest T_c element (atmospheric pressure)	d, e
Amorphous Nb	5.2	0.8	Amorphous d metal	f
Nb_3Sn	18	1.65–1.95	Best-studied A15 compound	g, h
Nb_3Ge (thin film)	23	1.64 ± 0.20	Highest known T_c	i, j
Amorphous Nb_3Ge	3–4		Amorphous A15 compound	k
$Gd_{.2}PbMo_6S_8$	14.3	—	Highest known critical field $H_{c2} \sim 60$ T	l
$ErRh_4B_4$	8.5	—	Superconducting order suppressed at 0.9°K by onset of ferromagnetic order	m
$ErMo_6S_8$	2.2	—	Superconducting order coexists with long-range antiferromagnetic order below 0.2°K	m
$Gd_xTh_{1-x}Ru_2$	2–4	—	Coexistence of superconductivity and "spin glass" magnetic freezing	n, o
$CeCu_2Si_2$	0.5	—	Unstable 4f-shell behavior	p
Au	<0.001°K	0.1	Very weak coupler	q
$SrTiO_{3-x}$	0.05–0.5°K	—	Doped semiconductor $n < 10^{20}$ cm^{-3}	r
(2H)–$NbSe_2$	7.1	—	Layer structure metal (atmospheric pressure)	s
$NbSe_3$	3.5	—	Quasi-1d metal (under pressure of 5.5 kbar)	t
$(TMTSF)_2ClO_4$	1.2–1.4	—	Quasi-1d "organic" material (atmospheric pressure)	u

(*Continued*)

TABLE I (*Continued*)

Material	T_c	λ	Comments	Reference
$BaPb_{1-x}Bi_xO_3$	~13	—	Highest T_c sp metal, metal–insulator transition suppresses superconductivity as x decreases	v

[a] W. L. McMillan, *Phys. Rev.* **167,** 331 (1968).
[b] W. L. McMillan and J. M. Rowell, *Phys. Rev. Lett.* **14,** 108 (1965).
[c] J. D. Leslie, J. T. Chen, and T. T. Chen, *Can. J. Phys.* **48,** 2783 (1970).
[d] E. L. Wolf, J. Zasadzinski, J. N. Osmun, and G. B. Arnold, *Solid State Commun.* **31,** 321 (1979).
[e] Yu. F. Revenko, A. I. Diachenko, V. M. Svistunov, and B. Schöneich, *Fiz. Nizk. Temp. (Kiev)* **6,** 1304 (1980), *Sov. Phys.—Low Temp. (Engl. Transl.)* **6,** 635 (1980).
[f] D. B. Kimhi and T. H. Geballe, *Phys. Rev. Lett.* **45,** 1039 (1980).
[g] L. Y. Shen, *Phys. Rev. Lett.* **29,** 1082 (1972).
[h] E. L. Wolf, J. Zasadzinski, G. B. Arnold, D. F. Moore, J. M. Rowell, and M. R. Beasley, *Phys. Rev. B* **22,** 1214 (1980).
[i] K. E. Kihlstrom and T. H. Geballe, *Phys. Rev. B* **24,** 4101 (1981).
[j] A. I. Golovashkin, E. V. Pechen', A. I. Skvortsov, and N. E. Khlebova, *Fiz. Tverd. Tela* **23,** 1324 (1981); *Sov. Phys.—Solid State (Engl. Transl.)* **23,** 774 (1981).
[k] C. C. Tsuei, S. von Molnar, and J. M. Coey, *Phys. Rev. Lett.* **41,** 664 (1978).
[l] Ø. Fischer, *Appl. Phys.* **16,** 1 (1978).
[m] G. Shirane, W. Thomlinson, and D. E. Moncton, *in* "Superconductivity in d- and f-Band Metals" (H. Suhl and M. B. Maple, eds.), p. 381. Academic Press, New York, 1980.
[n] D. Davidov, K. Baberschke, J. A. Mydosh, and G. J. Nieuwenhuys, *J. Phys. F* **7,** L47 (1977).
[o] S. Roth, *Appl. Phys.* **15,** 1 (1978).
[p] F. Steglich, J. Aarts, C. D. Bredl, W. Lieke, D. Meschede, W. Franz, and H. Schäfer, *Phys. Rev. Lett.* **43,** 1892 (1979).
[q] R. F. Hoyt, H. N. Scholz, and D. O. Edwards, *Phys. Lett. A* **84,** 145 (1981).
[r] J. F. Schooley, W. R. Hosler, and M. L. Cohen, *Phys. Rev. Lett.* **12,** 474 (1964).
[s] T. H. Geballe, *AIP, Conf. Proc.* **4,** 237 (1972).
[t] A Briggs, P. Moncleau, M. Nuñez Regueiro, J. Peyrard, M. Ribault, and J. Richard, *J. Phys. C* **13,** 2117 (1980).
[u] K. Bechgaard, K. Carneiro, M. Olsen, F. B. Rasmussen, and C. S. Jacobsen, *Phys. Rev. Lett.* **46,** 852 (1981).
[v] C. W. Chu, S. Huang, and A. W. Sleight, *Solid State Commun.* **18,** 977 (1976).

metals[11] (e.g., Nb_3Sn and Nb_3Ge) have such high values of T_c, and this poses an important challenge to ME theory. In cases such as Al and Nb,[12,13] satisfactory explanations for T_c exist.

[11] Several reviews of A15 metals have been written. Yu. A. Izyumov and E. Z. Kurmaev, *Usp. Fiz. Nauk* **113,** 193 (1974); *Sov. Phys.—Usp. (Engl. Transl.)* **19,** 26 (1974).
[12] H. K. Leung, J. P. Carbotte, D. W. Taylor, and C. R. Leavens, *J. Low Temp. Phys.* **24,** 2534 (1976).
[13] W. H. Butler, F. J. Pinski, and P. B. Allen, *Phys. Rev. B* **19,** 3708 (1979).

At a crude level, the explanation consists in the calculation of the electron–phonon interaction parameter λ, to be defined in Section 3, and the estimation of T_c from McMillan's equation,[14]

$$T_c = \frac{\omega_{\log}}{1.2} \exp\left(\frac{-1.04(1 + \lambda)}{\lambda - \mu^*(1 + 0.62\lambda)} \right). \tag{1.1}$$

Here μ^* is an adjustable parameter of order 0.1 which represents Coulomb repulsion, and $\omega_{\log}$ is a logarithmic average phonon frequency, discussed in Section 14. This article aims to clarify the underlying theory to which Eq. (1.1) is a popular approximation. In Section 14, Eq. (1.1) will be critically discussed.

In other cases, such as $ErRh_4B_4$, where superconductivity and magnetism compete,[14a] ME theory in its present form is not very relevant. A theory at the BCS level, introducing new magnetic-order parameters, is needed first.[15] It remains a task for the future to generalize to the ME level if T_c is to be explained in microscopic detail. Similar comments may apply to other systems given in Table I, where superconductivity competes with other kinds of order. The pseudo-one-dimensional compounds may demand new theories which do not invoke the Migdal theorem[2] and which go beyond the BCS mean-field level.

This article attempts to describe primarily those systems and effects which are firmly in the grasp of ME theory, and, secondarily, certain extensions of ME theory which have gained some credence or seem likely to develop in the near future. Thus we include anisotropy effects and dilute impurities, either nonmagnetic or magnetic, as examples of the primary aim, and a brief discussion of possible "paramagnon" effects as an example of the secondary aim.

2. BCS Theory

For a proper review of BCS theory,[1] the reader may consult many texts.[16–18] This section derives the BCS integral equation by Gor'kov's method.[19] This introduces the Green's function needed in later sections,

[14] W. L. McMillan, *Phys. Rev.* **167,** 331 (1968).

[14a] G. Shirane, W. Thomlinson, and D. E. Moncton, *in* "Superconductivity in d- and f-Band Metals" (H. Suhl and M. B. Maple, eds.), p. 381. Academic Press, New York, 1980.

[15] M. J. Nass, K. Levin, and G. S. Grest, *Phys. Rev. Lett.* **46,** 614 (1981).

[16] P. G. de Gennes, "Superconductivity of Metals and Alloys." Benjamin, New York, 1966.

[17] G. Rickayzen, *in* "Superconductivity" (R. D. Parks, ed.), Vol. 1, Chapter 2. Dekker, New York, 1969.

[18] V. Ambegaokar, *in* "Superconductivity" (R. D. Parks, ed.), Vol. 1, Chapter 5. Dekker, New York, 1969.

[19] L. P. Gor'kov, *Zh. Eksp. Teor. Fiz.* **34,** 735 (1958); *Sov. Phys.—JETP (Engl. Transl.)* **7,** 505 (1958).

especially Gor'kov's "anomalous" F function, and provides simple examples of manipulations using Matsubara's imaginary frequencies.

The key feature of BCS theory is Cooper-pair condensation. The pair of states $(k\uparrow, -k\downarrow)$ is occupied coherently. The Cooper-pair amplitude, $\langle c_{k\uparrow}c_{-k\downarrow}\rangle$, which is zero by number conservation in the normal state, becomes finite below T_c. The simplest model[1] which permits such behavior is given by the BCS "reduced" Hamiltonian, $\mathscr{H}_0 + \mathscr{H}_{red}$, where

$$\mathscr{H}_0 = \sum_{k\sigma} \epsilon_k c^\dagger_{k\sigma} c_{k\sigma}, \tag{2.1}$$

$$\mathscr{H}_{red} = -\sum_{kk'} V_{k'k}(c^\dagger_{-k'\downarrow}c^\dagger_{k'\uparrow})(c_{k\uparrow}c_{-k\downarrow}). \tag{2.2}$$

The operator $c_{k\sigma}$ destroys an electron in a Bloch state $k\sigma$ of energy ϵ_k (measured relative to the chemical potential μ), where k is short for $\mathbf{k}n$ (wave number and band index). Interaction (2.2) only contains terms that scatter pairs of electrons from one pair state $(k\uparrow, -k\downarrow)$ to a different one $(k'\uparrow, -k'\downarrow)$. In reality, terms also exist that scatter only one of the paired electrons; these complicate the Cooper-pair phenomena in what is assumed to be an inessential way, and they have been omitted. The interaction matrix elements $V_{k'k}$ are at this stage unspecified. Bardeen, Cooper, and Schrieffer had in mind the Fröhlich[20] or Bardeen–Pines[21] effective phonon-induced interaction, a time-independent potential with some of the features of the correct time-dependent Eliashberg interaction. Because of the minus sign in Eq. (2.2), a positive $V_{k'k}$ corresponds to an attractive interaction.

The BCS solution of this Hamiltonian is exact in the thermodynamic limit and can be derived in many ways. Gor'kov's Green's function method[19] is the one that can be generalized most easily to more realistic models. The field operators $c_{k\sigma}$ are written in the Heisenberg picture, except that time t is replaced by an imaginary time $-i\tau$:

$$c_{k\sigma}(\tau) \equiv e^{\mathscr{H}\tau} c_{k\sigma} e^{-\mathscr{H}\tau}. \tag{2.3}$$

Thermodynamic averages are denoted by

$$\langle A \rangle = Z^{-1}\, \mathrm{tr}(e^{-\beta\mathscr{H}} A), \tag{2.4}$$

where Z is $\mathrm{tr}(e^{-\beta\mathscr{H}})$ and $\beta = 1/T$. This is actually a grand canonical average: The operator $\mathscr{H}$ implicitly contains the term $-\mu N$ and the trace runs over states with all values of N. Units are used where $\hbar = k_B = 1$. The reason for an imaginary time variable is the similarity of $e^{-\mathscr{H}\tau}$ and $e^{-\beta\mathscr{H}}$. The Green's function is defined for τ in the range $(-\beta, \beta)$ as

[20] H. Fröhlich, *Proc. R. Soc. London, Ser. A* **215,** 291 (1952).

[21] J. Bardeen and D. Pines, *Phys. Rev.* **99,** 1140 (1955).

$$G(k, \tau) \equiv -\langle c_{k\sigma}(\tau)c_{k\sigma}^\dagger(0)\rangle\theta(\tau) + \langle c_{k\sigma}^\dagger(0)c_{k\sigma}(\tau)\rangle\theta(-\tau), \tag{2.5}$$

where $\theta(\tau)$ is 1 for positive τ and 0 otherwise. This can be written in a concise notation:

$$G(k\tau) = -\langle T_\tau c_{k\sigma}(\tau)c_{k\sigma}^\dagger(0)\rangle, \tag{2.6}$$

where T_τ is the Wick operator, which reorders the operators following it in such a way that τ increases from right to left. Whenever two fermion operators are interchanged in this process, a factor of -1 is introduced. For noninteracting electrons (Hamiltonian = $\mathscr{H}_0$) Eq. (2.6) can be directly evaluated:

$$G_0(k, \tau) = e^{-\epsilon_k\tau}[f_k\theta(-\tau) - (1 - f_k)\theta(\tau)], \tag{2.7}$$

where $f_k = \langle c_{k\sigma}^\dagger c_{k\sigma}\rangle_0$ is the Fermi function, $(e^{\beta\epsilon_k} + 1)^{-1}$. Within the range $-\beta < \tau < \beta$ Eq. (2.6) has the antiperiodicity property

$$G(k, \tau + \beta) = -G(k, \tau). \tag{2.8}$$

This property is used as the definition for G outside the range $(-\beta, \beta)$. It follows that G has Fourier representation

$$G(k, \tau) = \frac{1}{\beta}\sum_{n=-\infty}^{\infty} e^{-i\omega_n\tau}G(k, i\omega_n), \tag{2.9}$$

$$\omega_n = (2n + 1)\pi T, \tag{2.10}$$

where only odd integers $2n + 1$ appear in the "Matsubara frequencies" $i\omega_n$. The Fourier coefficients $G(k, i\omega_n)$ are given by inverting Eq. (2.9),

$$G(k, i\omega_n) = \frac{1}{2}\int_{-\beta}^{\beta} d\tau e^{i\omega_n\tau}G(k, \tau). \tag{2.11}$$

For noninteracting electrons, using Eqs. (2.7) and (2.11),

$$G_0(k, i\omega_n) = (i\omega_n - \epsilon_k)^{-1}. \tag{2.12}$$

It is proved in textbooks[22–24] that the analytic continuation of $G(k, i\omega_n)$ to complex values of z, $G(k, z)$, contains full information about the time-dependent thermal Green's function. In particular, the poles z_0 of $G(k, z)$ can be interpreted as quasiparticle energies, provided that the damping, $1/2\tau = \operatorname{Im} z_0$, is not too large. For G_0 the pole $z_0 = \epsilon_k$ occurs on the real axis; that is, there is no damping. Furthermore, thermodynamic properties of systems can be evaluated once the Green's functions $G(k, i\omega_n)$ are known.

[22] A. L. Fetter and J. D. Walecka, "Quantum Theory of Many Particle Systems." McGraw-Hill, New York, 1971.

[23] A. A. Abrikosov, L. P. Gor'kov, and I. E. Dzyaloshinski, "Methods of Quantum Field Theory in Statistical Physics." Prentice-Hall, Englewood Cliffs, New Jersey, 1963.

[24] G. Baym and D. Mermin, *J. Math. Phys.* **2**, 232 (1961).

To evaluate $G(k, i\omega_n)$ for the BCS model, calculate the τ derivative of $G(k, \tau)$, using Eqs. (2.1)–(2.5):

$$\left(-\frac{d}{d\tau} - \epsilon_k\right)G(k, \tau) = \delta(\tau) - \sum_{k'} V_{kk'}\langle T_\tau c^\dagger_{-k\downarrow}(\tau)c_{k'\uparrow}(\tau)c_{-k'\downarrow}(\tau)c^\dagger_{k\uparrow}(0)\rangle. \quad (2.13)$$

The complicated thermal average on the right-hand side of Eq. (2.13) is evaluated using a mean-field argument. In Hartree–Fock mean-field theory, averages like $\langle c_1^\dagger c_2^\dagger c_3 c_4\rangle$ are approximated by $\langle c_1^\dagger c_4\rangle\langle c_2^\dagger c_3\rangle - \langle c_1^\dagger c_3\rangle\langle c_2^\dagger c_4\rangle$ in the hope that the fluctuations $\langle AB\rangle - \langle A\rangle\langle B\rangle$ are unimportant. It was mentioned that BCS theory requires a nonzero pair amplitude $\langle c_{k\uparrow}c_{-k\downarrow}\rangle$. Therefore the following treatment is suggested:

$$\langle T_\tau c^\dagger_{-k\downarrow}(\tau)c_{k'\uparrow}(\tau)c_{-k'\downarrow}(\tau)c^\dagger_{k\uparrow}(0)\rangle \rightarrow \langle T_\tau c_{k'\uparrow}(\tau)c_{-k'\downarrow}(\tau)\rangle\langle T_\tau c^\dagger_{-k\downarrow}(\tau)c^\dagger_{k\uparrow}(0)\rangle. \quad (2.14)$$

Other Hartree–Fock-like terms on the right-hand side of Eq. (2.14) are zero unless $k' = k$, which is an unimportant case. Also, the band energy ϵ_k already includes Hartree–Fock energies, so it would be incorrect to include them again on the right-hand side of Eq. (2.13).

The Gor'kov "anomalous" Green's functions are now defined:

$$F(k, \tau) = -\langle T_\tau c_{k\uparrow}(\tau)c_{-k\downarrow}(0)\rangle, \quad (2.15)$$

$$\bar{F}(k, \tau) = -\langle T_\tau c^\dagger_{-k\downarrow}(\tau)c^\dagger_{k\uparrow}(0)\rangle. \quad (2.16)$$

A careful inspection reveals that these are complex conjugates of each other:

$$\bar{F}(k, \tau) = F(k, \tau)^*. \quad (2.17)$$

Then, using Eqs. (2.14)–(2.16), Eq. (2.13) becomes

$$\left(-\frac{d}{d\tau} - \epsilon_k\right)G(k, \tau) = \delta(\tau) - \sum_{k'} V_{kk'}F(k', 0)\bar{F}(k, \tau). \quad (2.18)$$

An equation of motion for $\bar{F}(k, \tau)$ is next generated by similar procedures, giving

$$\left(-\frac{d}{d\tau} + \epsilon_k\right)\bar{F}(k, \tau) = -\sum_{k'} V_{kk'}\bar{F}(k', 0)G(k, \tau). \quad (2.19)$$

These are coupled nonlinear differential equations whose solution gives BCS theory. The solution is accomplished by extracting the factor

$$\Delta_k \equiv \sum_{k'} V_{kk'}F(k', 0). \quad (2.20)$$

This will turn out to be the BCS energy gap. It will be treated as an unknown complex parameter that must be constructed self-consistently at the end. Next Eqs. (2.18) and (2.19) are Fourier transformed using Eq. (2.11):

$$\begin{pmatrix} i\omega_n - \epsilon_k & \Delta_k \\ \Delta_k^* & i\omega_n + \epsilon_k \end{pmatrix} \begin{pmatrix} G(k, i\omega_n) \\ \bar{F}(k, i\omega_n) \end{pmatrix} = \begin{pmatrix} 1 \\ 0 \end{pmatrix}. \tag{2.21}$$

The determinant of the matrix is $-(\omega_n^2 + \epsilon_k^2 + |\Delta_k|^2)$, which never vanishes. The combination $\epsilon_k^2 + |\Delta_k|^2$ is denoted

$$E_k^2 \equiv \epsilon_k^2 + |\Delta_k|^2. \tag{2.22}$$

Inverting Eq. (2.21) gives the result

$$G = (-i\omega_n - \epsilon_k)/(\omega_n^2 + E_k^2), \tag{2.23}$$

$$\bar{F} = \Delta_k^*/(\omega_n^2 + E_k^2). \tag{2.24}$$

In the limit $\Delta_k \to 0$, the normal state solution $G = (i\omega_n - \epsilon_k)^{-1}$, $\bar{F} = 0$ is retrieved. The parameter E_k can be interpreted as a quasiparticle energy with Δ_k as an energy gap, because $G(k, z)$ given by Eq. (2.23) has poles at $z_0 = \pm E_k$.

Finally, Δ_k must be constructed self-consistently. The $\tau = 0$ F function is computed using the Fourier representation of Eq. (2.9):

$$F(k, \tau = 0) = \frac{1}{\beta} \sum_{n=-\infty}^{\infty} F(k, i\omega_n). \tag{2.25}$$

Next, Eq. (2.20) is written

$$\Delta_k = \frac{1}{\beta} \sum_{k'} \sum_{n=-\infty}^{\infty} \frac{V_{kk'}\Delta_{k'}}{\omega_n^2 + E_{k'}^2}. \tag{2.26}$$

This is the BCS gap equation[1] in a disguised form. The more familiar version,

$$\Delta_k = \sum_{k'} \frac{V_{kk'}\Delta_{k'}}{2E_{k'}} \tanh\left(\frac{E_{k'}}{2k_B T}\right), \tag{2.27}$$

follows from the identity

$$\sum_{n=-\infty}^{\infty} [(2n + 1)^2\pi^2 + a^2]^{-1} = (2a)^{-1} \tanh\left(\frac{a}{2}\right). \tag{2.28}$$

This identity can be derived from the Poisson sum formula.[5,25] An alternate derivation via the digamma function is given in Appendix A.

The gap Δ obtained from Eq. (2.27) decreases as T increases, going to zero at $T = T_c$ by the law $\Delta \propto (T_c - T)^{1/2}$. BCS theory gives a famous

[25] L. P. Kadanoff and G. Baym, "Quantum Statistical Mechanics." Benjamin, New York, 1962.

formula for T_c:

$$T_c = 1.13\theta_D \exp[-1/N(0)V]. \tag{2.29}$$

This result explains (to the first approximation) the isotope effect and the fact that T_c is always low. In the second approximation, the isotopic shift[26] $d \log T_c/d \log M$ is observed to depart from $-\frac{1}{2}$, and the parameter $N(0)V$ is significantly larger than Eq. (2.29) requires it to be to explain T_c. Improved theories of T_c, to be discussed in the remainder of this article, repair both of these defects.

The derivation of Eq. (2.29) from Eq. (2.27) is a standard exercise[17] in Western texts. The derivation from Eq. (2.26) is often found in Russian literature and is closely related to the treatment used later in this article. Following BCS, a model for $V_{kk'}$ is used which imitates the more complicated Fröhlich or Bardeen–Pines interaction:

$$V_{kk'} \longrightarrow V\theta(\theta_D - |\epsilon_k|)\theta(\theta_D - |\epsilon_{k'}|). \tag{2.30}$$

This interaction vanishes unless the states k and k' both lie within the Debye energy θ_D of the Fermi energy. The gap Δ_k is a constant, Δ, within this range and zero otherwise. At $T = T_c$, Δ goes to zero and E_k becomes ϵ_k. The nonlinear equations (2.26) and (2.27) become linear equations, with only the trivial solution $\Delta = 0$, except at the special temperature T_c, where $\Delta \neq 0$ becomes possible. Thus Eq. (2.26) becomes

$$\Delta = T_c V \Delta \sum_{k'}^{|\epsilon'_k|<\theta_D} \sum_{n=-\infty}^{\infty} [(2n+1)^2\pi^2 T_c^2 + \epsilon_{k'}^2]^{-1}, \tag{2.31}$$

$$1 = T_c N(0) V \int_{-\theta_D}^{\theta_D} d\epsilon' \sum_{n=-\infty}^{\infty} [(2n+1)^2\pi^2 T_c^2 + \epsilon'^2]^{-1}. \tag{2.32}$$

In Eq. (2.32) it has been assumed that the density of states $N(\epsilon)$ is approximately a constant $N(0)$ over the interval $(-\theta_D, \theta_D)$ around ϵ_F. The energy integral is done next, yielding

$$[N(0)V]^{-1} = \sum_{n=-\infty}^{\infty} \frac{2}{|2n+1|\pi} \tan^{-1}\left(\frac{\theta_D}{|2n+1|\pi T_c}\right). \tag{2.33}$$

The inverse tangent function takes the value $\pi/2$ when its argument is large (i.e., $|2n + 1|$ is small) and approaches zero when its argument is small (i.e., $|2n + 1|$ is large). This behavior can be approximated by $(\pi/2)\theta(\theta_D - |2n + 1|\pi T_c)$, yielding

[26] R. Meservey and B. B. Schwartz, *in* "Superconductivity" (R. D. Parks, ed.), Vol. 1, Chapter 3. Dekker, New York, 1969.

$$[N(0)V]^{-1} = \sum_{n=0}^{\theta_D/2\pi T_c - 1/2} \frac{1}{n+1/2}$$

$$= \psi(\theta_D/2\pi T_c + 1) - \psi(1/2) \tag{2.34}$$

$$\approx \ln(\theta_D/2\pi T_c) - \ln(e^{-\gamma}/4). \tag{2.35}$$

Equation (2.29) follows immediately from Eq. (2.35), using $\gamma = 0.577 \cdots$ (Euler's constant). The identity (2.34) follows from Eq. (A.7), and the approximation (2.35) follows from Eqs. (A.8) and (A.16).

II. The Normal State

3. Electron–Phonon Effects

The BCS T_c equation, Eq. (2.29), fails to account well for actual transition temperatures. The fault lies not with the approximation necessary to get Eq. (2.29) from Eq. (2.26) or (2.27) but, rather, in that the starting model, Eqs. (2.1) and (2.2), of electrons interacting via a time-independent pairwise interaction does not incorporate enough of the physics of the electron–phonon system.[2] For example, the electron–phonon interaction causes a mass enhancement of electron states near ϵ_F, seen in specific heat, and a finite lifetime of electron quasi-particle states. These effects are so strong that in most metals the damping $1/2\tau$ of quasi-particles is comparable to the energy ϵ_k when T or ω (temperature or external frequency) is raised to $\sim\theta_D/5$. Well-defined quasi-particles no longer exist.[27,28] Nevertheless, Migdal[2] showed that Feynman–Dyson perturbation theory can solve the electron–phonon problem to high accuracy, because the small parameter $N(0)\theta_D$ keeps higher order corrections small.

These normal effects are best derived by means of the thermal Green's functions (2.6). An accurate theory of T_c requires Cooper-pair condensation in a medium with strong electron–phonon effects and was given by Eliashberg[3] and extended by Scalapino, Schrieffer, and Wilkins.[29] This theory reduces in the normal state to Migdal's theory. The first step is to master Migdal theory, which is the task of this section.

In addition to the electron thermal Green's function, Eq. (2.6), we need the corresponding phonon function

$$D_{\alpha\beta}(\mathbf{Q}, \tau) = -\langle T_\tau u_{\mathbf{Q}\alpha}(\tau) u_{-\mathbf{Q}\beta}(0)\rangle. \tag{3.1}$$

[27] S. Engelsberg and J. R. Schrieffer, *Phys. Rev.* **131,** 993 (1963).

[28] G. Grimvall, *Sel. Top. Solid State Phys.* **16** (1981).

[29] D. J. Scalapino, J. R. Schrieffer, and J. W. Wilkins, *Phys. Rev.* **148,** 263 (1966).

The displacement operator $\mathbf{u}$ is a sum of operators with boson commutation relations

$$\mathbf{u}_Q = \sum_i \left(\frac{\hbar}{2NM\omega_{\mathbf{Q}i}}\right)^{1/2} \hat{\epsilon}_{\mathbf{Q}i}(a_{\mathbf{Q}i} + a^{\dagger}_{-\mathbf{Q}i}), \tag{3.2}$$

where $a_{\mathbf{Q}i}$ destroys a phonon with energy $\omega_{\mathbf{Q}i}$ of wave vector $\mathbf{Q}$, branch i, and polarization vector $\hat{\epsilon}_{\mathbf{Q}i}$. For simplicity, only the equation for primitive lattices with a single species of mass M is given. Later the composite index $Q = \mathbf{Q}i$ will be used. The Wick operator T_τ introduces no factors of -1 when two boson operators or one fermion and one boson operator are interchanged. If phonons are governed by a harmonic Hamiltonian, the corresponding free-phonon Green's function D^0 can be easily evaluated:

$$D^0_{\alpha\beta}(\mathbf{Q}\tau) = -\sum_i \left(\frac{\hbar}{2MN\omega_Q}\right)\epsilon_{Q\alpha}\epsilon_{-Q\beta}\{e^{-\omega_Q\tau}[(n_Q + 1)\theta(\tau) + n_Q\theta(-\tau)]$$
$$+ e^{\omega_Q\tau}[n_Q\theta(\tau) + (n_Q + 1)\theta(-\tau)]\}, \tag{3.3}$$

where n_Q is the Bose function $[\exp(\beta\omega_Q) - 1]^{-1}$ and the symmetry $\omega_{-Q} = \omega_Q$ has been used.

In close analogy with the electron Green's function, $D_{\alpha\beta}(\mathbf{Q}\tau)$ has the periodicity property (in the interval $-\beta < \tau < \beta$)

$$D_{\alpha\beta}(\mathbf{Q}; \tau + \beta) = D_{\alpha\beta}(\mathbf{Q}, \tau), \tag{3.4}$$

which is used to define D outside the range $-\beta < \tau < \beta$. The Fourier representation is

$$D_{\alpha\beta}(\mathbf{Q}, \tau) = \frac{1}{\beta}\sum_{\nu=-\infty}^{\infty} e^{-i\omega_\nu\tau} D_{\alpha\beta}(\mathbf{Q}, i\omega_\nu), \tag{3.5}$$

$$\omega_\nu = 2\nu\pi T; \tag{3.6}$$

that is, only even integers 2ν appear in the boson Matsubara frequencies ω_ν. The inverse relation is the same as Eq. (2.11), which gives for free phonons the result

$$D^0_{\alpha\beta}(Q, i\omega_\nu) = \sum_i \left(\frac{\hbar}{2MN\omega_Q}\right)\epsilon_{Q\alpha}\epsilon_{-Q\beta}\left(\frac{1}{i\omega_\nu - \omega_Q} - \frac{1}{i\omega_\nu + \omega_Q}\right). \tag{3.7}$$

Fortunately, there are textbook treatments of Migdal theory.[5,6,23,28,30] Therefore this article omits some complexities of logic and focuses instead on deriving results in a form which permits application to actual materials with complicated electron and phonon spectra. The aim is to compute the Green's functions of the interacting systems. There are two useful ways to

[30] T. Holstein, *Ann. Phys. (N.Y.)* **29**, 410 (1964).

express these functions: first, in terms of self-energies Σ and Π defined by

$$G^{-1}(k, i\omega_n) = G_0^{-1}(k, i\omega_n) - \Sigma(k, i\omega_n), \tag{3.8}$$

$$[D^{-1}(\mathbf{Q}, i\omega_\nu)]_{\alpha\beta} = [D_0^{-1}(\mathbf{Q}, i\omega_\nu)]_{\alpha\beta} - \Pi_{\alpha\beta}(\mathbf{Q}, i\omega_\nu), \tag{3.9}$$

and second, in terms of their spectral representations,

$$G(k, i\omega_n) = \int_{-\infty}^{\infty} d\epsilon \, \frac{C(k, \epsilon)}{i\omega_n - \epsilon}, \tag{3.10}$$

$$D_{\alpha\beta}(\mathbf{Q}, i\omega_\nu) = \int_0^{\infty} d\Omega \, B_{\alpha\beta}(\mathbf{Q}, \Omega)[(i\omega_\nu - \Omega)^{-1} - (i\omega_\nu + \Omega)^{-1}]. \tag{3.11}$$

The spectral functions C and $B_{\alpha\beta}$ can be shown[22,23] to be real and nonnegative and to obey sum rules. The spectral functions are related to the analytic continuation of the Green's function off the imaginary axis onto the real axis, approached from above:

$$C(k, \epsilon) = -(1/\pi)\mathrm{Im}\, G(k, \epsilon + i\eta), \tag{3.12}$$

$$B_{\alpha\beta}(Q, \Omega) = -(1/\pi)\mathrm{Im}\, D_{\alpha\beta}(Q, \Omega + i\eta). \tag{3.13}$$

For noninteracting electrons and phonons, the spectral functions are

$$C_0(k, \epsilon) = \delta(\epsilon - \epsilon_k), \tag{3.14}$$

$$B^0_{\alpha\beta}(\mathbf{Q}, \Omega) = \sum_i \left(\frac{\hbar}{2MN\omega_Q}\right)\epsilon_{Q\alpha}\epsilon_{-Q\beta}\delta(\Omega - \omega_Q). \tag{3.15}$$

A "quasi-particle approximation" is the assumption that even with interactions, C and B are approximately δ functions, with shifted energies. This turns out to be an excellent approximation for phonons in real crystalline metals. The phonon self-energy Π causes a large shift of energies of phonons (which is very difficult to evaluate numerically), but only a small damping. The spectral function $B_{\alpha\beta}$ is directly measurable by inelastic neutron scattering[31] and has narrow peaks. Therefore it is unnecessary to evaluate Π numerically. We shall take $B_{\alpha\beta}$ from experiment (or, more precisely, from a Born–von Karman interpolation[32] fitted to whatever data are available).

The electronic self-energy Σ has entirely different properties. No experiment measures directly the spectral function C, so we are compelled to calculate it. As already mentioned, the damping can be large and C need not look like a δ function. Fortunately, nature is kind and Σ is easier to evaluate than Π. A correct theory for Σ can be obtained from the "Fröhlich

[31] W. Marshall and S. W. Lovesey, "Theory of Thermal Neutron Scattering." Oxford Univ. Press (Clarendon), London and New York, 1971.

[32] B. N. Brockhouse, E. D. Hallman, and S. C. Ng, *Metall. Soc. Conf.* [*Proc.*] **43,** 161 (1967).

Hamiltonian"[28]

$$\mathscr{H} = \mathscr{H}_{\text{el}} + \mathscr{H}_{\text{L}} + \mathscr{H}_{\text{ep}}, \tag{3.16}$$

where $\mathscr{H}_{\text{el}}$ describes noninteracting electron quasi-particles, which are the quasi-stationary states of the rigid perfect lattice problem of band electrons plus Coulomb interactions:

$$\mathscr{H}_{\text{el}} = \sum_{k\sigma} \epsilon_k c_{k\sigma}^{\dagger} c_{k\sigma} . \tag{3.17}$$

Technically, ϵ_k is determined by the pole of an electron Green's function that includes Coulomb interactions.[33,34] Available experimental evidence suggests that this "quasi-particle band structure"[35] is surprisingly similar to the bands calculated by band theorists, even though the usual versions of band theory aim to be "density-functional" band theories. Density-functional theory[36] in principle gives a correct ground-state energy, but the bands need have little resemblance to the quasi-particle band structure used to describe low-lying excitations. We shall assume (without justification, but with considerable precedent) that the energies ϵ_k are available from band theorists.

The lattice Hamiltonian $\mathscr{H}_{\text{L}}$ describes noninteracting phonons whose energies derive from experiment:

$$\mathscr{H}_{\text{L}} = \sum_{Q} \hbar\omega_Q (a_Q^{\dagger} a_Q + 1/2). \tag{3.18}$$

Finally, the interaction $\mathscr{H}_{\text{ep}}$ is written

$$\mathscr{H}_{\text{ep}} = \sum_{kQ} \langle k + Q | \nabla_\alpha V | k \rangle u_{Q\alpha} c_{k+Q}^{\dagger} c_k , \tag{3.19}$$

where the displacement $u_{Q\alpha}$ is given by Eq. (3.2). The matrix element contains the operator ∇V, which scatters electrons from a single atom at the origin, displaced by an infinitesimal amount. Linear response theory shows[37] that this operator is $\epsilon^{-1}\nabla V^0$, where V^0 is a bare electron–ion potential, and ϵ^{-1} a screening operator. Only in the weak scattering limit can one expect ∇V to be the gradient of a screened potential (the gradient and the screening operator do not commute except in a homogeneous medium; the notation ∇V is not meant to be literal). The essence of the Fröhlich Hamiltonian is that Coulomb interactions have already renormalized ϵ_k and ∇V, and *both* Coulomb and electron–phonon interactions have renormalized ω_Q. This Hamiltonian is now used to construct Σ. It cannot be used to construct Π because ω_Q is already the observed renormalized spectrum.

[33] V. Heine, P. Nozières, and J. W. Wilkins, *Philos. Mag.* [8] **13,** 741 (1966).
[34] P. B. Allen, *Phys. Rev. B* **18,** 5217 (1978).
[35] L. J. Sham and W. Kohn, *Phys. Rev.* **145,** 561 (1966).
[36] W. Kohn, *in* "Many Body Theory" (R. Kubo, ed.), p. 73. Syokabō, Tokyo, 1966.
[37] S. K. Sinha, *Phys. Rev.* **169,** 477 (1968).

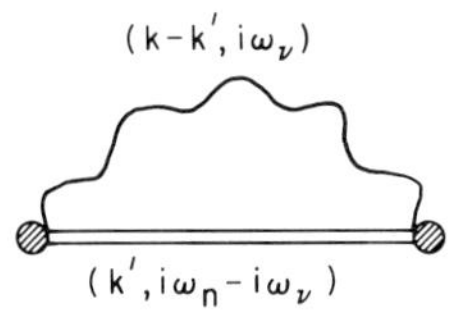

FIG. 1. The Feynman graph for the electronic self-energy $\Sigma_{ep}(k, i\omega_n)$ [Eq. (3.20)] according to Migdal theory. The wavy line is a phonon Green's function D, and the double solid line is a renormalized electron Green's function G.

The essence of Migdal theory is to use only the lowest order Feynman graph (Fig. 1) for Σ. The value of this graph is

$$\Sigma_{ep}(k, i\omega_n) = -\frac{1}{\beta}\sum_{k',\nu} \langle k|\nabla_\alpha V|k'\rangle D_{\alpha\beta}(k-k', i\omega_\nu) \times \langle k'|\nabla_\beta V|k\rangle G(k', i\omega_n - i\omega_\nu). \tag{3.20}$$

All relevant properties of the electron–phonon system are derived from this formula and Eq. (3.8), which form a coupled pair. The analogous formulas in the superconducting state will constitute Eliashberg theory. Nothing further will be said about the derivation of Eq. (3.20), except a brief discussion in Section 4 of the justification (Migdal's "theorem") for omitting higher graphs.

The first step in evaluating Eq. (3.20) is to introduce the spectral representation (3.11) for the phonon Green's function:

$$\Sigma_{ep}(k, i\omega_n) = T\sum_{k',\nu}\int_0^\infty d\Omega\, \frac{\alpha^2F(k, k', \Omega)}{N(0)}\left(\frac{2\Omega}{\omega_\nu^2+\Omega^2}\right)G(k', i\omega_n - i\omega_\nu), \tag{3.21}$$

$$\alpha^2F(k, k', \Omega) \equiv N(0)\langle k|\nabla_\alpha V|k'\rangle B_{\alpha\beta}(k-k', \Omega)\langle k'|\nabla_\beta V|k\rangle. \tag{3.22}$$

This defines the important "electron–phonon spectral function" α^2F. Frequently this is found with either one or both of the electron states (k, k') averaged over the Fermi surface (FS):

$$\alpha^2F(k, \Omega) = \frac{1}{N(0)}\sum_{k'} \alpha^2F(k, k', \Omega)\delta(\epsilon_{k'}), \tag{3.23}$$

$$\alpha^2F(\Omega) = \frac{1}{N(0)^2}\sum_{kk'} \alpha^2F(k, k', \Omega)\delta(\epsilon_k)\delta(\epsilon_{k'}) \tag{3.24a}$$

$$\equiv \langle \alpha^2F(k, \Omega)\rangle_{FS} \equiv \langle\langle \alpha^2F(k, k', \Omega)\rangle\rangle_{FS}, \tag{3.24b}$$

where $N(0) = \sum_k \delta(\epsilon_k)$ is the density of states at the Fermi surface. These functions are each dimensionless measures of the effectiveness of phonons of frequency Ω in scattering electrons from k to k' $[\alpha^2F(k, k', \Omega)]$, from k

to any state on the Fermi surface $[\alpha^2F(k, \Omega)]$, and from any state to any other state on the Fermi surface $[\alpha^2F(\Omega)]$. The last is the function that McMillan and Rowell[7] extracted from tunneling experiments. An alternate version of Eq. (3.22) is

$$\alpha^2F(k, k', \Omega) = \sum_j N(0)|g^j_{k,k'}|^2\delta(\Omega - \omega_{\mathbf{k}-\mathbf{k}',j}), \tag{3.25}$$

$$g^j_{kk'} = (\hbar/2M\omega_{\mathbf{Q}j})^{1/2}\langle k|\hat{\epsilon}_{\mathbf{Q}j}\cdot\nabla V|k'\rangle, \tag{3.26}$$

where $Q = k' - k$. Here Eq. (3.15) was used and $g^i_{kk'}$ can be called the electron–phonon matrix element. In all cases studied in detail so far, $\alpha^2F(\Omega)$ bears a close resemblance to the phonon density of states $F(\Omega)$, which motivates the awkward notation α^2F.

At this stage it is convenient to depart somewhat from the standard treatments. We wish to be careful in the evaluation of Eq. (3.21), allowing for the full complexities of d-band electronic structure and phonons. It is helpful to separate "energy" and "angular" parts of the k' summation. Let us assume the existence of a set of functions that is complete and orthonormal when integrated on constant-energy surfaces:

$$\sum_k F_J(k)F_{J'}(k)\delta(\epsilon_k - \epsilon) = \delta_{JJ'}N(\epsilon), \tag{3.27}$$

$$N(\epsilon) = \sum_k \delta(\epsilon_k - \epsilon). \tag{3.28}$$

For spherical energy surfaces, these are spherical harmonics: $F_J(k) = Y_{lm}(\hat{k})(4\pi)^{1/2}$. Procedures for general energy surfaces have been described by Allen[38] and Butler and Allen[39] and the functions are called "Fermi surface harmonics." In terms of this set we can write

$$\Sigma_{\text{ep}}(k, i\omega_n) = \sum_J \Sigma_{\text{ep},J}(\epsilon_k, i\omega_n)F_J(k), \tag{3.29}$$

$$G(k, i\omega_n) = \sum_J G_J(\epsilon_k, i\omega_n)F_J(k), \tag{3.30}$$

$$\alpha^2F(k, k', \Omega) = \sum_{JJ'} \alpha^2F(JJ', \epsilon_k, \epsilon_{k'}, \Omega)F_J(k)F_{J'}(k'). \tag{3.31}$$

Inserting these definitions into Eq. (3.21), the exact result is

$$\Sigma_{\text{ep},J}(\epsilon, i\omega_n) = T\sum_{J',\nu}\int_{-\infty}^{\infty} d\epsilon' \frac{N(\epsilon')}{N(0)}\int_0^{\infty} d\Omega\, \alpha^2F(JJ', \epsilon\epsilon', \Omega) \times \left(\frac{2\Omega}{\omega_\nu^2 + \Omega^2}\right)G_{J'}(\epsilon', i\omega_n - i\omega_\nu). \tag{3.32}$$

[38] P. B. Allen, *Phys. Rev. B* **13,** 1416 (1976).

[39] W. H. Butler and P. B. Allen, *in* "Superconductivity in d- and f-Band Metals" (D. H. Douglass, ed.), p. 73. Plenum, New York, 1976.

Now the "angular" part of the k' summation has become a discrete J' sum, and the "energy" part stands in isolation. It is possible to neglect the (ϵ, ϵ') dependence of $N(\epsilon')\alpha^2F(JJ', \epsilon, \epsilon', \Omega)$ due to the following long-winded argument:

1. We wish to know $\Sigma_J(\epsilon, i\omega_n)$ for ϵ, ω_n within $\pm\omega_D$ of $\epsilon_F = 0$.
2. Because $\Omega \leq \omega_D$, only values of ω_ν in the region $|\omega_\nu| \leq \omega_D$ contribute significantly to Eq. (3.32).
3. Thus $G_{J'}(\epsilon', i\omega_n - i\omega_\nu)$ enters Eq. (3.32) only with *small* values of $|\omega_n - \omega_\nu|$.
4. Thus $G_{J'}(\epsilon', i\omega_n - i\omega_\nu)$ varies *rapidly* as ϵ' varies in the range $\pm\omega_D$ and gets small for larger $|\epsilon'|$.
5. Nothing suggests that $N(\epsilon')\alpha^2F(JJ', \epsilon\epsilon', \Omega)$ should vary rapidly with energy ϵ, ϵ' in this range (barring extremely narrow electron bandwidths.)
6. Thus $\alpha^2F(JJ', \epsilon\epsilon', \Omega)$ can be replaced by $\alpha^2F(JJ', 00, \Omega)$, called $\alpha^2F(JJ', \Omega)$, and $N(\epsilon')/N(0)$ can be replaced by 1.

The result is that to a good approximation

$$\Sigma_{\text{ep},J}(i\omega_n) = T\sum_{J',\nu}\int_0^\infty d\Omega\, \alpha^2F(JJ', \Omega)\frac{2\Omega}{\omega_\nu^2 + \Omega^2} \times \int_{-\infty}^{\infty} d\epsilon'\, G_{J'}(\epsilon', i\omega_n - i\omega_\nu). \tag{3.33}$$

Because the right-hand side is now independent of ϵ (which is assumed to be not much larger than ω_D), the argument ϵ has been dropped from the left-hand side. It should be noted that the errors involved in the approximation should have little effect (i.e., on the order of ω_D/ϵ_F) on the ω_n dependence of Σ_{ep}. However, the contribution to Eq. (3.32) from larger values of $|\epsilon'|$ is not accurately accounted for. This contribution is largely independent of $i\omega_n$ and turns out to be irrelevant to superconductivity and other low-temperature properties. However, if a theory for high-T electron–phonon effects is desired (e.g., thermal shifts of energy bands and optical properties), then the approximation made in going from Eq. (3.32) to Eq. (3.33) is not warranted. Equation (3.33) correctly describes all effects which arise from the fact that electrons close to the Fermi surface are sensitive to the time dependence of the lattice vibrations, but omits certain adiabatic effects.

A complete description of the thermodynamics of electrons (including T_c) is contained in the Green's function $G(k, i\omega_n)$ at the "imaginary Matsubara frequencies" $i\omega_n$. However, additional dynamical information is contained in the analytic continuation $G(k, \omega + i\delta)$ to points just above the real-frequency axis, known as the "retarded" Green's function. It is therefore interesting to continue Σ analytically:

$$\Sigma(k, \omega + i\delta) \equiv \Sigma_1(k, \omega) + i\Sigma_2(k, \omega). \tag{3.34}$$

An important role is played by the poles of G, or the zeros of G^{-1}. Suppose that a pole occurs near $\omega = 0$:

$$\begin{aligned} G^{-1}(k, \omega + i\delta) &= \omega - \epsilon_k - \Sigma_1(k, \omega) - i\Sigma_2(k, \omega) \\ &\simeq \omega\left(1 - \left.\frac{\partial\Sigma_1(k, \omega)}{\partial\omega}\right|_{\omega=0}\right) - [\epsilon_k + \Sigma_1(k, 0)] - i\Sigma_2(k, \omega). \end{aligned} \tag{3.35}$$

Then the pole of G occurs at a frequency ω_0 given by

$$\omega_0 = E_k - i/2\tau_k\,, \tag{3.36}$$

$$E_k = \left(1 - \frac{\partial\Sigma_1}{\partial\omega}\right)^{-1} [\epsilon_k + \Sigma_1(k, 0)],$$

$$1/2\tau_k = -\left(1 - \frac{\partial\Sigma_1}{\partial\omega}\right)^{-1} \Sigma_2(k, E_k), \tag{3.37}$$

provided that $1/\tau_k$ is small compared to E_k. For electron–phonon problems at low temperature, $\Sigma_1(k, 0)$ is a small and uninteresting quantity [in fact, zero in the approximation of Eq. (3.33)], whereas $-\partial\Sigma_1/\partial\omega$ is a large (of the order of unity) and very interesting correction, called the mass-enhancement parameter λ_k:

$$\lambda_k \equiv -\partial\Sigma_{1,\mathrm{ep}}(k, \omega)/\partial\omega|_{\omega=0}\,. \tag{3.38}$$

The frequency and temperature dependence of λ_k and $1/\tau_k$ are also interesting and have been studied experimentally in the normal state by various Fermi surface probes. The subject has been reviewed by Grimvall.[28]

We can rewrite Eq. (3.33) in a form which facilitates the analytic continuation,

$$\begin{aligned} \Sigma_{\mathrm{ep},J}(i\omega_n) = \sum_{J'} \int_{-\infty}^{\infty} d\omega' \int_0^{\infty} d\Omega\, \alpha^2 F(JJ', \Omega) \\ \times \left(-\frac{1}{\pi}\int_{-\infty}^{\infty} d\epsilon' \,\mathrm{Im}\, G_{J'}(\epsilon', \omega' + i\delta)\right) I(i\omega_n, \Omega, \omega'), \end{aligned} \tag{3.39}$$

$$I(i\omega_n, \Omega, \omega') \equiv T \sum_{\nu} \frac{2\Omega}{\omega_\nu^2 + \Omega^2} \frac{1}{i\omega_n - i\omega_\nu - \omega'}\,. \tag{3.40}$$

In deriving Eq. (3.39) from Eq. (3.33), a spectral representation has been used for $G_{J'}(\epsilon', i\omega_n - i\omega_\nu)$ analogous to the spectral representation (3.10) for $G(k, i\omega_n)$. By analogy with Eq. (3.12), the spectral function for $G_J(\epsilon', i\omega_{n'})$ is $-(1/\pi)\mathrm{Im}\, G_J(\epsilon', \omega' + i\delta)$. The sum over Matsubara frequencies in the definition (3.40) of I can be performed explicitly, for example, by using the Poisson sum formula.[5,25] The result is

$N(0)g^2/\omega_D$, where g, the electron–phonon matrix element, has a typical size $(\epsilon_F\omega_D)^{1/2}$. Thus λ_k is of order unity, as has been found in experiments such as cyclotron resonance.[40]

The mathematical source of the large λ_k is the singular behavior of Eq. (3.48) when $\omega = \pm\Omega$, which can be traced to the ω' integral in Eq. (3.44). The integration encounters a simple pole at $\omega' = \omega \pm \Omega$, which generates a smooth result unless the sharp edge of the Fermi function $f(\omega')$ in the numerator coincides with the pole (which means that $\omega' = 0$, T small, and $\omega = \pm\Omega$.) Therefore at higher T, when the Fermi function no longer has a sharp edge, the rapid ω dependence of Σ must go away.

Inspecting Eq. (3.44) at high T, the real part involves principal-part integrals which give small answers. The imaginary part, at any temperature,

$$\operatorname{Im} R(\omega + i\delta, \Omega) = -\pi[2N(\Omega) + 1 + f(\omega + \Omega) - f(\omega - \Omega)]. \quad (3.50)$$

This will cause, at high T, a scattering rate $1/2\tau_k$ of order $\pi\lambda_k k_B T$, with only weak ω dependence.

An explicit result for the integral (3.44) for R is derived in Appendix B. The answer, valid for all T, z, and Ω, is

$$R(z, \Omega) = -2\pi i[N(\Omega) + 1/2] + \psi\left(\frac{1}{2} + i\frac{\Omega - z}{2\pi T}\right) - \psi\left(\frac{1}{2} - i\frac{\Omega + z}{2\pi T}\right), \quad (3.51)$$

where $\psi(z)$ is the digamma function, $d \ln \Gamma(z)/dz$. Using the properties of ψ given in Appendix A, it is easy to verify the limits (3.48) and (3.50), and to inspect other cases. For example, at small ω, using Eq. (A.12), R becomes

$$R(\omega + i\delta, \Omega) \cong -2\pi i \operatorname{csch}\left(\frac{\Omega}{T}\right) + \frac{2\omega}{\Omega}\left(\frac{\Omega}{2\pi T}\right) \operatorname{Im} \psi^{(1)}\left(\frac{1}{2} + i\frac{\Omega}{2\pi T}\right) + O(\omega^2), \quad (3.52)$$

where $\psi^{(1)}$ is the derivative of the digamma function. Explicit formulas for λ_k and $1/\tau_k$, as defined in Eqs. (3.38) and (3.37), valid at $\omega = 0$ and all T, are

$$\frac{1}{\tau_k(T)} = \frac{4\pi}{1 + \lambda_k(T)} \int_0^\infty d\Omega\, \alpha^2 F(k, \Omega) \operatorname{csch}\left(\frac{\Omega}{T}\right), \quad (3.53)$$

$$\lambda_k(T) = 2\int_0^\infty \frac{d\Omega}{\Omega}\, \alpha^2 F(k, \Omega) \frac{-\Omega}{2\pi T} \operatorname{Im} \psi^{(1)}\left(\frac{1}{2} + i\frac{\Omega}{2\pi T}\right). \quad (3.54)$$

To derive Eq. (3.49) for λ_k at $T = 0$ we need the large x limit of $(-x) \times \operatorname{Im} \psi^{(1)}(1/2 + ix)$, which is unity. The low-T behavior of $1/\tau_k$ is well known[28] to go as T^3. This follows from Eq. (3.53), which depends only on the small

[40] G. W. Crabtree, W. R. Johanson, S. A. Campbell, D. H. Dye, D. P. Karim, and J. B. Ketterson, *Conf. Ser.—Inst. Phys.* **55**, 79 (1980).

Ω limit of $\alpha^2F(k, \Omega)$. In this limit, acoustic phonons give $F(\Omega) \sim \Omega^2$. The same power law holds for $\alpha^2F(k, \Omega)$, since Eq. (3.26) requires $g_{k,k+Q} \sim |\mathbf{Q}|/\omega_Q{}^{1/2}$ in the small Q limit. At high temperatures ($2\pi T \gg \theta_D$), we need the small x behavior of $(-x)\psi^{(1)}(\frac{1}{2} + ix)$, namely $14\zeta(3)x^2 + O(x^4)$. This shows that the high-temperature mass-enhancement parameter $\lambda_k(T)$ diminishes to zero as

$$\lambda_k(T > \theta_D) \cong 14\zeta(3)\lambda_k\langle\Omega^2\rangle_k/(2\pi T)^2, \tag{3.55}$$

where $\langle\Omega^2\rangle_k$ is a mean-square phonon frequency defined by

$$\lambda_k\langle\Omega^2\rangle_k \equiv 2\int_0^\infty \frac{d\Omega}{\Omega}\, \alpha^2F(k, \Omega)\Omega^2. \tag{3.56}$$

The corresponding high-temperature result for the quasi-particle lifetime follows from Eq. (3.53),

$$1/\tau_k(T > \theta_D) = 2\pi\lambda_k T. \tag{3.57}$$

The temperature behavior of $\lambda_k(T)$ at intermediate temperatures shows initially a rise of the form $T^2 \ln T$, first found by Eliashberg[41] in a theory for specific heat. A detailed discussion of the full temperature range was first given by Grimvall.[42] Experimental confirmation of the effect was first found by Sabo[43] using cyclotron resonance in Zn. The current status of this subject is covered by Grimvall.[28]

4. Qualitative Discussion of Migdal's "Theorem"

The results derived in Section 3 all come from the single Feynman graph in Fig. 1. It is important to the success of the theory that higher order graphs can be neglected, as Migdal's "theorem" asserts. A discussion of this point is important, because the "theorem" can fail in certain cases of interest. Consider then the graphs of Fig. 2. Figure 2a is a higher order process which is implicitly included in the graph of Fig. 1 by use of a self-consistently calculated Green's function in the intermediate state. Figure 2b is not included in Fig. 1. This "vertex correction" graph is the first term omitted from Migdal theory. The justification comes from the observation in Section 3 that the rapid z dependence of $G(z)$ arises from regions of the intermediate ω' sums where the integrand of Eq. (3.44) has a vanishing denominator. Suppose that the intermediate energy ω' is greater in absolute magnitude than ω_D. Then the phonon energy Ω in the denominator becomes unimportant and can be set to zero. The two terms in Eq. (3.44) then have the

[41] G. M. Eliashberg, *Zh. Eksp. Teor. Fiz.* **43,** 1005 (1962); *Sov. Phys.—JETP (Engl. Transl.)* **16,** 780 (1963).

[42] G. Grimvall, *J. Phys. Chem. Solids* **29,** 1221 (1968).

[43] J. J. Sabo, Jr., *Phys. Rev. B* **1,** 1325 (1970).

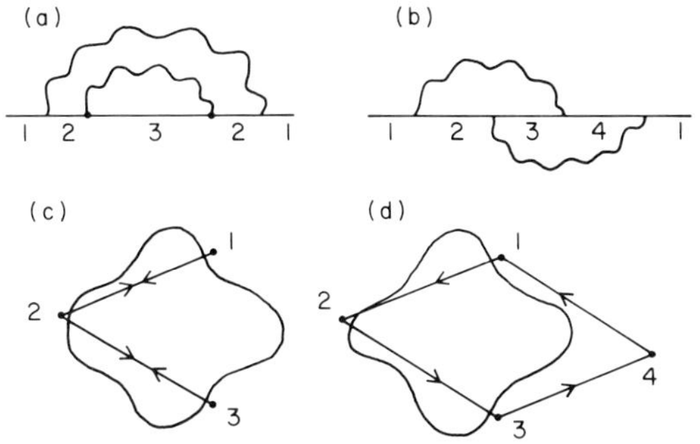

FIG. 2. Feynman graphs (a) and (b) are the next-order corrections to Σ_{ep}. Graph (a) is included in Migdal's theory and graph (b) is omitted. Diagrams (c) and (d) show a schematic Fermi surface and electron states 1–4 which occur in graphs (a) and (b). The smallness of (b) relative to (a) comes from the large denominator associated with state 4 (see text).

same denominator, and the Fermi factors cancel when the terms are added. The contribution to Σ from this part of ω' space is not necessarily smaller in absolute magnitude than $\sim\omega_D$, but it has no interesting z dependence and can be ignored, because it gives only a small and slowly varying shift to all electron energies. The interesting parts of $\Sigma(z)$ come from regions of ω' space, where the result is sensitive to the phonon energy Ω. The interesting processes violate the Born–Oppenheimer adiabatic theorem, which states that electrons are not sensitive to the time dependence of ion motions, because they instantly adjust, as if the ions were statically deformed.

Migdal pointed out that the Born–Oppenheimer-violating terms are unimportant in higher order graphs. This can be seen qualitatively from Fig. 2c and d, which shows schematic Fermi surfaces and particular k states which contribute to the Feynman graphs in Fig. 2a and b. Each of these graphs has three intermediate energy denominators, but in Fig. 2a only two of the three, ϵ_{12} and ϵ_{13}, are distinct, ϵ_{12} occurring twice, whereas in Fig. 2b ϵ_{12}, ϵ_{13}, and ϵ_{14} each occur once. Figure 2c shows a process contributing to Fig. 2a, where both ϵ_{12} and ϵ_{13} are small, suggesting that the corresponding contribution to $\Sigma(z)$ has rapid z dependence and is sensitive to the values of the phonon energies. Figure 2d shows that the corresponding process which contributes to Fig. 2b has one large denominator, ϵ_{14}. This suggests that the corresponding contribution to $\Sigma(z)$, although still strongly z dependent, will be smaller by ω_D/ϵ_F. The probability that all three denominators of the Feynman graph in Fig. 2b will be as small as in the Feynman graph in Fig. 2a is $\sim N(0)\omega_D$. Thus we expect to be able to neglect Fig. 2b relative to Fig. 2a, which is already included in the theory of Section 3. This is Migdal's "theorem." It is sometimes called an "adiabatic" theorem, but this is a misnomer; more accurately, it is a theorem about nonadiabatic processes. Superconductivity clearly arises from nonadiabatic processes (otherwise the isotope effect could not occur). Migdal's theorem underlies the Eliashberg theory of superconductivity.

Figure 2 enables us to see that Migdal's theorem will not hold in two special cases: first, if either phonon has $|\mathbf{Q}|$ small, and second, if the Fermi surface has a one-dimensional topology. Say that $|Q_{23}|$ is small. Then state 3 lies close to state 2, and state 4 close to state 1. The parallelogram of Fig. 2d becomes a long, skinny figure, and all three denominators are small. The Feynman graph in Fig. 2b is no longer small compared to that in Fig. 2a. However, in ordinary metallic superconductors, small $|\mathbf{Q}|$ phonons are a negligible region of phase space and contribute nothing of special significance. An exception would be a low-carrier-density system, such as a doped semiconductor,[44] for example, $SrTiO_3$,[45] or an inversion layer.[46] In such systems, the parameter ω_D/ϵ_F is not necessarily small. Alternatively, we can say that all phonons which span the Fermi surface have small $|\mathbf{Q}|$. Thus Migdal's theorem fails.

A one-dimensional Fermi surface consists of parallel planes at $k_x = \pm k_F$. By orienting the parallelogram of Fig. 2d so that two sides are parallel to the Fermi surface planes, it is easy to make all three denominators small, violating Migdal's theorem. It was once widely believed that the Fermi surface of A15 metals might have one-dimensional features, but this is not supported by recent evidence. However, there are now several quasi-one-dimensional superconductors, as mentioned in Section 1. Special theories are needed for 1d metals that avoid Migdal's approximation. Exact solutions are known for certain 1d models with nonretarded interactions,[47] but these models have no long-range superconducting order, even in the ground state. There is a large literature on the 1d electron–phonon problem, but so far no progress has been made beyond the Migdal level,[47a] and the subject will not be discussed here.

5. Coulomb Effects

There is no small parameter which enables a satisfactory perturbation theory to be constructed for the Coulomb interaction between electrons. Thus Coulomb contributions to the electron self-energy Σ cannot be reliably calculated. Fortunately this is not a serious problem in superconductivity theory. The dominant electron–phonon interaction can be fully accounted

[44] M. L. Cohen, *in* "Superconductivity" (R. D. Parks, ed.), Vol. 1, Chapter 12. Dekker, New York, 1969.

[45] J. F. Schooley, W. R. Hosler, and M. L. Cohen, *Phys. Rev. Lett.* **12,** 474 (1964).

[46] M. J. Kelly and W. Hanke, *Phys. Rev. B* **23,** 112 (1981).

[47] V. J. Emery, *in* "Highly Conducting One-Dimensional Solids" (J. T. Devreese, R. P. Evrard, and V. E. van Doren, eds.), Chapter 6. Plenum, New York, 1979.

[47a] See, however, J. E. Hirsch and E. Fradkin, *Phys. Rev. Lett.* **49,** 402 (1982).

for, as in Section 3, provided that the electron and phonon quasiparticle energies ϵ_k and ω_Q are known and that the matrix elements $g_{kk'}$ can be calculated. Coulomb effects enter in profound ways into ϵ_k, ω_Q, and $g_{kk'}$, but are presumed to be dealt with either by experiment (ω_Q) or band theory (ϵ_k, $g_{kk'}$). There remains a direct Coulomb repulsion between electrons that contributes negatively to the pairing interaction $V_{kk'}$. Fortunately, the Coulomb repulsion is greatly weakened by the fact that in forming Cooper pairs, electrons avoid meeting directly. The second electron is attracted by lattice polarization to the point where the first electron used to be one-quarter of a lattice vibration period earlier. The Coulomb parameter analogous to the electron–phonon parameter λ is called μ:

$$\mu = N(0)\langle\langle V^c_{kk'}\rangle\rangle_{\mathrm{FS}}\,, \tag{5.1}$$

where $V^c_{kk'}$ is the effective screened Coulomb repulsion and μ is of the order 1. Because of retardation, the number which actually enters the theory [e.g., Eq. (1.1)] is μ^*:

$$\mu^* = \mu/[1 + \log(\omega_{\mathrm{el}}/\omega_{\mathrm{D}})], \tag{5.2}$$

where ω_{el} measures Coulombic response frequencies, of order 10 eV. Thus μ^* is of the order 0.1. The algebra by which Eq. (5.2) is derived will be explained in Section 9.

This section briefly describes a model for the Coulomb terms Σ_{C} in the electron self-energy for the normal state. The model *cannot* be justified and is probably not an acceptable approximation to reality. The reasons for discussing it at all are, first, to provide a basis for the algebra in Section 9, where μ^* is introduced, and, second, because the model seems to have some intrinsic interest as a starting point for examining speculations about exciton and plasmon effects in superconductivity. The Coulomb term in the Hamiltonian is

$$\mathscr{H}_{\mathrm{C}} = \sum_{Q,G} \frac{4\pi e^2}{|Q+G|^2}\,\rho^\dagger_{Q+G}\rho_{Q+G}\,, \tag{5.3}$$

$$\rho_{Q+G} \equiv \sum_{kk'} \langle k'|e^{-i(Q+G)\cdot r}|k\rangle c^\dagger_{k'}c_k\,. \tag{5.4}$$

The operator ρ_{Q+G} is the Fourier transform of the electron density operator. As usual, the symbol k is short for $\mathbf{k}n$, and the sum in Eq. (5.4) includes both band-diagonal and nondiagonal contributions.

A few of the Feynman graphs for Σ_{C} are shown in Fig. 3. Figure 3a and b are the Hartree and Fock terms Σ_{H} and Σ_{C}, respectively, and have the forms[48]

[48] In Ref. 22, p. 123ff., it is shown how the Hartree–Fock Schrödinger equation follows from the zero-temperature version of Eqs. (5.5) and (5.6).

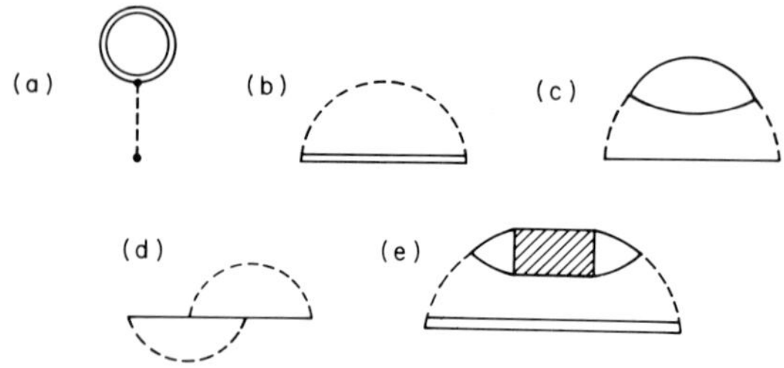

FIG. 3. Feynman graphs (a) and (b) are the direct and exchange parts of Hartree–Fock theory [Eqs. (5.5) and (5.6)]. Graphs (c) and (d) are the second-order corrections to Hartree–Fock theory, and graph (e) when added to (b) is the "screened-exchange" self-energy. Graph (e) is denoted Σ_{SC} [Eq. (5.11)] and contains graph (c) as its lowest-order part.

$$\Sigma_{\mathrm{H}}(k, i\omega_n) = 2 \sum_{Q,G} \frac{4\pi e^2}{G^2} \langle k|e^{-i\mathbf{G}\cdot\mathbf{r}}|k\rangle\langle k-Q|e^{i\mathbf{G}\cdot\mathbf{r}}|k-Q\rangle \times T \sum_{\nu} G(k-Q, i\omega_n - i\omega_\nu), \tag{5.5}$$

$$\Sigma_{\mathrm{F}}(k, i\omega_n) = -\sum_{Q,G} \frac{4\pi e^2}{|Q+G|^2} |\langle k-Q|e^{-i(\mathbf{Q}+\mathbf{G})\cdot r}|k\rangle|^2 \times T \sum_{\nu} G(k-Q, i\omega_n - i\omega_\nu), \tag{5.6}$$

where the extra factor of 2 in Eq. (5.5) comes from a sum over spins. The double lines in Fig. 3a and b indicate that the complete (self-consistent) electron Green's function G is used, rather than G_0. These two pieces of Σ are real and ω_n independent. In order to deal properly with terms like Eqs. (5.5) and (5.6), it becomes essential to recognize that the Green's function $G(k, i\omega_n)$ defined in Eq. (2.6) is implicitly a matrix $G(\mathbf{k}nn'; i\omega_n)$ in band-index space.[34,49] The operators $c_{k\sigma}$ are actually $c_{\mathbf{k}n\sigma}$, and there is no need for the two operators c and $c\dagger$ to have the same band index n. As long as self-energies Σ are small compared to energies ϵ_k (as is true for electron–phonon problems), it is reasonable to neglect interband effects, that is, force $n' = n$. For Coulomb problems, Σ is not small. Fortunately, we can leave these complexities to the band theorists, who are in principle supposed to solve the Dyson equation correctly and provide us with basis functions $\psi_{kn}(r)$ which diagonalize $\mathscr{H}_0 + \Sigma_{\mathrm{C}}$. Sections 8 and 9 examine the *difference* between Σ_{C} in the normal and in the superconducting states. This difference will be small, permitting the band-diagonal approximation.

The only second-order Feynman graphs which are not implicitly included

[49] J. M. Luttinger, *Phys. Rev.* **119,** 1153 (1960).

in Fig. 3a and b are shown in Fig. 3c and d. Having no analog of Migdal's theorem, there is no reason to believe that Fig. 3d is less important than Fig. 3c. Without justification, the model to be discussed drops Fig. 3d, but keeps an infinite series of graphs shown in Fig. 3e, of which Fig. 3c is the simplest example. These graphs contain the screening by electronic polarization of the Coulomb interaction of the Fock graph, Fig. 3b, and will be denoted Σ_{SC}.

It is convenient to define an electron–hole pair propagator or polarization propagator analogous to the Green's functions (2.6) and (3.1):

$$\chi_{GG'}(Q\tau) \equiv -\langle T_\tau \rho_{Q+G}(\tau)\rho^\dagger_{Q+G'}(0)\rangle \tag{5.7}$$

$$= T \sum_\nu e^{-i\omega_\nu \tau}\chi_{GG'}(Q, i\omega_\nu). \tag{5.8}$$

The analytic continuation of $\chi_{GG'}(Q, i\omega_\nu)$ to frequencies $\omega + i\delta$ just above the real axis yields the electron density–density retarded susceptibility which gives the charge density response at wave vector $Q + G$ to an external potential with Fourier components $(Q + G', \omega)$:

$$\delta\rho_{\text{ind}}(Q + G, t) = \sum_{G'} \chi_{GG'}(Q, \omega)\delta V_{\text{ext}}(Q + G')e^{-i\omega t}. \tag{5.9}$$

This connection can be proved using the retarded-commutator formula for $\chi(Q, \omega)$ and by working out spectral representations.[50] The nonlocal dielectric function[51,52] $\epsilon(r, r')$ has a simple relation to χ:

$$\epsilon^{-1}(Q + G, Q + G', \omega) = \delta_{GG'} - \frac{4\pi e^2}{|Q + G|^2}\chi_{GG'}(Q, \omega). \tag{5.10}$$

The polarization χ is analogous to the phonon Green's function (3.1), and Σ_{SC} is analogous to Σ_{ep}, Eq. (3.20):

$$\begin{aligned}\Sigma_{SC}(k, i\omega_n) = -T \sum_{QGG'} \sum_\nu &\left\langle k \left| \frac{4\pi e^2}{|Q + G|^2} e^{+i(Q+G')\cdot r} \right| k - Q \right\rangle \\ &\times \chi_{GG'}(k - k', i\omega_\nu)\left\langle k - Q \left| \frac{4\pi e^2}{|Q + G'|^2} e^{-i(Q+G')\cdot r} \right| k \right\rangle \\ &\times G(k - Q, i\omega_n - i\omega_\nu).\end{aligned} \tag{5.11}$$

Just as in the phonon problem, Eqs. (3.13) and (3.22), it is possible to define an electron Coulomb spectral function $\mathcal{S}(k, k', \Omega)$:

[50] S. Doniach and E. H. Sondheimer, "Green's Functions for Solid State Physicists," Appendix 2. Benjamin, Reading, Massachusetts, 1974.

[51] S. L. Adler, *Phys. Rev.* **126,** 413 (1962).

[52] N. Wiser, *Phys. Rev.* **129,** 62 (1963).

$$\mathcal{S}(k, k', \Omega) \equiv \left(-\frac{1}{\pi}\right) N(0) \operatorname{Im}\left(\sum_{GG'} \left\langle k \left| \frac{4\pi e^2}{|Q+G|^2} e^{+i(Q+G)\cdot r} \right| k' \right\rangle \right.$$
$$\left. \times \chi_{GG'}(Q, \omega + i\delta) \left\langle k' \left| \frac{4\pi e^2}{|Q+G'|^2} e^{-i(Q+G')\cdot r} \right| k \right\rangle \right), \quad (5.12)$$

where Q is $k - k'$ reduced to the first Brillouin zone. The simplest sensible approximation to χ is the random phase approximation (RPA).[51,52] After tedious manipulations, it can be shown that this spectral function has quite a simple form in RPA theory:

$$\mathcal{S}_{\mathrm{RPA}}(kk', \Omega) = N(0) \sum_{k_1, k_2} |\langle k', k_2 | v_s(r, r', \Omega) | k, k_1 \rangle|^2$$
$$\times [f(k_1) - f(k_2)] \delta(\epsilon_{k_1} - \epsilon_{k_2} + \Omega), \quad (5.13)$$

where $v_s(r, r', \Omega)$ is the screened Coulomb interaction, that is, $e^2/|r - r'|$ screened by $\epsilon_{\mathrm{RPA}}^{-1}(r, r', \Omega)$. Unfortunately this latter quantity is extremely hard to calculate for real materials, although significant progress has been made.[53] In terms of this spectral function, the self-energy (5.11) becomes

$$\Sigma_{\mathrm{SC}}(k, i\omega_n) = T \sum_{k', \nu} \int_0^\infty d\Omega \, \frac{\mathcal{S}(k, k', \Omega)}{N(0)} \, \frac{2\Omega}{\omega_\nu^2 + \Omega^2} \, G(k', i\omega_n - i\omega_\nu), \quad (5.14)$$

a formula entirely analogous to Eq. (3.21) for Σ_{ep}. It is also possible to separate the "angular" and "energy" parts of the k' integral, as in Eqs. (3.31) and (3.32):

$$\mathcal{S}(kk'\Omega) = \sum_{JJ'} \mathcal{S}(JJ', \epsilon\epsilon', \Omega) F_J(k) F_{J'}(k'), \quad (5.15)$$

$$\Sigma_{\mathrm{SC},J}(\epsilon, i\omega_n) = T \sum_{J'} \int_{-\infty}^{\infty} d\epsilon' \, \frac{N(\epsilon')}{N(0)}$$
$$\times \int_0^\infty d\Omega \, \mathcal{S}(JJ', \epsilon\epsilon', \Omega) \, \frac{2\Omega}{\omega_\nu^2 + \Omega^2} \, G(\epsilon', i\omega_n - i\omega_\nu). \quad (5.16)$$

This formula is identical to Eq. (3.32), except that $\alpha^2 F$ is replaced by $\mathcal{S}$. At this stage it is no longer a simple process. Unlike $\alpha^2 F$, $\mathcal{S}$ has no upper characteristic frequency or cutoff which is small on the scale of ϵ_{F}. We can expect $\mathcal{S}$ to remain large for energies up to perhaps 100 eV in d-band elements. Thus the approximations which converted Eq. (3.32) into Eq. (3.33) have no analog here. It is similarly unclear to what extent the band quasi-particle approximation holds for G in Eq. (3.32). We have reached the end of the development and must leave the story unfinished. Further developments in the special case of a free-electron gas have been described

[53] W. Hanke, *Adv. Phys.* **27**, 287 (1978).

by Hedin and Lundqvist,[54] but the extension to band electrons has only recently been attempted for the first time.[55]

For later reference we now write the Fock self-energy, Eq. (5.6), in the form of Eq. (5.16), with energy and angle dependence separated. First define the bare Coulomb matrix element $V_{JJ'}$:

$$\sum_{\mathbf{G}} \frac{4\pi e^2}{|\mathbf{k}-\mathbf{k}'+\mathbf{G}|^2} |\langle k'|e^{-i(\mathbf{k}-\mathbf{k}'+\mathbf{G})\cdot\mathbf{r}}|k\rangle|^2 = \sum_{JJ'} V_{JJ'}(\epsilon_k, \epsilon_{k'})F_J(k)F_{J'}(k'). \quad (5.17)$$

With this definition, Eq. (5.6) can be written as

$$\Sigma_{\mathrm{F},J}(\epsilon, i\omega_n) = -T\sum_{J',\nu} \int_{-\infty}^{\infty} d\epsilon'\, N(\epsilon')V_{JJ'}(\epsilon, \epsilon')G_{J'}(\epsilon', i\omega_n - i\omega_\nu). \quad (5.18)$$

6. Impurity Effects

As a preparation for dealing with impurity effects in superconductors, this section summarizes the relevant aspects of impurity effects in the normal state. Several notable efforts[56,57] have recently been made to calculate T_c in concentrated alloys and highly disordered systems, but no systematic general approach is available. Therefore only dilute impurities will be discussed. It will also be assumed that no bound state is created by the impurity near the Fermi level. Then perturbation theory looks well behaved, with the concentration n_i of impurities as a small parameter. There is an important difference between nonmagnetic (N) and magnetic (P for paramagnetic) impurities. The subscript i will assume the value N or P. The terms in the Hamiltonian are

$$\mathscr{H}_{\mathrm{N}} = \sum_{\mu}\sum_{kk'} \langle k'|V_{\mathrm{N}}|k\rangle S_{\mathrm{N}}(k'-k)c^{\dagger}_{k'\mu}c_{k\mu}, \quad (6.1)$$

$$\mathscr{H}_{\mathrm{P}} = \sum_{\mu,\nu}\sum_{kk'} \langle k'|V_{\mathrm{P}}|k\rangle \mathbf{S}_{\mathrm{P}}(k'-k)\cdot(c^{\dagger}_{k'\mu}\boldsymbol{\sigma}_{\mu\nu}c_{k\nu}), \quad (6.2)$$

$$S_{\mathrm{N}}(Q) = \sum_a e^{-i\mathbf{Q}\cdot\mathbf{R}_a}, \quad (6.3)$$

$$\mathbf{S}_{\mathrm{P}}(Q) = \sum_a \mathbf{S}_a e^{-i\mathbf{Q}\cdot\mathbf{R}_a}, \quad (6.4)$$

where R_a is the impurity location, $\mathbf{S}_a$ the impurity spin, $S_i(Q)$ impurity structure factors, and V_i the alteration of the crystal potential due to the

[54] L. Hedin and S. Lundqvist, *Solid State Phys.* **23,** 1 (1969).

[55] G. Strinati, H. J. Mattausch, and W. Hanke, *Phys. Rev. Lett.* **45,** 290 (1980).

[56] B. L. Gyorffy, A. Pindor, and W. L. Temmerman, *Phys. Rev. Lett.* **43,** 1343 (1979).

[57] G. Zwicknagl, *Z. Phys. B* **40,** 23, 31 (1980).

impurity, and μ and ν run over the spin orientations. Magnetic impurities can cause spin-flip scattering of conduction electrons. The structure factors have the following average properties:

$$\langle S_{\mathrm{N}}(Q)\rangle = n_{\mathrm{N}}\delta_{\mathbf{Q},0}\,, \tag{6.5}$$

$$\langle S_{\mathrm{N}}(Q)S_{\mathrm{N}}(-Q')\rangle = n_{\mathrm{N}}\delta_{QQ'} + \langle S_{\mathrm{N}}(Q)\rangle\langle S_{\mathrm{N}}(-Q')\rangle, \tag{6.6}$$

$$\langle \mathbf{S}_{\mathrm{P}}(Q)\rangle = 0, \tag{6.7}$$

$$\langle S_{\mathrm{P}}(Q)_{\alpha}S_{\mathrm{P}}(-Q')_{\beta}\rangle = n_{\mathrm{P}}\,\frac{S(S+1)}{3}\,\delta_{\alpha\beta}\delta_{QQ'}\,, \tag{6.8}$$

where S is the magnitude of the spin $\mathbf{S}_a$ and α and β are Cartesian components of $\mathbf{S}_a$.

The graphical perturbation theory[18,50] for the Green's function, averaged over impurity positions, has peculiar similarities and differences as compared with the theory of $\mathscr{H}_{\mathrm{ep}}$ and $\mathscr{H}_{\mathrm{C}}$. Some low-order graphs are shown in Fig. 4. Figure 4a vanishes for magnetic impurities, and for nonmagnetic ones gives

$$\Sigma_{\mathrm{N}}^{(\mathrm{a})}(k, i\omega_n) = n_{\mathrm{N}}\langle \mathbf{k}n|V_{\mathrm{N}}|\mathbf{k}n'\rangle. \tag{6.9}$$

In a free-electron gas, this is an uninteresting constant shift of all the electron energies. In real solids, there are interband matrix elements, and the Green's function is not diagonal in band index.[49] The best procedure is to get rid of this term by recalculating the energy bands, using the potential $V + n_{\mathrm{N}}V_{\mathrm{N}}$ at each atom site in place of the unperturbed electron–atom potential V. This is the "virtual crystal approximation."[58] For dilute impurities, only a small change of band structure is expected. We neglect non-band-diagonal effects in higher order graphs on the assumption that the self-energies are small.

Figure 4b contributes a finite amount for both magnetic and nonmagnetic cases,

$$\Sigma_{\mathrm{N}}^{(\mathrm{b})}(k, i\omega_n) = n_{\mathrm{N}}\sum_{k'} |\langle k|V_{\mathrm{N}}|k'\rangle|^2 G(k', i\omega_n), \tag{6.10}$$

$$\Sigma_{\mathrm{P}}^{(\mathrm{b})}(k, i\omega_n) = n_{\mathrm{P}}S(S+1)\sum_{k'} |\langle k|V_{\mathrm{P}}|k'\rangle|^2 G(k', i\omega_n). \tag{6.11}$$

Unlike the electron–phonon and Coulomb problems, there is no sum over intermediate frequencies $i\omega_\nu$. The reason is that a nonmagnetic impurity is time independent, so it makes no difference how much time elapses between the two scattering events. The same would be true for phonon scattering if the frequencies ω_Q were approximated by zero. In this case, $\alpha^2F(\Omega)$ would be $\sim\Omega\delta(\Omega)$ and only the term $i\omega_\nu = 0$ would contribute to Eq. (3.21).

[58] H. Ehrenreich and L. M. Schwartz, *Solid State Phys.* **31,** 149 (1976).

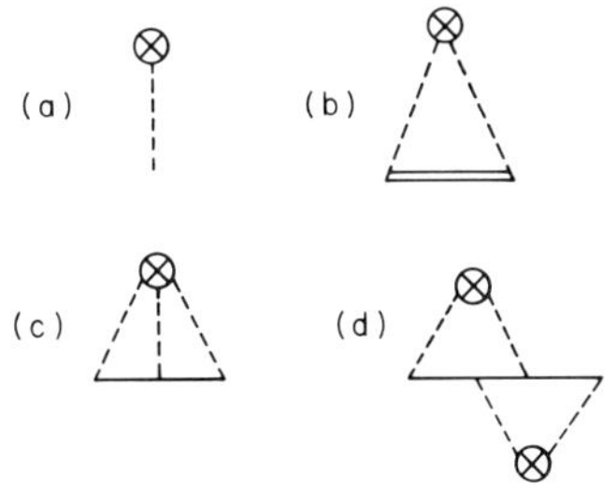

FIG. 4. Impurity graphs contributing to Σ_i (see text). Graph (b) corresponds to Eqs. (6.10) and (6.11).

A paramagnetic impurity is actually time dependent, because its spin occasionally changes by ± 1 owing to electron scattering. However, the time-dependent aspect does not enter until third order, Fig. 4c, which gives the Kondo divergence[59] as $T \to 0$. For nonmagnetic impurities, Fig. 4c and higher order graphs sum to a result which looks like Eq. (6.10), except that V_N is replaced by a t matrix. We therefore redefine V_N to be a t matrix. For magnetic impurities we do the same thing and ignore the Kondo problems which occur at higher orders. The reason is that the important effects in T_c come already in second order, Fig. 4b, which affects up-spin electrons differently from down-spin electrons. The Kondo divergences are not severe at $T = T_c$ unless T_c is quite low.

The graph of Fig. 4d is the first of an infinite set of graphs describing correlated scattering from more than one impurity. Such effects are small when n_i is small. However, these graphs also have an interesting pathology, as they contain the phenomenon of Anderson localization,[60] which, like the Kondo problem, requires a nonperturbative solution. In three dimensions, dilute impurities are safely treated by ignoring this effect, but in lower dimensions[61] it is possible that at very low temperatures localization-related effects can begin to appear even when impurities are dilute.[61a] We shall ignore this and assume that Eqs. (6.10) and (6.11) are an adequate answer. The superscript (b) will thus be dropped.

For later reference, we separate energy and angle variables in Σ_N and Σ_P. The following definitions are needed:

$$\pi N(0) n_N |\langle k'|V_N|k\rangle|^2 = \sum_{JJ'} \gamma_N(JJ', \epsilon\epsilon') F_J(k) F_{J'}(k'), \quad (6.12)$$

$$\pi N(0) n_P S(S+1) |\langle k'|V_P|k\rangle|^2 = \sum_{JJ'} \gamma_P(JJ', \epsilon\epsilon') F_J(k) F_{J'}(k'), \quad (6.13)$$

[59] J. Kondo, *Solid State Phys.* **23,** 183 (1969).

[60] P. W. Anderson, *Phys. Rev.* **109,** 1492 (1958).

[61] E. Abrahams, P. W. Anderson, D. C. Licciardello, and T. V. Ramakrishnan, *Phys. Rev. Lett.* **42,** 673 (1979).

[61a] S. Maekawa and H. Fukuyama, *J. Phys. Soc. Jpn.* **51,** 1380 (1982).

where γ_{N} and γ_{P} are scattering rates related to $1/2\tau$. Then the self-energies become

$$\Sigma_{\mathrm{N},J}(\epsilon, i\omega_n) = \frac{1}{\pi N(0)} \sum_{J'} \int_{-\infty}^{\infty} d\epsilon' \, N(\epsilon')\gamma_{\mathrm{N}}(JJ', \epsilon\epsilon')G_{J'}(\epsilon', i\omega_n), \qquad (6.14)$$

$$\Sigma_{\mathrm{P},J}(\epsilon, i\omega_n) = \frac{1}{\pi N(0)} \sum_{J'} \int_{-\infty}^{\infty} d\epsilon' \, N(\epsilon')\gamma_{\mathrm{P}}(JJ', \epsilon\epsilon')G_{J'}(\epsilon', i\omega_n). \qquad (6.15)$$

If we make the noninteracting approximation (3.42) for the Green's function $G_{J'}$ and neglect the (ϵ, ϵ') dependence of $N(\epsilon')$, γ_{N} and γ_{P}, then Eqs. (6.14) and (6.15) can be evaluated with the help of Eqs. (B.1)–(B.3) of Appendix B:

$$\Sigma_{(\mathrm{N,P})J} = -i\gamma_{J0}^{(\mathrm{N,P})} \operatorname{sign} \omega_n , \qquad (6.16)$$

$$\gamma_{JJ'}^{(\mathrm{N,P})} \equiv \gamma_{(\mathrm{N,P})}(JJ', 00). \qquad (6.17)$$

When analytically continued to ω just above the real axis, sign ω_n becomes 1. These formulas can then be written in the standard form

$$\Sigma_{(\mathrm{N,P})}(k, \omega) = -i/2\tau_k^{(\mathrm{N,P})} , \qquad (6.18)$$

$$1/2\tau_k^{(\mathrm{N,P})} \equiv \sum_J \gamma_{(\mathrm{N,P})}(J0, 00)F_J(k). \qquad (6.19)$$

The impurity self-energy simply broadens the electron quasi-particle by the impurity scattering rate. The reason for the factor of ½ is that $G(k, t) \sim e^{-t/2\tau}$ gives the amplitude decay; the lifetime τ is for probability decay, related to $|G(k, t)|^2$.

III. The Superconducting State

7. Nambu Matrix Notation

In Section 2 BCS theory was derived by Gor'kov's methods.[19] Although the derivation used equations of motion, a diagram method works equally well. The Dyson equations corresponding to Eqs. (2.18) and (2.19) are shown in Fig. 5. The double lines are renormalized Green's functions. The "normal" Green's functions G have two arrows pointing in the same direction, while the "anomalous" Green's functions F have two arrows pointing in opposite directions. The meaning of the notation is the following: The left-hand arrow of a line segment points left or right, depending on whether $c\dagger$ or c appears for c^b on the right of the expression $-\langle T_\tau c^a(\tau)c^b(0)\rangle$, which defines the Green's function; the right-hand arrow uses the opposite convention for the c^a on the left of the definition. A pair interaction V is denoted by a dotted line, which conserves the directions

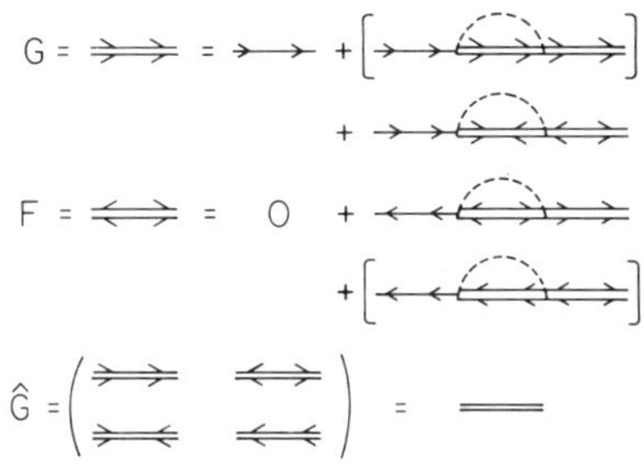

FIG. 5. The graphical version of the Dyson equation which gives BCS theory. The pieces enclosed in square brackets are omitted from Eqs. (2.18) and (2.19).

of the arrow. The bracketed terms of Fig. 5 are omitted in Section 2, because they contain effects assumed to be already in the bare Green's function G_0. Eliashberg theory[3] generalizes BCS theory to incorporate the Migdal theory[2] of the time-dependent electron–phonon interaction. This is done simply by replacing the dotted line for the pair interaction V in Fig. 5 by the wiggly line of Fig. 1 for the phonon propagator and the filled circles for the electron–phonon matrix element. The bracketed terms are no longer omitted.

Nambu[62] invented a condensed notation which has been widely used to simplify the algebra and graphs of Fig. 5. He introduced a two-component field operator Ψ_k,

$$\Psi_k = \begin{pmatrix} c_{k\uparrow} \\ c^{\dagger}_{-k\downarrow} \end{pmatrix}, \qquad \psi^{\dagger}_k = (c^{\dagger}_{k\uparrow}, c_{-k\downarrow}). \tag{7.1}$$

The generalized Green's function $\hat{G}$ is a 2×2 matrix, which is depicted schematically in Fig. 5c:

$$\begin{aligned} \hat{G}(k, \tau) &= -\langle T_\tau \Psi_k(\tau) \Psi^{\dagger}_k(0) \rangle \\ &= -\begin{pmatrix} \langle T_\tau c_{k\uparrow}(\tau) c^{\dagger}_{k\uparrow}(0) \rangle & \langle T_\tau c_{k\uparrow}(\tau) c_{-k\downarrow}(0) \rangle \\ \langle T_\tau c^{\dagger}_{-k\downarrow}(\tau) c^{\dagger}_{k\uparrow}(0) \rangle & \langle T_\tau c^{\dagger}_{-k\downarrow}(\tau) c_{-k\downarrow}(0) \rangle \end{pmatrix}, \end{aligned} \tag{7.2}$$

$$\hat{G}(k, \tau) = T \sum_{i\omega_n} e^{-i\omega_n \tau} \hat{G}(k, i\omega_n). \tag{7.3}$$

The upper left element is the "normal" G of Eq. (2.6). The upper right and lower left (off-diagonal) elements are Gor'kov's F and $\bar{F}$, respectively, Eqs. (2.15) and (2.16). For noninteracting band electrons, the off-diagonal elements vanish, and the (1, 1) element has the usual form $(i\omega_n - \epsilon_k)^{-1}$. It is easy to show (using the properties of the T_τ operator and assuming no spin-dependent forces) that

[62] Y. Nambu, *Phys. Rev.* **117**, 648 (1960).

$$\begin{aligned} G_{22}(k, \tau) &= -G_{11}(-k, -\tau), \\ G_{22}(k, i\omega_n) &= -G_{11}(-k, -\omega_n). \end{aligned} \tag{7.4}$$

From this it follows that for band electrons

$$\hat{G}_0(k, i\omega_n) = \begin{pmatrix} (i\omega_n - \epsilon_k)^{-1} & 0 \\ 0 & (i\omega_n + \epsilon_k)^{-1} \end{pmatrix}. \tag{7.5}$$

It is convenient to introduce the Pauli matrices

$$\begin{aligned} \hat{\tau}_0 &= \begin{pmatrix} 1 & 0 \\ 0 & 1 \end{pmatrix}, & \hat{\tau}_1 &= \begin{pmatrix} 0 & 1 \\ 1 & 0 \end{pmatrix}, \\ \hat{\tau}_2 &= \begin{pmatrix} 0 & -i \\ i & 0 \end{pmatrix}, & \hat{\tau}_3 &= \begin{pmatrix} 1 & 0 \\ 0 & -1 \end{pmatrix}. \end{aligned} \tag{7.6}$$

Then Eq. (7.5) can be written as

$$\begin{aligned} \hat{G}_0(k, i\omega_n) &= (i\omega_n \hat{\tau}_0 - \epsilon_k \hat{\tau}_3)^{-1} \\ &= (i\omega_n \hat{\tau}_0 + \epsilon_k \hat{\tau}_3)/[(i\omega_n)^2 - \epsilon_k^2]. \end{aligned} \tag{7.7}$$

Just as in the normal state, the self-energy is defined by Dyson's equation,

$$\hat{G}(k, i\omega_n)^{-1} = \hat{G}_0(k, i\omega_n)^{-1} - \hat{\Sigma}(k, i\omega_n). \tag{7.8}$$

This formula encompasses Fig. 5a and b. In this notation, the most general form that the self-energy could take is

$$\begin{aligned} &\hat{\Sigma}(k, i\omega_n) \\ &\quad = i\omega_n[1 - Z(k, i\omega_n)]\hat{\tau}_0 + \chi(k, i\omega_n)\hat{\tau}_3 + \phi(k, i\omega_n)\hat{\tau}_1 + \bar{\phi}(k, i\omega_n)\hat{\tau}_2, \end{aligned} \tag{7.9}$$

where Z, χ, ϕ, $\bar{\phi}$ are four independent and, so far, arbitrary functions. The corresponding Green's function is

$$\hat{G}^{-1}(k, i\omega_n) = i\omega_n Z\hat{\tau}_0 - (\epsilon_k + \chi)\hat{\tau}_3 - \phi\hat{\tau}_1 - \bar{\phi}\hat{\tau}_2. \tag{7.10}$$

This matrix can be inverted, yielding

$$\hat{G}(k, i\omega_n) = [i\omega_n Z\hat{\tau}_0 + (\epsilon_k + \chi)\hat{\tau}_3 + \phi\hat{\tau}_1 + \bar{\phi}\hat{\tau}_2]/\det(\hat{G}^{-1}), \tag{7.11}$$

$$\det(\hat{G}^{-1}) = (i\omega_n Z)^2 - (\epsilon_k + \chi)^2 - \phi^2 - \bar{\phi}^2. \tag{7.12}$$

In the normal state, since $\hat{G}$ is diagonal, it is clear that ϕ and $\bar{\phi}$ must both vanish, and Z and χ must be determined by the normal-state self-energy. It follows from Eq. (7.4) that the coefficient of $\hat{\tau}_0$ in Eq. (7.11) is an odd function of $i\omega_n$, while the coefficient of $\hat{\tau}_3$ is even. Thus the precise identification of Z and χ in the *normal state* is

$$i\omega_n[1 - Z(k, i\omega_n)] = \tfrac{1}{2}[\Sigma(k, i\omega_n) - \Sigma(k, -i\omega_n)], \tag{7.13}$$

$$\chi(k, i\omega_n) = \tfrac{1}{2}[\Sigma(k, i\omega_n) + \Sigma(k, -i\omega_n)], \tag{7.14}$$

where it has been assumed that everything is even in k. Thus Z and χ are both even functions of $i\omega_n$. To maintain property (7.4) in the superconducting state, it is necessary for $\phi^2 + \bar{\phi}^2$ also to be even in $i\omega_n$.

The theory of the superconductivity state now simply consists of reevaluating the graphs of Figs. 1, 3, and 4, using the definitions (7.3) and (7.8) of $\hat{G}$ and $\hat{\Sigma}$. It is first necessary to rewrite the Hamiltonian in terms of the operators Ψ_k instead of c_k:

$$\mathscr{H}_0 = \sum_k \epsilon_k(\Psi_k^\dagger, \hat{\tau}_3\Psi_k), \tag{7.15}$$

$$\mathscr{H}_{\text{ep}} = \sum_{kk'} \langle k'|\nabla_\alpha V|k\rangle(\Psi_{k'}^\dagger, \hat{\tau}_3\Psi_k)u_{k-k'\alpha}\,. \tag{7.16}$$

$\mathscr{H}_\text{C}$ = Eq. (5.3), but ρ_{Q+G} is redefined as

$$\rho_{Q+G} = \sum_{kk'} \langle k'|e^{-i(Q+G)\cdot r}|k\rangle(\Psi_{k'}^\dagger, \hat{\tau}_3\Psi_k), \tag{7.17}$$

$$\mathscr{H}_\text{N} = \sum_{kk'} \langle k'|\nabla_\text{N}|k\rangle S_\text{N}(k' - k)(\Psi_{k'}, \hat{\tau}_3\Psi_k), \tag{7.18}$$

$$\mathscr{H}_\text{P} = \sum_{kk'} \langle k'|V_\text{P}|k\rangle[S_{\text{P}z}(k' - k)(\Psi_{k'}^\dagger, \hat{\tau}_0\Psi_k)$$

$$+ S_{\text{P}+}(k' - k)(\Psi_{k'}^\dagger, \hat{\tau}_-\Psi_k) + S_{\text{P}-}(k' - k)(\Psi_{k'}^\dagger, \hat{\tau}_+\Psi_k)]. \tag{7.19}$$

The paramagnetic impurity Hamiltonian is the only one which involves Pauli matrices other than $\hat{\tau}_3$. The matrices $\hat{\tau}_\pm$ are $\tfrac{1}{2}(\hat{\tau}_1 \pm i\hat{\tau}_2)$, with similar definitions for $S_{\text{P}\pm}(Q)$. The different form of $\mathscr{H}_\text{P}$ is connected with the fact that it is not time-reversal invariant. This leads to the pair-breaking effect and gapless superconductivity.[63]

The Feynman–Dyson perturbation series for $\hat{G}$ turns out to be identical to that for G. The only practical difference is that $\hat{G}$ is a matrix and that factors of $\hat{\tau}_i$ (i = 0, 1, 2, 3) enter the expressions attached to the various interaction matrix elements, as in Eqs. (7.15)–(7.19). This means that the formal results (3.21) for Σ_ep, (5.6) for Σ_F, (5.14) for Σ_SC, and (6.10) and (6.11) for Σ_N and Σ_P remain valid for $\hat{\Sigma}$ as well, provided that G on the right-hand side is replaced by $\hat{\tau}_3\hat{G}\hat{\tau}_3$ in $\hat{\Sigma}_\text{ep}$, $\hat{\Sigma}_\text{F}$, $\hat{\Sigma}_\text{SC}$, and $\hat{\Sigma}_\text{N}$, and by $\hat{\tau}_0\hat{G}\hat{\tau}_0$ in $\hat{\Sigma}_\text{P}$.

[63] A. A. Abrikosov and L. P. Gor'kov, *Zh. Eksp. Teor. Fiz.* **39,** 1781 (1960); *Sov. Phys.—JETP (Engl. Transl.)* **12,** 1243 (1961).

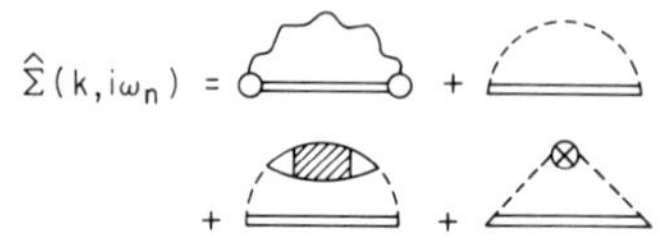

FIG. 6. The Feynman graphs kept in the discussion of Eliashberg theory given here [Eqs. (8.2)–(8.6)].

8. Complete Self-Energy

As just described, Eliashberg theory consists of generalizing the normal-state self-energy Σ. The Feynman graphs which are kept are summarized in Fig. 6. Additional Feynman graphs are needed in some cases, for example, for a more realistic treatment of the Coulomb problem or to include concentrated alloys or Kondo effects. However, the most important graph is the electron–phonon graph, and Migdal's theorem operates in the superconducting state, just as in the normal state, to justify omitting higher order phonon graphs.

The different pieces of the superconducting self-energy $\hat{\Sigma}$ will now be given. Just as in the normal state, we separate "angle" and "energy" dependence, using the complete set of "angular" functions $F_J(k)$ [see Section 3, Eq. (3.27)] and we can write

$$\hat{\Sigma}(k, i\omega_n) = \sum_J \hat{\Sigma}_J(\epsilon_k, i\omega_n) F_J(k). \tag{8.1}$$

A similar equation defines $\hat{G}_J(\epsilon_k, i\omega_n)$. The various species of $\hat{\Sigma}_J$ are

$$\hat{\Sigma}_{\mathrm{ep},J}(\epsilon, i\omega_n) = T \sum_{J',\nu} \int_{-\infty}^{\infty} d\epsilon' \frac{N(\epsilon')}{N(0)} \times \int_0^{\infty} d\Omega\, \alpha^2 F(JJ', \epsilon\epsilon', \Omega) \frac{2\Omega}{\omega_\nu^2 + \Omega^2} \hat{\tau}_3 \hat{G}_{J'}(\epsilon', i\omega_n - i\omega_\nu)\hat{\tau}_3 , \tag{8.2}$$

$$\hat{\Sigma}_{\mathrm{F},J}(\epsilon, i\omega_n) = -T \sum_{J',\nu} \int_{-\infty}^{\infty} d\epsilon'\, N(\epsilon') V_{JJ'}(\epsilon, \epsilon') \times \hat{\tau}_3[\hat{G}_{J'}(\epsilon', i\omega_n - i\omega_\nu) - \hat{G}_{J'}^{\mathrm{N}}(\epsilon', i\omega_n - i\omega_\nu)]\hat{\tau}_3 , \tag{8.3}$$

$$\hat{\Sigma}_{\mathrm{SC},J}(\epsilon, i\omega_n) = T \sum_{J',\nu} \int_{-\infty}^{\infty} d\epsilon' \frac{N(\epsilon')}{N(0)} \int_0^{\infty} d\Omega\, \mathcal{S}(JJ', \epsilon\epsilon', \Omega) \times \frac{2\Omega}{\omega_\nu^2 + \Omega^2} \hat{\tau}_3[\hat{G}_{J'}(\epsilon', i\omega_n - i\omega_\nu) - \hat{G}_{J'}^{\mathrm{N}}(\epsilon', i\omega_n - i\omega_\nu)]\hat{\tau}_3 , \tag{8.4}$$

$$\hat{\Sigma}_{\mathrm{N},J}(\epsilon, i\omega_n) = \frac{1}{\pi N(0)} \sum_{J'} \int_{-\infty}^{\infty} d\epsilon'\, N(\epsilon') \gamma_{\mathrm{N}}(JJ', \epsilon\epsilon') \hat{\tau}_3 \hat{G}_{J'}(\epsilon', i\omega_n) \hat{\tau}_3 , \tag{8.5}$$

$$\hat{\Sigma}_{\mathrm{P},J}(\epsilon, i\omega_n) = \frac{1}{\pi N(0)} \sum_{J'} \int_{-\infty}^{\infty} d\epsilon' \, N(\epsilon')\gamma_{\mathrm{P}}(JJ', \epsilon\epsilon')\hat{\tau}_0 G_{J'}(\epsilon', i\omega_n)\hat{\tau}_0 \, . \quad (8.6)$$

Equations (8.2)–(8.5) are direct generalizations of the corresponding normal state formulas (3.32), (5.18), (5.16), (6.14), and (6.15). Only Eqs. (8.3) and (8.4) contain any surprise, namely, the subtraction of the normal-state Green's function $\hat{G}^{\mathrm{N}}$. The reason for this is that the Coulomb pieces of $\hat{\Sigma}$ in the normal state are included in the band structure ϵ_k, which occurs in $\hat{G}_0$. Thus they must not also occur in $\hat{\Sigma}$.

The theory of T_c is now reduced to solving these equations self-consistently by using Eq. (7.11) for $\hat{G}$. The solution $\phi = \bar{\phi} = 0$ always exists and yields the normal-state results of Section II. However, if a solution with $\phi \neq 0$ or $\bar{\phi} \neq 0$ (or both) exists, then it can be shown[64] to have a lower free energy and to describe a state with Cooper-pair condensation. The definition of T_c is the highest temperature at which there exists a solution with nonzero ϕ and $\bar{\phi}$. The effect of impurities on T_c is contained in Eqs. (8.5) and (8.6). By building pair correlations simultaneously with impurity averaging, the impurity graphs become no different from any other self-energy graphs, greatly simplifying the algebra. The treatment of paramagnetic impurity scattering using Nambu 2-matrices is a little unnatural. A full discussion is more easily done using 4-matrices, as described by Ambegaokar and Griffin.[65] However, the net result is the same as Eq. (8.6).

A careful examination of Eqs. (7.9), (7.11), (7.12), and (8.1)–(8.6), which determine $\hat{\Sigma}$, reveals that ϕ and $\bar{\phi}$ satisfy identical nonlinear equations. The solution is expected to have ϕ and $\bar{\phi}$ equal, except for a proportionality factor. Any scaling of ϕ and $\bar{\phi}$ that preserves $\phi^2 + \bar{\phi}^2$ will yield an equally good solution. Thus there is a solution with $(\phi, \bar{\phi}) = (\phi_0, 0)$, as well as a family of solutions $(\phi, \bar{\phi}) = [\phi_0 \cos(2\alpha), \phi_0 \sin(2\alpha)]$. The arbitrary phase 2α comes from the phase $e^{i\alpha}$ of the one-electron state used to define the operator c_k. The usual gauge symmetry of quantum mechanics shows that physical observables cannot depend on this phase. However, the phase α is measured by Josephson tunneling.[66] Thus BCS theory exhibits broken gauge symmetry. For purposes of calculating thermodynamic properties such as T_c, we can set α to any convenient value, and we choose $\alpha = 0$, which makes $\bar{\phi} = 0$.

[64] J. Bardeen and M. Stephen, *Phys. Rev.* **136,** A1485 (1964).

[65] V. Ambegaokar and A. Griffin, *Phys. Rev.* **137,** A1151 (1965).

[66] B. D. Josephson, *in* "Superconductivity" (R. D. Parks, ed.), Vol. 1, Chapter 9. Dekker, New York, 1969.

9. Rescaling the Coulomb Interaction

This section describes how the large Coulomb effects $\hat{\Sigma}_F + \hat{\Sigma}_{SC}$ of Eqs. (8.3) and (8.4) are replaced by the harmless adjustable number $\mu^* \sim 0.1$ of Eq. (1.1). The process is done in more mathematical detail than is warranted by the inherent approximations of Eqs. (8.3) and (8.4). However, omitting the details might create more confusion than it would dispel.

Based on Migdal's work, the long-winded argument of Section 3 shows how to eliminate the energy (ϵ') integrals from Σ_{ep}, giving accurate results for $\Sigma_J(\epsilon, i\omega_n)$ when ϵ is at the Fermi surface, $\epsilon = 0$. In Section 5 it is shown that no similar simplification exists for Coulomb terms in the *normal* state. However, the superconducting state is easier, because we assume that the large *normal-state* Coulomb effects contained in $\Sigma_F + \Sigma_{SC}$ are already included in the band structure ϵ_k. The remaining off-diagonal parts of the superconducting parts $\hat{\Sigma}_F + \hat{\Sigma}_{SC}$ turn out to have only a small effect on superconductivity, which is treated phenomenologically.

The Fock piece $\hat{\Sigma}_F$ has been written separately from the screened Coulomb piece $\hat{\Sigma}_{SC}$, because $\hat{\Sigma}_F$ alone of all the pieces of $\hat{\Sigma}$ is independent of $i\omega_n$. This creates a problem, because we would prefer to be able to truncate the $i\omega_\nu$ summation at some finite (although possibly large) cutoff ω_c. A rigorous procedure can be constructed as follows. First, note that as T approaches T_c, $\hat{G}$ approaches $\hat{G}^N$, and $\hat{G} - \hat{G}^N$ can be accurately approximated by $\hat{G}^{od}$, the off-diagonal part of G. Even well below T_c, this remains an excellent approximation. Also the following relations hold:

$$\hat{\tau}_3 \hat{G}^d \hat{\tau}_3 = \hat{G}^d, \qquad \hat{\tau}_3 \hat{G}^{od} \hat{\tau}_3 = -\hat{G}^{od} = -\hat{\Sigma}^{od}/\det(\hat{G}^{-1}); \tag{9.1}$$

that is, the effect of the $\hat{\tau}_3$ matrices on either side of $\hat{G}$ is simply to change the signs of the off-diagonal parts. Next, we will split the $i\omega_\nu$ sum in Eq. (8.3) into two pieces, first with $|\omega_{n'}| < \omega_c$, and second with $|\omega_{n'}| > \omega_c$, where $\omega_{n'} = \omega_n - \omega_\nu$. The second part simplifies, because when $|\omega_n|$ exceeds ω_c, the only part of $\hat{\Sigma}$ which is not small is $\hat{\Sigma}_F$:

$$\hat{\tau}_3 \hat{G}^{od}(i\omega_{n'}) \hat{\tau}_3 \xrightarrow[|\omega_{n'}|>\omega_c]{} -\hat{\Sigma}_F^{od}/[(i\omega_{n'})^2 - \epsilon_{k'}^2]. \tag{9.2}$$

This also assumes that the anomalous part ϕ in det G^{-1}, as well as Z and χ, can be ignored relative to $i\omega_{n'}$. Putting the $|\omega_{n'}| > \omega_c$ piece onto the left-hand side, Eq. (8.3) becomes

$$\sum_{J'} \int_{-\infty}^{\infty} d\epsilon' \left(\delta_{JJ'} \delta(\epsilon - \epsilon') - T \sum_{n'}^{|\omega_{n'}|>\omega_c} N(\epsilon') \frac{V_{JJ'}(\epsilon, \epsilon')}{(i\omega_{n'})^2 - \epsilon'^2} \right) \hat{\Sigma}_{F,J'}(\epsilon')$$
$$= -T \sum_{n'}^{|\omega_{n'}|<\omega_c} \sum_{J'} \int_{-\infty}^{\infty} d\epsilon' \, N(\epsilon') V_{JJ'}(\epsilon, \epsilon') \hat{\tau}_3 \hat{G}^{od}(\epsilon', i\omega_{n'}) \hat{\tau}_3 . \tag{9.3}$$

The integral operator on the left-hand side is positive and thus can be formally inverted. Its inverse, denoted R, satisfies

$$\sum_{J'} \int_{-\infty}^{\infty} d\epsilon' \left(\delta_{JJ'}\delta(\epsilon - \epsilon') - T \sum_{n'}^{|\omega_{n'}|>\omega_c} N(\epsilon') \frac{V_{JJ'}(\epsilon, \epsilon')}{(i\omega_{n'})^2 - \epsilon'^2} \right) R_{J'J''}(\epsilon', \epsilon''; \omega_c)$$
$$= \delta_{JJ''}\delta(\epsilon - \epsilon''). \quad (9.4)$$

Then we can define a new, dimensionless Coulomb "pseudopotential" U:

$$U_{JJ'}(\epsilon, \epsilon'; \omega_c) \equiv N(0) \sum_{J''} \int_{-\infty}^{\infty} d\epsilon''\, R_{JJ''}(\epsilon, \epsilon''; \omega_c) V_{J''J'}(\epsilon'', \epsilon'). \quad (9.5)$$

In terms of U, Eq. (8.3) becomes

$$\hat{\Sigma}_{F,J}(\epsilon') = -T \sum_{n',J'}^{|\omega_{n'}|<\omega_c} \int_{-\infty}^{\infty} d\epsilon' \frac{N(\epsilon')}{N(0)} U_{JJ'}(\epsilon, \epsilon'; \omega_c) \hat{\tau}_3 \hat{G}^{\mathrm{od}}(\epsilon', i\omega_{n'}) \hat{\tau}_3 \,. \quad (9.6)$$

This equation now replaces Eq. (8.3) and contains the desired cutoff. An integral equation for U follows by combining Eqs. (9.4) and (9.5) to eliminate R:

$$U_{JJ'}(\epsilon, \epsilon'; \omega_c) = N(0) V_{JJ'}(\epsilon, \epsilon')$$
$$+ T \sum_{n',J''}^{|\omega_{n'}|>\omega_c} \int_{-\infty}^{\infty} d\epsilon''\, N(\epsilon'') \frac{V_{JJ'}(\epsilon, \epsilon'')}{(i\omega_{n'})^2 - \epsilon''^2} U_{J''J'}(\epsilon'', \epsilon'; \omega_c). \quad (9.7)$$

A model solution can be found for U by making a model for the (ϵ, ϵ') dependence of $N(\epsilon)$ and $V_{JJ'}$. If both are assumed to be independent of $\epsilon\epsilon'$ for $|\epsilon|$ and $|\epsilon'|$ less than ϵ_0, and zero otherwise, then $U_{JJ'}$ will acquire the same dependence on (ϵ, ϵ') and the ϵ' integral can be performed:

$$U_{J,J'}(\omega_c) \approx N(0) \left(V_{JJ'} - l(\omega_c, \epsilon_0) \sum_{J''} V_{JJ''} U_{J'',J'}(\omega_c) \right)$$
$$= \sum_{J''} [\hat{1} + l(\omega_c, \epsilon_0) N(0) \hat{\mathrm{V}}]^{-1}_{JJ''} N(0) V_{J''J'} \,, \quad (9.8)$$

where $l(\omega_c, \epsilon_0)$ is given by

$$l(\omega_c, \epsilon_0) = T \sum_{n'}^{|\omega_{n'}|>\omega_c} \int_{-\epsilon_0}^{\epsilon_0} d\epsilon'' \frac{1}{\omega_{n'}^2 + \epsilon''^2} \approx \ln\left(\frac{\epsilon_0}{\omega_c}\right). \quad (9.9)$$

The cutoff ϵ_0 is the energy at which the average matrix element of $e^2/|r - r'|$ gets small, while ω_0 is the frequency at which $\mathcal{S}(\Omega)$ gets small (forcing Σ_{SC} to get small for $\omega_n > \omega_c$.) In an electron gas the former is a few times ϵ_F, whereas the latter is the plasma energy ω_p, a very similar number. This means that l is a fairly small number, and we have succeeded only in showing that $U_{JJ'} \approx N(0)V_{JJ'}$. This is a useful result—it says that introducing a cutoff ω_c artificially into the $i\omega_\nu$ sum in Eq. (8.3) does not affect the answer much, provided that ω_c is about the same as the *natural* cutoff of $\mathscr{S}(\Omega)$ in Eq. (8.4).

Solutions of Eliashberg equations become realistically possible if the ϵ' integrals can be approximately eliminated, and if a cutoff of the order of a few times ω_D can be introduced in the ω_ν sums. Both of these can be accomplished in a realistic way, *provided that we believe that T_c is not sensitive to the details of the electronic screening in* $\mathscr{S}(\Omega)$. In other words, we neglect the possibility of excitons or plasmons enhancing T_c. Since there is still no experimental evidence for such an effect, perhaps this is not an important restriction. However, since these effects are permitted in principle, it is desirable to make realistic estimates of the possible magnitudes. In spite of a number of efforts, it is not clear that such a realistic estimate exists yet. The reasons for this feeling are (1) as mentioned in Section 5, the vertex correction diagrams are not counted, (2) realistic calculations of $\mathscr{S}(kk'\Omega)$ are very difficult because of the well-known technical difficulties of dielectric theory, and (3) solving the equations with the full (ϵ, ϵ') dependence *and* no small cutoff on $i\omega_\nu$ can exhaust all current computers. The possibility of Coulombic enhancement of T_c remains an unsolved problem in the theory of T_c.

It is not too hard to justify the neglect of Coulomb enhancement effects in the simpler real materials. In order to affect T_c, they should also be able to affect the normal-state properties near ϵ_F at low T, for example, by contributing some mass enhancement, $-\partial\Sigma/\partial\omega$, beyond the part due to phonons, which band theory would not contain unless time-dependent potentials were used.[35] It is difficult to tell to what degree experiments can rule out such Coulombic enhancement. Data for transition elements (such as Nb and Ta) seem consistent with the idea of no Coulomb enhancement effect, whereas Pd seems to show an enhanced Coulombic repulsion associated with spin fluctuations.

The hypothesis of no Coulomb enhancement permits a simplified treatment of the problem. First, define $\hat{\Sigma}_C$ to be the sum of $\hat{\Sigma}_F$ and $\hat{\Sigma}_{SC}$. We are concerned with $\Sigma_J(\epsilon, i\omega_n)$ only at small values of ϵ and ω_n and expect the ω_n dependence of the Coulomb part to be weak in this range. Thus we can write in this range

$$\hat{\Sigma}_{C,J}(\epsilon, i\omega_n) = -T \sum_{n',J'}^{|\omega_{n'}|<\omega_c} \int_{-\infty}^{\infty} d\epsilon' \frac{N(\epsilon')}{N(0)} U_{JJ'}^{\text{tot}}(\epsilon, \epsilon'; \omega_c)\hat{\tau}_3 \hat{G}^{\text{od}}(\epsilon', i\omega_{n'})\hat{\tau}_3 \,, \quad (9.10)$$

$$U_{JJ'}^{\text{tot}}(\epsilon, \epsilon'; \omega_c) = U_{JJ'}(\epsilon, \epsilon', \omega_c) + 2 \int_0^\infty \frac{d\Omega \, \mathscr{S}(JJ', \epsilon\epsilon', \Omega)}{\Omega} \,, \quad (9.11)$$

where the static screening approximation is made; that is, ω_ν in the denominator of Eq. (8.4) is neglected relative to Ω, which is of the order of ω_p.

This causes $\hat{\Sigma}_C$ to be independent of ω_n. Next, use approximation (9.2) with Σ_F^{od} replaced by Σ_C^{od} to simplify $\hat{G}^{od}(i\omega_{n'})$ provided that ω_n significantly exceeds the phonon frequencies. A new lower cutoff ω_{co} is chosen, typically $\sim 5\omega_D$. The scaling procedure (9.3)–(9.7) is applied again to Eq. (9.10). The result is a modified version of Eq. (9.10), with $U_{JJ'}^{tot}(\epsilon, \epsilon'; \omega_c)$ replaced by $U_{JJ'}^{*}(\epsilon, \epsilon'; \omega_{co})$ and the cutoff ω_c replaced by ω_{co}. The new Coulomb pseudopotential U^* satisfies the analog of Eq. (9.7):

$$U_{JJ'}^{*}(\epsilon, \epsilon'; \omega_{co}) = U_{JJ'}^{tot}(\epsilon, \epsilon'; \omega_c) + T \sum_{n',J''}^{\omega_c > |\omega_{n'}| > \omega_{co}} \int_{-\infty}^{+\infty} d\epsilon'' \frac{U_{JJ''}^{tot}(\epsilon, \epsilon''; \omega_c)}{(i\omega_{n'})^2 - \epsilon''^2} U_{J''J'}^{*}(\epsilon, \epsilon'; \omega_{co}). \quad (9.12)$$

Next make the model $U_{JJ'}^{tot}(\epsilon, \epsilon'; \omega_c) = \mu_{JJ'}$; that is, neglect the (ϵ, ϵ') dependence of the Coulomb interaction. The reason is the expectation that the energy variation will be dominated by the rapid ϵ' dependence of $\hat{G}^{od}(\epsilon', i\omega_{n'})$ when $\omega_{n'}$ is small. By the same reasoning, $N(\epsilon')$ is replaced by $N(0)$ in the rescaled version of Eq. (9.10). The result is that Eq. (9.10) is replaced by

$$\hat{\Sigma}_{C,J}(i\omega_n) = -T \sum_{J',n'}^{|\omega_{n'}| < \omega_{co}} \mu_{JJ'}^{*}(\omega_{co}) \int_{-\infty}^{\infty} d\epsilon' \, \hat{\tau}_3 \hat{G}^{od}(\epsilon', i\omega_{n'}) \hat{\tau}_3 \,, \quad (9.13)$$

$$\mu_{JJ'}^{*}(\omega_{co}) = \sum_{J''} \left[\hat{1} + \hat{\mu} \log\left(\frac{\omega_c}{\omega_{co}}\right) \right]_{JJ''}^{-1} \mu_{J''J'} \,, \quad (9.14)$$

where Eq. (9.14) is the solution of Eq. (9.12) in the energy-independent model.

The ultimate justification for having dealt so crudely with Coulomb effects is that the value of $\log(\omega_p/5\omega_D)$ is large enough ($\sim$5–10) that $\mu_{JJ'}^{*}(5\omega_D)$ is fairly small. The anisotropy of $\mu_{JJ'}^{*}$ should be even smaller, and thus negligible. Only a single number, $\mu_{00}^{*}(5\omega_D) \equiv \mu^{*}(5\omega_D)$, is left and can be fitted to experiment, for example, by tunneling.[7,67] This is very fortunate, given our inability to calculate Coulomb effects reliably. The number μ^* is often referred to as the Morel–Anderson[68] pseudopotential.

10. Anisotropic Equations

Equation (9.13) gives the Coulomb self-energy $\hat{\Sigma}_C(\epsilon = 0, i\omega_n)$ at the Fermi energy in a form such that the ϵ' integration involves only $\hat{G}$. The electron–

[67] A. A. Galkin, A. I. D'yachenko, and V. M. Svistunov, *Zh. Eksp. Teor. Fiz.* **66,** 2262 (1974); *Sov. Phys.—JETP (Engl. Transl.)* **39,** 1115 (1974).

[68] P. Morel and P. W. Anderson, *Phys. Rev.* **125,** 1263 (1962).

phonon part $\hat{\Sigma}_{\rm ep}$ and the impurity part $\hat{\Sigma}_{\rm N} + \hat{\Sigma}_{\rm P}$ can be written in the same way. In the case of $\hat{\Sigma}_{\rm ep}$, the justification is the Migdal argument. The steps by which Eq. (3.33) was derived from Eq. (3.32) remain equally valid in the superconducting state. The justification in the case of impurity terms is that ϵ will be within $\omega_{\rm co} \approx 5\omega_{\rm D}$ of $\epsilon_{\rm F}$, and the important imaginary part of $\hat{\Sigma}_{\rm i}$ has an energy-conserving δ function $\delta(\epsilon - \epsilon')$. We can neglect the unimportant real part of $\hat{\Sigma}_{\rm i}$ when impurities are dilute. Equations (8.2)–(8.6) become

$$\hat{\Sigma}_{{\rm ep},J}(i\omega_n) = T \sum_{J',\nu} \int_0^\infty d\Omega\, \alpha^2 F(JJ',\Omega) \frac{2\Omega}{\omega_\nu^2 + \Omega^2} \times \int_{-\infty}^\infty d\epsilon'\, \hat{\tau}_3 \hat{G}_{J'}(\epsilon', i\omega_n - i\omega_\nu)\hat{\tau}_3\,, \tag{10.1}$$

$$\hat{\Sigma}_{{\rm C},J}(i\omega_n) = -T \sum_{J',n'}^{|\omega_{n'}|<\omega_{\rm co}} \mu^*_{JJ'}(\omega_{\rm co}) \int_{-\infty}^\infty d\epsilon'\, \hat{\tau}_3 \hat{G}^{\rm od}_{J'}(\epsilon', i\omega_{n'})\hat{\tau}_3\,, \tag{10.2}$$

$$\hat{\Sigma}_{{\rm N},J}(i\omega_n) = \frac{1}{\pi}\sum_{J'} \gamma^{\rm N}_{JJ'} \int_{-\infty}^\infty d\epsilon'\, \hat{\tau}_3 \hat{G}_{J'}(\epsilon', i\omega_n)\hat{\tau}_3\,, \tag{10.3}$$

$$\Sigma_{{\rm P},J}(i\omega_n) = \frac{1}{\pi}\sum_{J'} \gamma^{\rm P}_{JJ'} \int_{-\infty}^\infty d\epsilon'\, \hat{\tau}_0 \hat{G}_{J'}(\epsilon', i\omega_n)\hat{\tau}_0\,. \tag{10.4}$$

The ϵ' integral appearing in these equations can be evaluated to good approximation by using Eqs. (7.11) and (7.12) for $\hat{G}$. The dominant ϵ' dependence of $\hat{G}(\epsilon', i\omega_{n'})$ comes from the explicit ϵ' appearing as the $(\epsilon_k + \chi)\hat{\tau}_3$ term of $\hat{G}^{-1}$ in Eq. (7.10). Because $\omega_n Z$ and ϕ are small, the denominator varies rapidly with ϵ' from this source. All other ϵ' variations, that is, the implicit ϵ' dependence of Z, χ, and ϕ, are negligible by comparison. Therefore the integral is elementary, with the result

$$\int_{-\infty}^\infty d\epsilon'\, \hat{G}_{J'}(\epsilon', i\omega_{n'}) = -\pi \left\{ \frac{i\omega_{n'} Z(i\omega_{n'})\hat{\tau}_0 + \phi(i\omega_{n'})\hat{\tau}_1}{\{[\omega_{n'} Z(i\omega_{n'})]^2 + \phi(i\omega_{n'})^2\}^{1/2}} \right\}_{J'}. \tag{10.5}$$

Here the choice of gauge $\alpha = \bar{\phi} = 0$ has been made explicit. The curly brackets with subscript J mean that the enclosed expression is k dependent and that the Jth expansion parameter $\langle\{\ \ \}F_J(k)\rangle_{\rm FS}$ in Fermi surface harmonics $F_J(k)$ is needed. Note that only two unknown functions, Z and ϕ, remain to be determined self-consistently in Eq. (10.5). The coefficient of $\hat{\tau}_3$, which gives χ, vanishes in performing integral (10.5).

A further simplification comes about because we are interested in $T = T_{\rm c}$, where ϕ becomes vanishingly small. We need to keep the infinitesimal function ϕ in the numerator, because the condition for $T_{\rm c}$ is that a solution $\phi \neq 0$ should exist. However, the ϕ^2 term in the denominator can be dropped. $Z(i\omega_n)$ turns out to be positive, and Eq. (10.5) at T_c becomes

$$\int_{-\infty}^{\infty} d\epsilon' \, \hat{G}_{J'}(\epsilon', i\omega_{n'}) = -\pi \text{ sign } \omega_{n'} \left(i\delta_{J'0}\hat{\tau}_0 - \frac{\Delta_{J'}(i\omega_{n'})}{\omega_{n'}} \hat{\tau}_1 \right), \quad (10.6)$$

$$\Delta_J(i\omega_n) \equiv \{\phi(i\omega_n)/Z(i\omega_n)\}_J \, . \quad (10.7)$$

This defines a function $\Delta(i\omega_n)$ which gives the gap in the energy spectrum for $T < T_c$. In k-space notation, Eq. (10.7) is

$$\Delta(k, i\omega_n) \equiv \sum_J \Delta_J(i\omega_n) F_J(k) = \frac{\phi(k, i\omega_n)}{Z(k, i\omega_n)} \, . \quad (10.8)$$

The fact that Δ is the gap can be seen from Eq. (7.12), where after analytic continuation from $i\omega_n$ to z, it can be seen that the pole of $\hat{G}$ gives the quasi-particle energy as $(\epsilon_k^2/Z^2 + \Delta^2)^{1/2}$.

Now we can write out the explicit self-consistent equations for the self-energy, using the definition (7.9) of the matrix components of $\hat{\Sigma}$:

$$i\omega_n[\delta_{J0} - Z_J(i\omega_n)]\hat{\tau}_0 + \phi_J(i\omega_n)\hat{\tau}_1 = \hat{\Sigma}_{\text{ep},J} + \hat{\Sigma}_{\text{C},J} + \hat{\Sigma}_{\text{N},J} + \hat{\Sigma}_{\text{P},J} \, . \quad (10.9)$$

After inserting Eq. (10.6) into Eqs. (10.1)–(10.4), the coefficient of $\hat{\tau}_0$ can be picked out:

$$i\omega_n[\delta_{J0} - Z_J(i\omega_n)] = -i\pi T \sum_{n'} \int_0^{\infty} d\Omega \, \alpha^2 F(J0, \Omega) \frac{2\Omega \text{ sign } \omega_{n'}}{(\omega_n - \omega_{n'})^2 + \Omega^2}$$
$$- i(\gamma_{J0}^{\text{N}} + \gamma_{J0}^{\text{P}}) \text{ sign } \omega_n \, . \quad (10.10)$$

The first term on the right of Eq. (10.10) agrees with the normal-state electron–phonon result derived in Section 3, namely, Eq. (3.43) with Eq. (B.3) for $R(i\omega_n, \Omega)$. This provides a belated justification for the use of the approximation $G(k, i\omega_n) \to (i\omega_n - \epsilon_k)^{-1}$ in Eq. (3.42). No such approximation was used here, but the result is unchanged. The second term of Eq. (10.10) is exactly the normal-state impurity result derived in Section 6. It is convenient to define an electron–phonon interaction function $\lambda(i\omega_\nu)$, denoted $\lambda(\nu)$ for short:

$$\lambda_{JJ'}(\nu) \equiv \int_0^{\infty} d\Omega \, \alpha^2 F(JJ', \Omega) \frac{2\Omega}{\omega_\nu^2 + \Omega^2} \, . \quad (10.11)$$

Then Eq. (10.10) can be rewritten as

$$Z_J(i\omega_n) = \delta_{J0} + \frac{\pi T}{|\omega_n|} \sum_{\omega_{n'}} \lambda_{J0}(n - n') s_n s_{n'} + \frac{1}{|\omega_n|} (\gamma_{J0}^{\text{N}} + \gamma_{J0}^{\text{P}}), \quad (10.12)$$

$$s_n \equiv \text{sign } \omega_n = (n + \tfrac{1}{2})/|n + \tfrac{1}{2}| \, . \quad (10.13)$$

The coefficient of $\hat{\tau}_1$ is the pair field ϕ which determines the superconducting properties. Again, using Eq. (10.9) and inserting Eq. (10.6) into Eqs. (10.1)–(10.4), the result is

$$\phi_J(i\omega_n) = \pi T \sum_{J',n'} [\lambda_{JJ'}(n-n') - \mu^*_{JJ'}(\omega_{co})] \frac{\Delta_{J'}(i\omega_{n'})}{|\omega_{n'}|} + \sum_{J'} (\gamma^{N}_{JJ'} - \gamma^{P}_{JJ'}) \frac{\Delta_{J'}(i\omega_n)}{|\omega_n|}. \tag{10.14}$$

Note the important change of sign of γ_P, but not γ_N, in ϕ relative to Z. The cutoff ω_{co} of the $\omega_{n'}$ sum in μ^* is to be chosen large enough ($\approx 5\omega_D$) so that $\lambda_{JJ'}(\omega_n - \omega_{n'})$ has fallen effectively to zero. The pair field ϕ_J must now be related back to the "gap" Δ_J and the "renormalization" Z_J via Eq. (10.7) or (10.8). At this point further manipulation of these equations requires introducing "Clebsch–Gordan coefficients" for Fermi surface harmonics. To avoid this technicality, we can switch back to k space. The Fermi surface harmonic or J-space description will be reintroduced in Section 17 for the discussion of "p-wave" superconductivity.

The k-space versions of Eqs. (10.12) and (10.14) follow from the J-space versions and definitions (3.29) and (3.30). It is also necessary to remember that $F_0 = 1$. The results are

$$Z(k, i\omega_n) = 1 + \frac{\pi T}{|\omega_n|} \sum_{k'n'} \frac{\delta(\epsilon_{k'})}{N(0)} \times \{\lambda(k, k'; n-n') + \delta_{nn'}[\gamma_N(k, k') + \gamma_P(k, k')]\} s_n s_{n'}, \tag{10.15}$$

$$Z(k, i\omega_n)\Delta(k, i\omega_n) = \pi T \sum_{k'n'}^{|\omega_{n'}|<\omega_c} \frac{\delta(\epsilon_{k'})}{N(0)|\omega_{n'}|} \{\lambda(k, k'; n-n') - \mu^*(\omega_c) + \delta_{nn'}[\gamma_N(k, k') - \gamma_P(k, k')]\} \Delta(k', i\omega_{n'}). \tag{10.16}$$

The states k are on the Fermi surface and, because of the δ functions, the states k' are also. The definition of $\lambda(k, k'; \nu)$ is

$$\lambda(k, k'; \nu) = \sum_{JJ'} \lambda_{JJ'}(\nu) F_J(k) F_{J'}(k'), \tag{10.17}$$

with similar formulas for $\gamma_i(k, k')$ except that a factor $(\pi T)^{-1}$ is introduced to make $\gamma_i(k, k')$ dimensionless:

$$\gamma_i(k, k') = (\pi T)^{-1} \sum_{JJ'} \gamma^i_{JJ'} F_J(k) F_{J'}(k'). \tag{10.18}$$

Using Eqs. (3.31), (6.12), and (6.13), these formulas can be written as

$$\lambda(k, k'; \nu) = \int_0^\infty d\Omega\, \alpha^2 F(k, k'; \Omega) \frac{2\Omega}{\omega_\nu^2 + \Omega^2} \tag{10.19}$$

$$= 2N(0)|g_{kk'}|^2\omega_{k-k'}/[\omega_\nu^2 + \omega_{k-k'}^2], \tag{10.20}$$

$$\gamma_i(k, k') = [\tfrac{1}{2}\tau_i(k, k')]/\pi T_c . \tag{10.21}$$

Equations (10.15) and (10.16) are the fundamental equations that determine T_c. No approximations beyond Eqs. (10.1)–(10.4) have been made. The possibility of exotic Coulomb effects has been discarded, as has energy (ϵ, ϵ') dependence of the parameters. However, a full account of the details of band structure, Fermi surface, gap anisotropy, complicated phonon spectra, and impurity scattering are all included. The J-space versions are simple enough to be completely solved on a modest computer (if only one or two J's are kept) or a large computer (if many J's are needed.) An alternative derivation of these equations was given by Allen.[69]

11. Isotropic Equations

Theory and experiment[70] both indicate now that known metallic superconductors have surprisingly isotropic gaps. Thus we can write

$$\Delta(k, i\omega_n) = \Delta(i\omega_n) + \delta\Delta(k, i\omega_n), \tag{11.1}$$

$$Z(k, i\omega_n) = Z(i\omega_n) + \delta Z(k, i\omega_n), \tag{11.2}$$

$$\langle\delta\Delta(k, i\omega_n)\rangle_{FS} = 0 = \langle\delta Z(k, i\omega_n)\rangle_{FS} . \tag{11.3}$$

The rms anisotropy $\langle\delta\Delta(k, i\omega_n)^2\rangle_{FS}^{1/2}$ seldom exceeds 10% of $\Delta(i\omega_n)$. The fractional rms anisotropy can be taken as an expansion parameter, the size of which will be denoted as a. To the lowest order, Eqs. (10.15) and (10.16) can be replaced by simpler isotropic equations for $\Delta(i\omega_n)$:

$$Z(i\omega_n) = 1 + \frac{1}{|2n+1|}\sum_{n'} [\lambda(n-n') + \delta_{nn'}(\gamma_N + \gamma_P)]s_n s_{n'} , \tag{11.4}$$

$$Z(i\omega_n)\Delta(i\omega_n) = \sum_{n'}^{|\omega_{n'}|<\omega_c} \frac{1}{|2n'+1|}[\lambda(n-n') - \mu^* + \delta_{nn'}(\gamma_N - \gamma_P)]\Delta(i\omega_{n'}). \tag{11.5}$$

These equations follow from Eqs. (10.15) and (10.16) by substitution of Eqs. (11.1) and (11.2), averaging over k on the Fermi surface and keeping only terms of order a^0. The terms of order a^1 will be studied in Section 16.

[69] P. B. Allen, *Lect. Notes Phys.* **115,** 388 (1980).

[70] A. G. Shepelev, *Usp. Fiz. Nauk* **96,** 217 (1968); *Sov. Phys.—Usp. (Engl. Transl.)* **11,** 690 (1969).

The effect on T_c will be shown to be of order a^2. It will also be shown that impurities make Δ more isotropic. Thus the isotropic equations are correct for a "dirty" superconductor. The definitions of $\lambda(\nu)$ and γ_i are

$$\lambda(\nu) = \langle\langle\lambda(k, k'; \nu)\rangle\rangle_{FS} = \int_0^\infty d\Omega\, \alpha^2F(\Omega)\, \frac{2\Omega}{\omega_\nu^2 + \Omega^2}, \tag{11.6}$$

$$\gamma_i = (\tfrac{1}{2}\tau_i)/\pi T_c, \tag{11.7}$$

$$1/\tau_N = (2\pi/\hbar)n_N N(0)\langle\langle|V_N|^2\rangle\rangle_{FS} = \langle 1/\tau_N(k)\rangle_{FS}, \tag{11.8}$$

$$1/\tau_P = (2\pi/\hbar)n_P S(S+1)N(0)\langle\langle|V_P|^2\rangle\rangle_{FS} = \langle 1/\tau_P(k)\rangle_{FS}, \tag{11.9}$$

which follows from Eqs. (3.24), (6.12), and (6.13). This version of the T_c equations, except with $\gamma_i = 0$, was first studied by Owen and Scalapino[71] and Bergmann and Rainer.[72] An alternate way of extracting the isotropic approximation starts with the J-space equations, Eqs. (10.12) and (10.14). Weak anisotropy implies that $\lambda_{JJ'}(\nu)$ and similar matrices are nearly diagonal. The isotropic gap Δ and renormalization Z are the $J = 0$ values of Δ_J and Z_J, which obey Eqs. (11.4) and (11.5) exactly if $\lambda_{JJ'}(\nu)$ and so on are diagonal.

The renormalization function Z can be eliminated from Eqs. (11.4) and (11.5), yielding a linear matrix equation for $\Delta(i\omega_n)$:

$$\Delta(i\omega_n) = \sum_{n'}^{|\omega_{n'}|<\omega_{co}} \frac{1}{|2n'+1|} S(n, n')\Delta(i\omega_{n'}), \tag{11.10}$$

$$S(n, n') = \lambda(n - n') - \mu^* - \delta_{nn'} \sum_{n''} \lambda(n - n'')s_n s_{n''} - 2\gamma_P\delta_{nn'}. \tag{11.11}$$

Note that the nonmagnetic impurity term γ_N has canceled out of Eq. (11.11), whereas the magnetic term γ_P remains, because of the sign inversion between Eqs. (11.4) and (11.5). Thus, as discussed qualitatively by Anderson[73,73a] and mathematically by Abrikosov and Gor'kov,[74] dilute nonmagnetic impurities have no effect on the T_c of an isotropic superconductor. This is known as "Anderson's theorem." Magnetic impurities have a large effect. As discussed by Anderson, this has to do with the fact that Cooper pairs involve states $k\uparrow$ and $-k\downarrow$ which are time-reversed images of each other. Magnetic impurities cause spin-flip scattering and thus destroy the

[71] C. S. Owen and D. J. Scalapino, *Physica (Amsterdam)* **55**, 691 (1971).
[72] G. Bergmann and D. Rainer, *Z. Phys.* **263**, 445 (1974).
[73] P. W. Anderson, *J. Phys. Chem. Solids* **11**, 26 (1959).
[73a] P. W. Anderson, *Proc. Int. Conf. Low Temp. Phys., 7th, 1960* p. 298 (1961).
[74] A. A. Abrikosov and L. P. Gor'kov, *Zh. Eksp. Teor. Fiz.* **35**, 1558 (1958); *Sov. Phys.—JETP (Engl. Transl.)* **8**, 1090 (1959).

time-reversal symmetry which guaranteed the degeneracy of $k\uparrow$ and $-k\downarrow$. Nonmagnetic impurities preserve time-reversal symmetry.

The matrix $S(n, n')$ in Eq. (11.11) is Hermitian, and Eq. (11.10) can be cast in the form of a Hermitian eigenvalue problem

$$\hat{S}|\Delta\rangle = \rho|\Delta\rangle, \tag{11.12}$$

where ρ is an eigenvalue that depends on temperature T and equals 1 by definition when $T = T_{co}$, the isotropic approximation to T_c. In order for Eq. (11.12) to represent Eq. (11.10), we have to make an unconventional definition of inner products and matrix multiplication, namely,

$$\langle\phi|\psi\rangle = \sum_n \frac{1}{|2n+1|} \phi^*(n)\psi(n), \tag{11.13}$$

$$\langle n|\hat{A}|\phi\rangle = \sum_{n'} \frac{1}{|2n'+1|} A(n, n')\phi(n'). \tag{11.14}$$

This is a well-defined inner product because the weight factor, $|2n+1|^{-1}$, is positive.

The dimension of the matrix S used in Eq. (11.10) is determined by the value of N such that $\omega_N = \omega_{co} \simeq 5\omega_D$. Thus the dimension is $\sim 5\omega_D/2\pi T_c$. For a strong-coupling material like Pb, a 10×10 matrix is sufficient, whereas for Al, 200×200 is required. Thus Eq. (11.10) is easily manageable on a computer and yields solutions for T_c quite rapidly unless T_c/ω_D is very small.

12. Continuation to Real Frequencies

The Green's function $G(k, i\omega_n)$ acquires much additional significance when analytically continued from the imaginary frequencies $i\omega_n$ onto the real-frequency axis from above, $\omega + i\delta$, where δ is an infinitesimal. As discussed in Section 3 for the normal state, the poles of $G(k, z)$ give the quasi-particle spectrum. The density of states measured by tunneling[7] is contained in Im $G(k, \omega + i\delta)$. In the superconducting state, this allows a measurement of the complex gap function $\Delta(\omega)$, which is the analytic continuation of $\Delta(i\omega_n)$. The function $\Delta(\omega)$ obeys an integral equation which can be obtained from the equations for $\Delta(i\omega_n)$ by analytic continuation. This latter equation is far harder to solve than Eq. (11.10). Therefore a desirable procedure would be to continue numerically the computed values of $\Delta(i\omega_n)$ to get $\Delta(\omega)$. A Padé method was first discussed by Vidberg and Serene.[75] Unfortunately this method works well only for $T \lesssim 0.1T_c$. Thus it is often necessary to solve for $\Delta(\omega)$ directly from the real-frequency equa-

[75] H. J. Vidberg and J. Serene, *J. Low Temp. Phys.* **29**, 179 (1977).

tions. Historically the real-ω equations were important to the theory of T_c, because they were used in the classic paper of McMillan.[14] Therefore, for completeness, the real-ω equations will now be derived.

The starting point is Eqs. (10.1)–(10.4). For simplicity we make them isotropic, which is accomplished by simply dropping all subscripts J, J'. The ϵ' integral in Eqs. (10.3) and (10.4) is evaluated in the style of Eq. (10.5); however, for Eqs. (10.1) and (10.2), we use the spectral representations (3.10) and (3.12):

$$\hat{G}(\epsilon', i\omega_n) = -\frac{1}{\pi}\int_{-\infty}^{\infty} \frac{d\omega' \,\mathrm{Im}\, \hat{G}(\epsilon', \omega' + i\delta)}{i\omega_n - \omega'} . \tag{12.1}$$

Then in place of Eq. (10.5) we have

$$\int_{-\infty}^{\infty} d\epsilon' \,\hat{G}(\epsilon', i\omega_{n'}) = \int_{-\infty}^{\infty} \frac{d\omega' \,Q_1(\omega')}{i\omega_{n'} - \omega'} , \tag{12.2}$$

$$\hat{Q}_1(\omega') \equiv -\frac{1}{\pi}\int_{-\infty}^{\infty} d\epsilon' \,\mathrm{Im}\, \hat{G}(\epsilon', \omega' + i\delta). \tag{12.3}$$

Next we perform the sums over $i\omega_{n'}$. For Eq. (10.1) the procedure is exactly as in Eq. (3.39) for the normal state. For Eq. (10.2) the sum uses the identity

$$T \sum_{n=-\infty}^{\infty} \frac{1}{\epsilon - i\omega_n} = \tfrac{1}{2} \tanh\left(\frac{\epsilon}{2T}\right) . \tag{12.4}$$

This follows from Eq. (2.28) and is the basis for the Poisson sum formula. At this stage it is safe to replace $i\omega_n$ by $\omega + i\delta$, yielding

$$\hat{\Sigma}_{\mathrm{ep}}(\omega) = \int_{-\infty}^{\infty} d\omega' \int_0^{\infty} d\Omega \, \alpha^2F(\Omega) I(\omega + i\delta, \Omega, \omega')\hat{\tau}_3\hat{Q}_1(\omega')\hat{\tau}_3 , \tag{12.5}$$

$$\hat{\Sigma}_{\mathrm{C}}(\omega) = -\tfrac{1}{2}\int_{-\omega_{\mathrm{co}}}^{\omega_{\mathrm{co}}} d\omega' \,\mu^*(\omega_c) \tanh\left(\frac{\omega'}{2T}\right)\hat{\tau}_3\hat{Q}_1^{\mathrm{od}}(\omega')\hat{\tau}_3 , \tag{12.6}$$

$$\Sigma_{\mathrm{i}}(\omega) = (-i/2\tau_i)\hat{\tau}_i\hat{Q}(\omega)\hat{\tau}_i , \tag{12.7}$$

where $\hat{\tau}_i$ in Eq. (12.7) is $\hat{\tau}_3$ if i = N and $\hat{\tau}_0$ if i = P. The function $I(\omega, \Omega, \omega')$ was defined in Eq. (3.41). The function $\hat{Q}(\omega)$ is $-i$ times the analytic continuation of the function in brackets in Eq. (10.5),

$$\begin{aligned}\hat{Q}(\omega) &= -i\,\frac{\omega Z(\omega)\hat{\tau}_0 + \phi(\omega)\hat{\tau}_1}{[-\omega^2 Z(\omega)^2 + \phi(\omega)^2]^{1/2}} \\ &= \frac{\omega\hat{\tau}_0 + \Delta(\omega)\hat{\tau}_1}{[\omega^2 - \Delta(\omega)^2]^{1/2}} .\end{aligned} \tag{12.8}$$

The first square root in Eq. (12.8) is defined to have a positive imaginary part when ω is in the upper half-plane. The second square root must be chosen to have a real part which has the same sign as ω. Now we need an explicit equation for $\hat{Q}_1(\omega')$, defined by Eq. (12.3). We use Eq. (7.11) for $\hat{G}(k, i\omega_n)$, replace $i\omega_n$ by $\omega + i\eta$, and ignore the ϵ' dependence of Z, χ, and ϕ, just as was done in deriving Eq. (10.5). The result is

$$\hat{Q}_1(\omega') = \mathrm{Re}\left(\frac{\omega' Z(\omega')\hat{\tau}_0 + \phi(\omega')\hat{\tau}_1}{[\omega'^2 Z(\omega')^2 - \phi(\omega')^2]^{1/2}}\right) = \mathrm{Re}\ \hat{Q}(\omega). \tag{12.9}$$

To complete the equations, we write the definition (7.9) of $\hat{\Sigma}$ as

$$\hat{\Sigma}(\omega) = \omega[1 - Z(\omega)]\hat{\tau}_0 + Z(\omega)\Delta(\omega)\hat{\tau}_1\ . \tag{12.10}$$

The system of equations (12.5)–(12.10) are the real-ω equations in isotropic approximation. It has not been assumed that $T = T_c$, so these nonlinear equations are valid at all T. In particular, the $T = 0$ equations are contained as a special case.

Finally, we separate out from Eqs. (12.5)–(12.10) the coefficients of $\hat{\tau}_0$ and $\hat{\tau}_1$. The resulting equations are

$$\left[[1 - Z(\omega)] + i\left(\frac{1}{2\tau_N} + \frac{1}{2\tau_P}\right)[\omega^2 - \Delta(\omega)^2]^{-1/2}\right]\omega$$
$$= \int_{-\infty}^{\infty} d\omega' \int_0^{\infty} d\Omega\ \alpha^2 F(\Omega) I(\omega + i\delta, \Omega, \omega')\ \mathrm{Re}\left(\frac{\omega'}{[\omega'^2 - \Delta(\omega')^2]^{1/2}}\right), \tag{12.11}$$

$$\left[Z(\omega) - i\left(\frac{1}{2\tau_N} - \frac{1}{2\tau_P}\right)[\omega^2 - \Delta(\omega)^2]^{-1/2}\right]\Delta(\omega)$$
$$= -\int_{-\infty}^{\infty} d\omega' \int_0^{\infty} d\Omega\ \alpha^2 F(\Omega) I(\omega + i\delta, \Omega, \omega')\ \mathrm{Re}\left(\frac{\Delta(\omega')}{[\omega'^2 - \Delta(\omega')^2]^{1/2}}\right)$$
$$- \tfrac{1}{2}\mu^*(\omega_c) \int_{-\omega_c}^{\omega_c} d\omega' \tanh\left(\frac{\omega'}{2T}\right) \mathrm{Re}\left(\frac{\Delta(\omega')}{[\omega'^2 - \Delta(\omega')^2]^{1/2}}\right), \tag{12.12}$$

$$I(\omega + i\delta, \Omega, \omega') = \frac{N(\Omega) + 1 - f(\omega')}{\omega + i\delta - \Omega - \omega'} + \frac{N(\Omega) + f(\omega')}{\omega + i\delta + \Omega - \omega'}. \tag{12.13}$$

These are the real-ω gap equations. As T approaches T_c, the denominator $[\omega'^2 - \Delta(\omega')^2]^{1/2}$ becomes ω', and Eq. (12.11) reproduces Eqs. (3.43) and (6.18) for the normal state. With $1/\tau_i = 0$, these are the equations solved by McMillan.[14]

IV. Solutions for T_c

13. Exact Results for Isotropic Superconductors

Many authors have solved the isotropic Eliashberg equations numerically on computers. The most famous are McMillan's solutions[14] of the "real-axis" equations (12.11) and (12.12). The "imaginary-axis" equations (11.10) are equivalent but easier; many solutions have been found,[71,72,76–81] and, in particular, Allen and Dynes[76] confirmed McMillan's numerical answers. Computer programs are available.[82] Some of the exact numerical solutions will be exhibited in Section 14.

In addition to numerical solutions, certain other exact results are known, in particular, some bounds on T_c. The Hermitian eigenvalue problem (11.12) has all eigenvalues ρ less than 1 when $T > T_c$. Exactly at T_c, the *largest* eigenvalue becomes 1. The largest eigenvalue exceeds any trial eigenvalue ρ_t:

$$\rho_{\max} \geq \rho_t = \langle \Delta_t | \hat{S} | \Delta_t \rangle / \langle \Delta_t | \Delta_t \rangle, \tag{13.1}$$

where $|\Delta_t\rangle$ is an arbitrary trial eigenvector. The temperature at which ρ_t equals 1 defines a trial transition temperature T_{ct} which is a lower bound on T_c.

It is convenient to exploit the symmetry of the kernel $\hat{S}$ under the interchange $(\omega_n, \omega_{n'}) \rightarrow (-\omega_n, -\omega_{n'})$. Thus eigenvectors can be chosen to be even or odd, $\Delta(-i\omega_n) = \pm\Delta(i\omega_n)$. Assuming that the maximum eigenvalue has an even eigenvector, it obeys the equation

$$\rho \Delta_n = \sum_{n'=0}^{N} \frac{\tilde{S}(n, n'; T)\Delta_{n'}}{2n' + 1}, \tag{13.2}$$

$$\tilde{S}(n, n'; T) = S(n, n') + S(n, -n' - 1), \tag{13.3}$$

where Δ_n is short for $\Delta(i\omega_n)$. A careful consideration of the sign $(n'' + \frac{1}{2})$ factor in Eq. (11.11) leads to the following equation for the kernel $\tilde{S}$:

[76] P. B. Allen and R. C. Dynes, *Phys. Rev. B* **12,** 905 (1975).

[77] C. R. Leavens, *Solid State Commun.* **15,** 1329 (1974).

[78] S. G. Louie and M. L. Cohen, *Solid State Commun.* **22,** 1 (1977).

[79] J. Cai, G. Ji, H. Wu, J. Cai, and C. Gong, *Sci. Sin. (Engl. Ed.)* **22,** 417 (1979).

[80] Z. Zhou, H. Wu, D. Mao, and Y. Gu, *Sci. Sin. (Engl. Ed.)* **23,** 1378 (1980).

[81] H. Wu, Z. Zhou, Z. Wang, and X. Zhang, *Acta Phys. Sin.* **29,** 409 (1980); *Chin. Phys. (Engl. Transl.)* **1,** 43 (1981); Z. Wang, Z. Zhou, D. Mao, X. Zhang, and H. Wu, *Acta Phys. Sin.* **29,** 843 (1980); *Chin. Phys. (Engl. Transl.)* **1,** 116 (1981).

[82] P. B. Allen, Technical Report No. 7 (TCM/4/1974). Theory of Condensed Matter Group, Cavendish Laboratory (available from the author at Dept. of Physics, SUNY, Stony Brook, NY 11794.) Other groups have equally accurate programs.

$$\tilde{S}(n, n') = \lambda(n - n') + \lambda(n + n' + 1) - 2\mu^*(\omega_{co}) - \delta_{nn'}\left(\lambda(0) + 2\sum_{m=1}^{n}\lambda(m) + 2\gamma_P\right). \tag{13.4}$$

The simplest trial vector for Eq. (13.2) is $\Delta_n = \delta_{n0}$. This gives the trial eigenvalue

$$\rho_t^0 = \tilde{S}(0, 0) = \lambda(1) - 2\mu^*(\omega_{co}) - 2\gamma_P\,. \tag{13.5}$$

At the actual T_c, ρ_t^0 is less than 1, giving the inequality

$$\lambda(1) \equiv \int_0^\infty d\Omega\, \alpha^2F(\Omega)\, \frac{2\Omega}{\Omega^2 + (2\pi T_c)^2} < 1 + 2\mu^*(\omega_{co}) + 2\gamma_P\,. \tag{13.6}$$

This lower bound on T_c was given in Ref. 76 for the case $\gamma_P = 0$. Wu and Ji[83] give the inequality

$$\int_0^\infty d\Omega\, \alpha^2F(\Omega)\, \frac{2\Omega}{\Omega^2 + a^2} \geq \frac{\lambda\langle\omega^2\rangle}{\langle\omega^4\rangle/\langle\omega^2\rangle + a^2}, \tag{13.7}$$

$$\lambda = \lambda(0) = 2\int_0^\infty \frac{d\Omega\, \alpha^2F(\Omega)}{\Omega}, \tag{13.8}$$

$$\langle\omega^n\rangle \equiv \frac{2}{\lambda}\int_0^\infty \frac{d\Omega}{\Omega}\, \alpha^2F(\Omega)\Omega^n, \tag{13.9}$$

where $\lambda = \langle\lambda_k\rangle_{FS}$, λ_k being defined in Eq. (3.49). The moments $\langle\omega^n\rangle$ of the distribution function $2\alpha^2F(\Omega)/\lambda\Omega$ have been widely used since McMillan's work.[14] The second moment was already defined in Eq. (3.56). Equations (13.6) and (13.7) can be combined to give a rigorous lower bound on T_c:

$$T_c > (2\pi)^{-1}(\lambda\langle\omega^2\rangle)^{1/2}\left(\frac{1}{1 + 2\mu^* + 2\gamma_P} - \frac{\langle\omega^4\rangle}{\lambda\langle\omega^2\rangle^2}\right)^{1/2}, \tag{13.10}$$

where the factor in brackets is assumed to be positive; otherwise the lower bound is simply zero. For $\lambda \gg 1$, this lower bound far exceeds the McMillan equation (1.1) and suggests a more optimistic outlook[76] for raising T_c than the one McMillan gave. Actually this bound is fairly weak; the largest eigenvalue always exceeds ρ_t^0 by a finite amount, and better bounds can be constructed using more complicated trial eigenvectors.

Wu and collaborators[79–81,83–85] have studied the mathematical behavior of the solution of Eq. (13.2) in great detail for the case $\gamma_P = 0$ with no magnetic impurities. They proved that the solution has the form

[83] H. Wu and G. Ji, *Sci. Sin. (Engl. Ed.)* **22,** 514 (1979).

[84] H. Wu, C. Tsai, C. Kung, K. Chi, and C. Tsai, *Sci. Sin. (Engl. Ed.)* **20,** 583 (1977).

[85] C. Kung, H. Wu, C. Tsai, C. Tsai, and K. Chi, *Sci. Sin. (Engl. Ed.)* **21,** 62 (1978).

$$T_c = \alpha_0(\lambda\langle\omega^2\rangle)^{1/2}\left[1 + \lambda^{-1}\alpha_1 \frac{\langle\omega^4\rangle}{\langle\omega^2\rangle^2} + \lambda^{-2}\left(\alpha_{21}\frac{\langle\omega^6\rangle}{\langle\omega^2\rangle^3} + \alpha_{22}\frac{\langle\omega^4\rangle^2}{\langle\omega^2\rangle^4}\right) + \cdots\right], \quad (13.11)$$

where the coefficients α_{ij} depend only on μ^* and not on the shape of $\alpha^2F(\Omega)$. Wu and collaborators numerically determined[79] the coefficients out to α_{33} for $\mu^* = 0$ and for μ^* small. When $\mu^* = 0$, they found $\alpha_0 = 0.18273$, agreeing with other workers.[76] For $\mu^* \neq 0$, they suggested that $\alpha_0 = 0.180(1 + 2.60\mu^*)^{-1/2}$ should be accurate to about 1% over a reasonable range of μ^*. This is the same as the first term of Eq. (13.10), except that $(2\pi)^{-1} = 0.159$ has been increased to 0.180 and $2\mu^*$ increased to $2.60\mu^*$. The first term of Eq. (13.11) adequately represents T_c in the extreme (and so far unphysical) regime $\lambda\langle\omega^2\rangle \gtrsim 5\omega_{max}^2$, where ω_{max} is the maximum phonon frequency.[86] Note that McMillan's equation (1.1) is completely wrong in this regime. Unlike Eq. (1.1), Eq. (13.11) predicts no upper bound on T_c as λ increases. The series (13.11) converges only for $\lambda > \Lambda$, where Λ depends on the shape of α^2F and varies from 1 to 3 for different materials investigated.[80] Unfortunately, materials have not yet been found with $\lambda > 3$. In principle,[85] the series could be analytically continued to span the whole range of $T_c > 0$, but in practice this has not been accomplished. The mathematical behavior found in Refs. 81 and 85 for very low T_c is consistent with the exponential behavior given by McMillan's Eq. (1.1).

Another interesting inequality was established by Leavens[87]:

$$T_c \leq c(\mu^*)A = \tfrac{1}{2}c(\mu^*)\lambda\langle\omega\rangle, \quad (13.12)$$

where $A = \tfrac{1}{2}\lambda\langle\omega\rangle$ is, according to Eq. (13.8), the integral of $\alpha^2F(\Omega)$ or the "area" under $\alpha^2F(\Omega)$. For $\mu^* = 0$, c has the value 0.2309, while for $\mu^*(\omega_{co}) = 0.1$ the value is 0.176 (using $\omega_{co} = 10\langle\omega^2\rangle^{1/2}$). The upper bound (13.12) is actually achieved if α^2F corresponds to an Einstein spectrum with $\omega_E = 1.75A$.

Bergmann and Rainer[72] devised an algorithm for calculating $\delta T_c/\delta\alpha^2F(\Omega)$, the functional derivative of T_c by $\alpha^2F(\Omega)$. This provides an exact answer to the question of how T_c is altered by an arbitrary small change $\Delta\alpha^2F(\Omega)$ of α^2F, namely,

$$\Delta T_c = \int_0^\infty d\Omega \, \frac{\delta T_c}{\delta\alpha^2F(\Omega)} \, \Delta\alpha^2F(\Omega). \quad (13.13)$$

Figure 7 shows examples of this functional derivative, calculated for $\alpha^2F(\Omega)$ having the shape measured for Pb by tunneling,[7] with the strength λ scaled to various values. The values of T_c are shown normalized to $\langle\omega^2\rangle^{1/2}$, which

[86] C. R. Leavens, P. B. Allen, and R. C. Dynes, *Solid State Commun.* **30**, 595 (1979).
[87] C. R. Leavens, *Solid State Commun.* **17**, 1499 (1975).

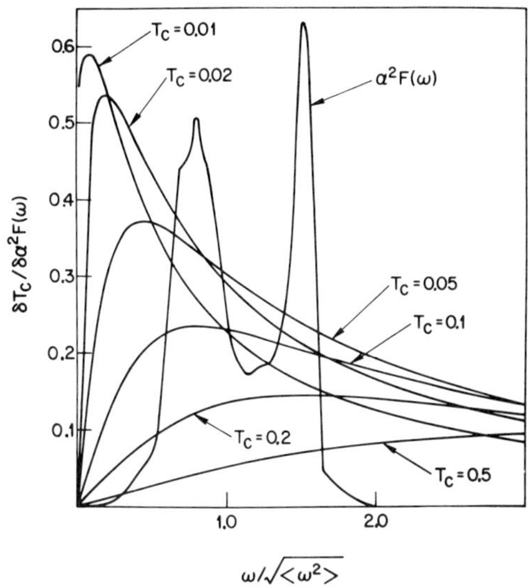

FIG. 7. The functional derivative of T_c by $\alpha^2F(\Omega)$ calculated using $\alpha^2F(\Omega)$ for Pb (also shown) and $\mu^* = 0.1$. The "strength" λ of $\alpha^2F(\Omega)$ has been varied to give six hypothetical values of T_c, denoted on the figure in units of $\langle\omega^2\rangle^{1/2}$. Reprinted from P. B. Allen and R. C. Dynes, *Phys. Rev. B* **12,** 905 (1975).

is 65°K for Pb. Thus $T_c = 7.2$°K corresponds to $T_c/\langle\omega^2\rangle^{1/2} = 0.11$ for actual Pb, with $\lambda = 1.55$. The functional derivatives have a maximum at a frequency $8T_c$, which corresponds in actual Pb to the frequency of the transverse phonon peak in α^2F. Thus reducing the phonon frequencies ("softening" the phonons) would not benefit T_c for Pb, as is confirmed by the fact that amorphous alloys of Pb with Bi have softer phonons and larger λ, but a T_c that is hardly affected. The functional derivatives rise as ω^1 for $\omega \ll 10T_c$ and fall as ω^{-1} for $\omega \gg T_c$. The latter effect occurs because $\delta\lambda/\delta\alpha^2F(\Omega) = 2/\Omega$ according to Eq. (13.8), and because the effect of high-Ω phonons on T_c is governed by λ. For low-Ω phonons, $\lambda\langle\omega^2\rangle$ governs the effect on T_c, and according to Eq. (13.9), $\delta(\lambda\langle\omega^2\rangle)/\delta\alpha^2F(\Omega) = 2\Omega$. The vanishing of $\delta T_c/\delta\alpha^2F(\Omega)$ at $\Omega = 0$ is understood[72] to be a consequence of Anderson's theorem[73]: A zero-frequency phonon would be a static defect and time-reversal invariant, and thus would not affect the T_c of an isotropic superconductor. The magnitude of the peak in $\delta T_c/\delta\alpha^2F(\Omega)$ decreases as $T_c/\langle\omega^2\rangle^{1/2}$ increases, showing that it is harder to increase T_c for a high-T_c material than for a low-T_c material. It is reassuring that $\delta T_c/\delta\alpha^2F(\Omega)$ is always positive. There are no regions of Ω where phonons are harmful to T_c. Mitrović and Carbotte[87a] have noticed that $\delta T_c/\delta\alpha^2F(\Omega)$ equals approximately $(1 + \lambda)^{-1}$ times a universal function of Ω/T_c.

[87a] B. Mitrović and J. P. Carbotte, *Solid State Commun.* **37,** 1009 (1981).

14. Approximate T_c Equations

This section will summarize a few of the more popular "T_c equations" available, explain the "derivation" of McMillan's equation from Eliashberg theory, and mention two alternatives, the BCS equation and the Kirzhnits–Maksimov–Khomskii[88] (KMK) equation. The main justification for having approximate T_c equations is that for most materials, less than perfect information is available. Perfect information would mean that $\alpha^2F(\Omega)$ and μ^* are known, for example, from tunneling,[7,88a] or else can be calculated theoretically. Then there is no need of a "T_c equation," because Eq. (11.10) or (13.2) is easily solvable on a computer. If $\alpha^2F(\Omega)$ is not known, but $F(\Omega)$ is (from neutron scattering probably), then it is often a good approximation to assume that α^2F is proportional to F. The Eliashberg equations can then determine λ from T_c if a reasonable value of μ^* is assumed. Unfortunately, often nothing is known except T_c.

a. *The Simplest Possible T_c Equation*

The usual point of view requires a minimum of three parameters to explain T_c, namely, λ, μ^*, and some measure of the phonon frequencies, ideally $\omega_{\log}$ (to be defined below), but more often only θ_D is available. A surprisingly good picture of T_c uses only one variable, the area $A = \frac{1}{2}\lambda\langle\omega\rangle$ of $\alpha^2F(\Omega)$:

$$T_c \cong 0.1477A. \tag{14.1}$$

This empirical relation was found by Leavens and Carbotte[89] to fit experiment well for a range of materials with $1.2 < \lambda < 2.4$ and $0.10 < \mu^* < 0.15$, where λ, μ^*, and A are determined from tunneling. A similar observation was made by Rowell.[90] Although A is a useful parameter for explaining T_c, it does not seem to occur in other situations, and one might prefer to think of Eq. (14.1) as a two-parameter equation involving λ and $\langle\omega\rangle$ (plus the observation that μ^* does not vary much and T_c is not sensitive to μ^* in this regime.) For low-T_c materials [where tunneling is not sensitive enough to give $\alpha^2F(\Omega)$] the relation (14.1) is completely wrong, and Eq. (1.1) is good. For materials with λ in the range $1.6 < \lambda < 2.4$, Eq. (14.1) is probably better than Eq. (1.1), because McMillan's equation begins to fail in this range. For larger λ, Eq. (14.1) is also wrong.

[88] D. A. Kirzhnits, E. G. Maksimov, and D. I. Khomskii, *J. Low Temp. Phys.* **10,** 79 (1973).
[88a] E. L. Wolf, *Rep. Prog. Phys.* **41,** 1439 (1978).
[89] C. R. Leavens and J. P. Carbotte, *J. Low Temp. Phys.* **14,** 195 (1974).
[90] J. M. Rowell, *Solid State Commun.* **19,** 1131 (1976).

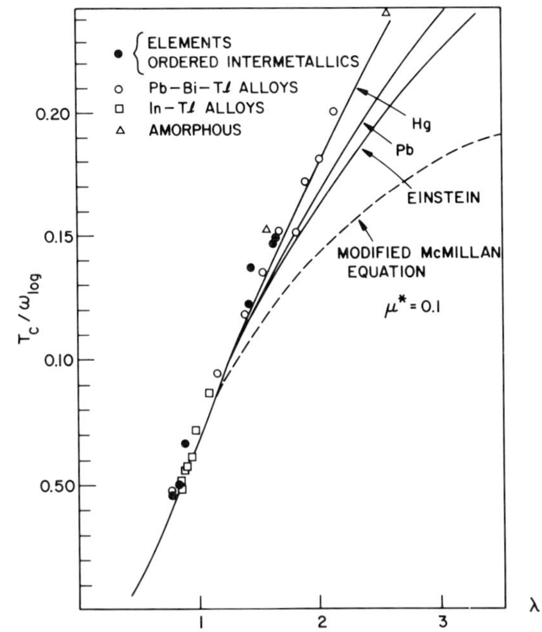

FIG. 8. $T_c/\omega_{\log}$ plotted versus λ for $\mu^* = 0.1$. The solid curves are exact numerical solutions of Eq. (13.2) using $\gamma_P = 0$, variable λ, and for the shape of α^2F an Einstein spectrum $\delta(\omega - \omega_E)$ and the measured spectra for Pd and Hg. The experimental points are from tunneling and are tabulated in P. B. Allen and R. C. Dynes, *Phys. Rev. B* **12**, 905 (1975), from which this figure is reprinted. The modified McMillan equation is Eq. (1.1).

b. The Most Popular T_c Equation

McMillan originally stated Eq. (1.1) with the prefactor $\theta_D/1.45$ in place of $\omega_{\log}/1.2$. The virtue of Eq. (1.1) is that it correctly fits exact solutions of Eliashberg theory provided that (1) $\alpha^2F(\Omega)$ has the shape of $F(\Omega)$ for Nb and (2) $T_c/\theta_D \lesssim 0.075$. Also, it is relatively simple and familiar. It is found empirically[76] that restriction (1) about the shape of α^2F can be relaxed by using $\omega_{\log}/1.2$ in place of $\theta_D/1.45$. The definition of $\omega_{\log}$ is

$$\omega_{\log} = \lim_{n\to 0} \langle \omega^n \rangle^{1/n} = \exp\langle \ln \Omega \rangle, \tag{14.2}$$

$$\langle \ln \Omega \rangle = \frac{2}{\lambda} \int_0^\infty d\Omega \, \alpha^2F(\Omega) \frac{\ln \Omega}{\Omega}. \tag{14.3}$$

Figure 8 illustrates this point. The three solid curves were computed from Eq. (13.2), using quite different shapes of $\alpha^2F(\Omega)$, namely, an Einstein spectrum and the Pb and Hg spectra derived from tunneling. When plotted versus λ with T_c normalized to $\omega_{\log}$, the three curves line up accurately for $\lambda < 1.5$. For extreme cases,[81] even $\omega_{\log}$ is not accurate and more parameters are needed. If T_c had been normalized to $\langle\omega\rangle$, the slopes of the curves would differ by the factor $\langle\omega\rangle/\omega_{\log} = 1, 1.07, 1.31$ (for Einstein, Pb, and Hg, respectively). Even worse discrepancies would have occurred had $T_c/\langle\omega^2\rangle^{1/2}$ or T_c/θ_D been used. Figure 8 also illustrates restriction (2). The McMillan result (2) is plotted as the dashed curve. It agrees very well with exact solutions for $\lambda < 1.2$, but underestimates T_c increasingly badly for larger values of λ. Allen and Dynes attempted to correct for this by introducing correction factors $f_1 f_2$ which multiply Eq. (1.1):

$$f_1 = \{1 + [\lambda/2.46(1 + 3.8\mu^*)]^{3/2}\}^{1/3}, \tag{14.4}$$

$$f_2 = \frac{1 + (\langle\omega^2\rangle^{1/2}/\omega_{\log} - 1)\lambda^2}{[\lambda^2 + 3.31(1 + 6.3\mu^*)^2\langle\omega^2\rangle/\omega_{\log}^2]}. \tag{14.5}$$

These formulas represent an *ad hoc* attempt to reconcile Eq. (1.1) with the correct large λ limit given by the first term of Eq. (13.11). The factor f_1 is shape independent and introduces the $\lambda^{1/2}$ factor for large λ, while f_2 is a shape factor which replaces $\omega_{\log}$ by $\langle\omega^2\rangle^{1/2}$ for large λ. The accuracy of these factors is not outstanding, as can be seen from the work of Cai *et al.*[79] One reason may be that the functional forms could have been chosen with more skill. Another reason is that it is inherently impossible to fit all the numerical results of Ref. 79 with a single shape parameter $\langle\omega^2\rangle^{1/2}/\omega_{\log}$ or any other choice. However, it seems superfluous to introduce more than one extra parameter to improve Eq. (1.1), because if more detailed shape information exists, Eq. (13.2) can be solved exactly.

c. *The Square-Well Model*

McMillan used an approximate solution[91,92] of the T_c equations (12.11) and (12.12) to guide the choice of the functional form of Eq. (1.1). The precise numerical coefficients were then adjusted to fit exact calculations. The method of McMillan is a version of a square-well model. It will now be shown how the same approximate solution emerges from the imaginary frequency equations (11.4) and (11.5) by an analogous approximation scheme. The idea is to replace $\lambda(n - n')$ by a constant, $\lambda(0) = \lambda$, for small n and n', and zero beyond a cutoff N, determined by $\omega_N = 2\pi(N + ½)T_c = \omega_D$. For the first term of S [Eq. (11.11)] the model is

$$\lambda(n - n') = \lambda\theta(\omega_D - |\omega_n|)\theta(\omega_D - |\omega_{n'}|), \tag{14.6a}$$

[91] N. N. Bogoliubov, V. V. Tolmachev, and D. V. Shirkov, "New Method in the Theory of Superconductivity." Acad. Sci. USSR, Moscow, 1958.

[92] S. Nakajima and M. Watabe, *Prog. Theor. Phys.* **29**, 341 (1963).

but for the second term [which derives from $Z(i\omega_n)$], a different version is used,

$$\lambda(\nu) = \lambda\theta(\omega_D - |\omega_\nu|). \tag{14.6b}$$

With this model, the T_c equations become

$$\left(1 + \lambda + \frac{2\gamma_P}{|2n+1|}\right)\Delta_n = (\lambda - \mu^*) \sum_{n'}^{|\omega_{n'}|<\omega_0} \frac{\Delta_{n'}}{|2n'+1|}. \tag{14.7}$$

This equation is valid for $|n| < N$. The Coulomb pseudopotential μ^* is scaled to the same cutoff that is used for $\lambda(m)$; thus μ^* is short for $\mu^*(\omega_D)$. If no magnetic impurities are present, $\gamma_P = 0$, the solution is $\Delta_n = \Delta\theta(\omega_D - |\omega_n|)$, and Eq. (14.7) becomes

$$\frac{1+\lambda}{\lambda - \mu^*} = \sum_{n=0}^{N-1} \frac{1}{n + 1/2} = \psi\left(\frac{\omega_D}{2\pi T_c} + 1\right) - \psi(\tfrac{1}{2}). \tag{14.8}$$

This is exactly Eq. (2.34) with $N(0)V$ replaced by $(\lambda - \mu^*)/(1 + \lambda)$. The result for T_c is therefore

$$T_c = 1.13\omega_D \exp[-(1+\lambda)/(\lambda - \mu^*)]. \tag{14.9}$$

McMillan's version was slightly more subtle and gave a factor $(1 + \lambda\langle\omega\rangle/\omega_D)$ multiplying μ^*. The result (1.1) then evolved in due course. Obviously, refinements can be attempted in this procedure, but experience shows that this does not necessarily lead to improved model T_c equations.[93] The square-well model will be used in Sections 15–17 to obtain approximate answers. However, the weakness of the model should be kept in mind.

d. The Most Rigorous T_c Equation

The most rigorous T_c equation is surely Eq. (13.11), which has only two defects: (1) a large number of input parameters $\langle\omega^{2n}\rangle$ and (2) the restriction to $\lambda > \Lambda \sim 1$–3. Thus Eq. (13.11) is primarily useful for the formal understanding it gives, rather than for its practical applications. For the region $\lambda < \Lambda$, Wu *et al.*[81] have given correction factors involving several new parameters which improve McMillan's equation (1.1). For most known materials, the corrections are very small.

e. Other T_c Equations

These are too numerous to list, some of them are quite useful, but the proliferation of T_c equations has gone much too far.

[93] P. B. Allen, *Solid State Commun.* **14**, 937 (1974).

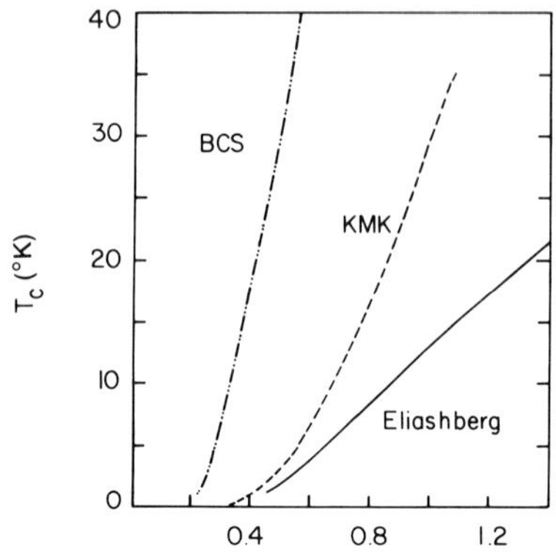

FIG. 9. Exact numerical solutions of T_c versus λ for $\mu^* = 0.1$ and α^2F proportional to $F(\omega)$ for Nb. The curve labeled "Eliashberg" is Eq. (13.2). The curve labeled "BCS" is the isotropic version of Eq. (2.27) with a Bardeen–Pines interaction for $V_{kk'}$. The curve labeled "KMK" is discussed in the text. Reprinted with permission from F. S. Khan and P. B. Allen, *Solid State Commun.* **36,** 481 (1980).

f. The BCS Equation

By the BCS equation we mean Eq. (2.27) with some model for $V_{kk'}$, often the Bardeen–Pines or Fröhlich interaction. Thus it is not a solution for T_c in the sense of Sections 14,a–d above but, instead, an alternative to Eliashberg theory as given in Eqs. (10.15) and (10.16). Unfortunately BCS theory cannot be used to compute T_c, although if T_c is used as an input parameter, Eq. (2.27) (or some generalization of it) can give a fairly accurate picture of how T_c is affected by circumstances such as anisotropy, magnetic impurities, and so forth. Khan and Allen[94] solved the BCS equation numerically and compared it with a corresponding solution of the Eliashberg equation, using an isotropic model with the shape of $\alpha^2F(\Omega)$ of niobium and $\mu^* = 0$ and 0.11. The results for $\mu^* = 0.11$ are shown in Fig. 9. When $\lambda = 0.4$, the BCS equation overestimates T_c by a factor of 18. Part of this comes from the "strong-coupling correction" $1 + \lambda$ in Eq. (1.1). However, if $Z(i\omega_n)$ is set to 1 in Eliashberg theory, a factor of 7 discrepancy still remains between the two theories.

g. The KMK Equation

Like the BCS equation, the KMK equation[88] is not a solution for T_c but, instead, an alternative to the standard Eliashberg theory of Eq. (10.15), (10.16), or (13.2). BCS theory emerges from Eliashberg theory if retardation

[94] F. S. Khan and P. B. Allen, *Solid State Commun.* **36,** 481 (1980).

is ignored. KMK theory emerges in a less drastic way. The aim was to deal with the full Eliashberg equations (8.2)–(8.6) without the subsequent approximations of ignoring the ϵ, ϵ', and $i\omega_\nu$ dependence of the Coulomb interaction which occurred in Section 9 where the μ^* rescaling was made. Since the full four-dimensional equations (8.2)–(8.6) are hopelessly difficult to solve, an approximation scheme was sought which would simplify the equations without losing the possible physics contained in the frequency-dependent electronic screening (i.e., plasmon or exciton attraction.) Unfortunately there may not exist a reliable scheme which does this. KMK made a great simplification of the equations by using the quasi-particle approximation (3.42) for electronic Green's functions. This is a weak-coupling approximation which seems possibly inconsistent with the expectation that interesting physics will emerge from electronic screening. The KMK equation formally resembles the BCS equation. Khan and Allen[94] solved it for the same phonon model previously described. The result shown in Fig. 9 is much better than BCS theory. The equation should be treated with caution, however.[94a]

15. Paramagnetic Impurities

First let us return to the square-well model equation (14.7) and solve it for the case $\gamma_P \neq 0$. The solution has the form

$$\Delta_n = \frac{C}{1 + \lambda + 2\gamma_P/|2n + 1|}\,. \tag{15.1}$$

Substituting this back into Eq. (14.7), the constant C cancels, leaving

$$\frac{1 + \lambda}{\lambda - \mu^*} = \sum_{n=0}^{N'-1} \frac{1}{n + 1/2 + \alpha} = \psi\left(\frac{\omega_D}{2\pi T'_c} + 1 + \alpha\right) - \psi(\tfrac{1}{2} + \alpha), \tag{15.2}$$

$$\alpha = \gamma_P/(1 + \lambda) = (\tfrac{1}{2}\tau_P)/[\pi T'_c(1 + \lambda)], \tag{15.3}$$

where α is the pair-breaking parameter. The digamma function expression for the sum is Eq. (A.7). A prime has been put on N and T_c to emphasize that they are altered by magnetic impurities from the values for pure isotropic materials used previously in Eq. (14.8). The left-hand sides of Eqs. (15.2) and (14.8) are identical, but on the right-hand sides, Eq. (15.2) has a larger denominator. This means that $N' = \omega_c/2\pi T'_c + \tfrac{1}{2}$ must exceed N, and therefore that T'_c is less than T_c. According to Eq. (15.3), there is a feedback effect: Decreasing T'_c makes α increase, which makes T'_c decrease. At a critical concentration n_c of impurities, $T'_c = 0$ and superconductivity disappears.

To simplify these equations, we use the approximate formula

[94a] B. Schuh and L. J. Sham, *J. Low Temp. Phys.* (in press).

$$\ln\left(\frac{T_c}{T'_c}\right) = \sum_{n=0}^{N'-1} \frac{1}{n+1/2} - \sum_{n=0}^{N-1} \frac{1}{n+1/2}$$

$$= \sum_{n=0}^{N'-1} \frac{1}{n+1/2} - \frac{1+\lambda}{\lambda-\mu^*}. \qquad (15.4)$$

When Eq. (15.2) is subtracted from Eq. (15.4), the result is

$$\ln\left(\frac{T_c}{T'_c}\right) = \sum_{n=0}^{N'-1} \frac{1}{n+1/2} - \frac{1}{n+1/2+\alpha}$$

$$= [\psi(\alpha+½) - \psi(½)] - [\psi(N'+\alpha+½) - \psi(N'+½)]$$

$$\cong \psi(\alpha+½) - \psi(½). \qquad (15.5)$$

It is a good approximation over the whole range of physical variation of α and N' to neglect the second term of the middle line of Eq. (15.5) relative to the first. The result is the famous Abrikosov–Gor'kov[63] equation for the reduction of T_c by magnetic impurities. The original Abrikosov–Gor'kov treatment was based on BCS theory, but since T_c is an input parameter in Eq. (15.5), the BCS prediction is quite accurate. The strong-coupling factor $1 + \lambda$ in Eq. (15.3) was missing from the original version of α, but since $1/\tau_P$ is not directly measurable, the factor is absorbed into an arbitrary scale factor. The function $\psi(\alpha + ½) - \psi(½)$ has the following limiting behavior:

$$\psi(\alpha+½) - \psi(½) \cong \pi^2\alpha/2 + \cdots \qquad (\alpha \ll 1) \qquad (15.6a)$$

$$\cong \ln(4e^\gamma\alpha) + 1/24\alpha^2 + \cdots \qquad (\alpha \gg 1). \qquad (15.6b)$$

According to Eq. (15.6b), T'_c goes to zero as $(n_c - n)^{1/2}$. The critical concentration n_c occurs when the scattering rate $1/\tau_P$ reaches the critical value

$$1/\tau_c = (1+\lambda)T_c/1.13. \qquad (15.7)$$

The parameter α can be written as

$$\alpha = 0.14(n/n_c)(T_c/T'_c), \qquad (15.8)$$

where n/n_c is defined as $(1/\tau_P)/(1/\tau_c)$. Thus Eq. (15.5) defines a universal relation between T'_c/T_c and n/n_c. For small n/n_c, the behavior is, according to Eq. (15.6a),

$$T'_c/T_c \cong 1 - 0.69n/n_c. \qquad (15.9)$$

The complete relation is shown in Fig. 10, along with data for Gd impurities in $LaAl_2$ as studied by Maple.[95] Perhaps an even more dramatic confirmation of Abrikosov–Gor'kov theory was the verification of their prediction

[95] M. B. Maple, *Phys. Lett. A* **26A**, 513 (1968).

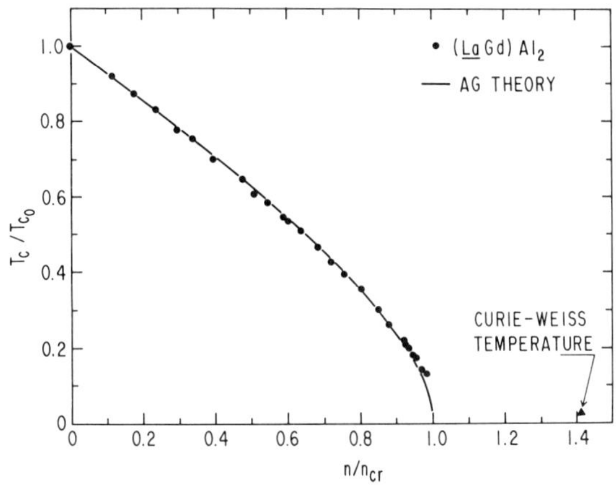

FIG. 10. Abrikosov–Gor'kov theory for the depression of T_c by magnetic impurities [Eq. (15.5)], compared with data for Gd impurities in $LaAl_2$. Reprinted with permission from M. B. Maple, *Phys. Lett.* **26A,** 513 (1968).

that close to the superconducting to normal transition line in Fig. 10, the superconducting state is gapless.[96]

Exact numerical solutions of Eq. (11.10) by Schachinger and Carbotte[97] have verified the accuracy of the square-well model equation (15.5) for a weak-coupling superconductor, Al (with parameters $T_c = 1.18°K$, $\lambda = 0.44$, $\mu^* = 0.15$, and α^2F calculated by Leung *et al.*[12]). For Pb with $\lambda = 1.55$, a calculation by Schachinger *et al.*[98] shows noticeable strong-coupling deviations from Abrikosov–Gor'kov theory.

A more serious problem is the reluctance of magnetic impurity spins in metals to be either simple or stable enough so that the Kondo type of Hamiltonian (6.2) can describe them adequately. For example, the moment often fluctuates at a rate $1/\tau_{sf}$ which is not negligible. Also, there may be complicated low-lying crystal field splittings, or spin–spin interactions. Finally, even when Eq. (6.2) is adequate, if the sign of the interaction $\langle k'|V|k\rangle$ is positive (antiferromagnetic coupling), higher than second-order diagrams of the type shown in Fig. 4b and c diverge as $T \to 0$ (the Kondo effect), invalidating the treatment used by Abrikosov and Gor'kov (and used here) at low T. These problems have attracted much interest, because the sensitivity of superconducting parameters (especially T_c and specific heat) to localized spins makes this a valuable probe of the magnetic state of im-

[96] K. Maki, *in* "Superconductivity" (R. D. Parks, ed.), Vol. 2, Chapter 18. Dekker, New York, 1969.

[97] E. Schachinger and J. P. Carbotte, *Phys. Rev. B* **24,** 6723 (1981).

[98] E. Schachinger, J. M. Daams, and J. P. Carbotte, *Phys. Rev. B* **22,** 3194 (1980).

purities in metals. The subject has been reviewed by Maple[99,100] and will not be discussed further here, except to mention the prediction of a "reentrant" superconducting transition by Müller-Hartmann and Zittartz.[101] They suggested that the "pair-breaking parameter" α [Eq. (15.3)] of Abrikosov–Gor'kov theory should acquire extra T dependence because of the Kondo effect, increasing as T decreases (for $T > T_k$, the Kondo temperature). For the case where $T_c > 8T_k$, and a range of concentrations $n < n_c$, the sample is superconducting for $T < T'_c$, but at still lower temperatures α has increased and drives the sample normal again. This interesting effect was seen[102] in $LaAl_2$ ($T_c \sim 3.3°K$) with Ce impurities ($T_k \sim 0.1°K$), for a Ce concentration of about 0.6%.

16. Anisotropic Superconductors

It is well known[103] that the energy gap $\Delta(k, \omega)$ varies from point to point on the Fermi surface. Direct evidence comes from such experiments as tunneling and ultrasonic attenuation. Indirect evidence is available in thermodynamic properties, such as T_c. Figure 11 shows T_c for Zn-based alloys as a function of the electron to atom ratio e/a. In "dirty" alloys, T_c displays a nearly linear variation with e/a, while the cleaner alloys show a cusplike behavior[104] centered on pure Zn. The interpretation, due first to Anderson,[73] is that pure Zn has a slightly enhanced value of T_c because of gap anisotropy. The "dirty" alloys have an isotropic gap and a nearly linear T_c versus e/a variation because of the "valence effect," that is, variation in electron and phonon behavior as e/a changes. As the alloys become cleaner, near pure Zn, gap anisotropy appears, and T_c rises above the simple valence behavior. An impurity scattering rate $1/\tau_k \ll \Delta$ will not affect gap anisotropy, but if $1/\tau_k \gg \Delta_k$, the gap anisotropy is washed out. This section neglects the valence effect, which requires sophisticated alloy theory, and focuses on the anisotropy effects.

The first theories were two-band models[105,106] in which $\Delta(k)$ was allowed to assume only two values, presumed to be on different sheets of the Fermi surface. The essential physics is already contained in this model, but the strict interpretation of the two different sheets, as well as the suggestion of possible extreme variation in $\Delta(k)$, should probably be abandoned.

[99] M. B. Maple, *in* "Magnetism: A Treatise on Modern Theory and Materials" (G. Rado and H. Suhl, eds.), Vol. 5, Chapter 10. Academic Press, New York, 1973.
[100] M. B. Maple, *Appl. Phys.* **9,** 179 (1976).
[101] E. Müller-Hartmann and J. Zittartz, *Phys. Rev. Lett.* **26,** 428 (1971).
[102] G. Riblet and K. Winzer, *Solid State Commun.* **9,** 1663 (1971).
[103] H. W. Weber, ed., "Anisotropy Effects in Superconductors." Plenum, New York, 1977.
[104] B. Serin, *Proc. Int. Conf. Low Temp. Phys., 7th, 1960,* p. 391 (1961).
[105] H. Suhl, B. T. Matthias, and L. R. Walker, *Phys. Rev. Lett.* **3,** 552 (1959).
[106] V. A. Moskalenko, *Fiz. Met. Metalloved.* **8,** 503 (1959).

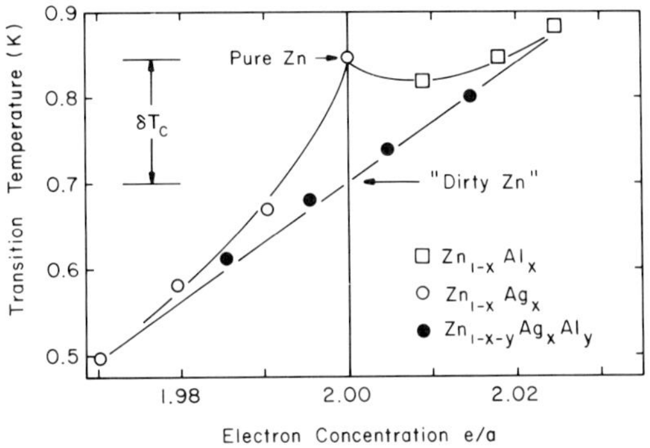

FIG. 11. T_c of Zn alloys plotted versus electron concentration. The solid lines are merely visual aids. The point labeled "dirty Zn" is probably not completely isotropic. The data show that pure Zn has T_c enhanced by $\delta T_c > 0.13$°K because of anisotropy, which is washed out in the dirty alloys. The data are from D. Farrell, Ph.D. Thesis, University of London, 1964 (unpublished); in the published version [D. Farrell, J. G. Park, and B. R. Coles, *Phys. Rev. Lett.* **13,** 328 (1964)] the horizontal scale was wrong.

Pokrovskii[107] made a nice theory, but left out impurities. Numerous groups[108–113] around 1962 added the impurity effects, following the work of Abrikosov and Gor'kov.[74] These were all BCS theories. The extension to Eliashberg theories began perhaps with Leavens and Carbotte.[114] The discussion given here is similar to Refs. 109 and 112. A different procedure was given by Allen.[115]

We begin by writing the anisotropic Eliashberg equations, Eqs. (10.15) and (10.16), in a more compact notation:

$$Z(k, i\omega_n) = 1 + \sum_{k'n'} w_{k'} w_n V^{(+)}(k, k'; n - n') s_n s_{n'}\,, \tag{16.1}$$

$$Z(k, i\omega_n)\Delta(k, i\omega_n) = \sum_{k'n'} w_{k'} w_{n'} V^{(-)}(k, k'; n - n')\Delta(k', i\omega_{n'}), \tag{16.2}$$

$$w_k = \delta(\epsilon_k)/N(0), \tag{16.3}$$

[107] V. L. Pokrovskii, *Zh. Eksp. Teor. Fiz.* **40,** 641 (1961); *Sov. Phys.—JETP (Engl. Transl.)* **13,** 447 (1961).

[108] T. Tsuneto, *Prog. Theor. Phys.* **28,** 857 (1962).

[109] C. Caroli, P. G. deGennes, and J. Matricon, *J. Phys. Radium* **23,** 707 (1962).

[110] D. Markowitz and L. P. Kadanoff, *Phys. Rev.* **131,** 563 (1963).

[111] M. J. Zuckermann and D. M. Brink, *Phys. Lett.* **4,** 76 (1963); D. M. Brink and M. J. Zuckermann, *Proc. Phys. Soc., London* **85,** 329 (1965).

[112] P. Hohenberg, *Zh. Eksp. Teor. Fiz.* **45,** 1208 (1963); *Sov. Phys.—JETP (Engl. Transl.)* **18,** 834 (1964).

[113] J. R. Clem, *Phys. Rev.* **148,** 392 (1966).

[114] C. R. Leavens and J. P. Carbotte, *Ann. Phys.* (N.Y.) **70,** 338 (1972).

[115] P. B. Allen, *Z. Phys. B* **47,** 45 (1982).

$$w_n = |2n + 1|^{-1}, \tag{16.4}$$

$$V^{(+)}(k, k'; n - n') = \lambda(k, k'; n - n') + \gamma_N \delta_{nn'}, \tag{16.5}$$

$$V^{(-)}(k, k'; n - n') = \lambda(k, k'; n - n') - \mu^* + \gamma_N \delta_{nn'}. \tag{16.6}$$

For simplicity, μ^* and γ_N have been assumed to be isotropic, and γ_P has been neglected, because usually magnetic impurity effects, when present, will swamp the anisotropy effect. Just as was done in the isotropic theory (11.10), $Z(i\omega_n)$ can be eliminated from Eq. (16.2), yielding

$$\rho\Delta(k, i\omega_n) = \sum_{k'n'} w_{k'} w_{n'} K(k, k'; n, n')\Delta(k'n'). \tag{16.7}$$

The formula for the kernel K will be given below. This is a Hermitian eigenvalue problem, if inner products are defined with the (positive) weight $w_k w_n$, analogous to Eqs. (11.13) and (11.14). The temperature at which the maximum eigenvalue ρ equals 1 gives T_c. A lower bound on the maximum eigenvalue ρ can be estimated as $\langle\Delta|\hat{K}|\Delta\rangle$ for any normalized trial eigenvector $|\Delta\rangle$. In particular, the trial eigenvector can be chosen to be the solution $\Delta(i\omega_n)$ of the isotropic problem (11.10). The value of $\langle\Delta|\hat{K}|\Delta\rangle$ is just $\langle\Delta|\hat{S}|\Delta\rangle$, the isotropic eigenvalue. The true eigenvalue and thus the true T_c exceed the estimate from isotropic theory. Thus anisotropy increases T_c above the isotropic value. This argument was formulated by Rainer.[116]

Let us represent $\Delta(k, i\omega_n)$ as its isotropic average plus its fluctuation around the average:

$$\Delta(k, i\omega_n) = \Delta(i\omega_n) + \delta(k, i\omega_n), \tag{16.8}$$

$$\Delta(i\omega_n) = \langle\Delta(k, i\omega_n)\rangle_{FS} = \sum_k w_k \Delta(k, i\omega_n), \tag{16.9}$$

$$0 = \sum_k w_k \delta(k, i\omega_n). \tag{16.10}$$

Note that the weight w_k gives the Fermi surface average, defined originally in Eq. (3.24a). Similarly, we represent the kernel K by its average plus fluctuations:

$$S(n, n') \equiv \langle\langle K(k, k'; n, n')\rangle\rangle_{FS}, \tag{16.11a}$$

$$K(k, k'; n, n') = S(n, n') + \delta K(k, k'; n, n'). \tag{16.11b}$$

An explicit formula for the kernel K is

$$K(k, k'; n, n') = V^{(-)}(k, k'; n - n') - \delta^{(2)}_{kk'}\delta_{nn'} \sum_{n''} V^{(+)}(k; n - n'')s_n s_{n''}, \tag{16.12}$$

where $\delta^{(2)}_{kk'}$ and $V^{(+)}(k; m)$ are defined by

[116] D. Rainer, *Solid State Commun.* **6**, 111 (1968).

$$\sum_{k'} w_{k'} \delta^{(2)}_{kk'} h(k') = h(k), \tag{16.13}$$

$$V^{(\pm)}(k; m) = \sum_{k'} w_{k'} V^{(\pm)}(k, k'; m). \tag{16.14}$$

The operator $\delta^{(2)}_{kk'}$ is a Fermi surface unit operator. From Eqs. (16.11), (16.12), (16.5), and (16.6) it can be seen that $S(n, n')$ is the isotropic kernel, Eq. (11.11). We also need to compute the average of δK when one variable, k or k', is averaged:

$$\delta K(k; n, n') = \delta\lambda(k, n - n') - \delta_{nn'} \sum_{n''} \delta\lambda(k, n - n'') s_n s_{n''}, \tag{16.15}$$

where $\delta\lambda(k; m)$ is $\lambda(k, m) - \lambda(m)$.

From here on, we assume that the fluctuations $\delta\Delta(k, i\omega_m)$ and $\delta\lambda(km)$ are small; both experiment and theory have found this to be generally true. Zinc, shown in Fig. 11, has a comparatively large anisotropy, but the rms value $\langle \delta\lambda_k^2 \rangle_{\mathrm{FS}}^{1/2}$ is no larger than 20% of λ. The variational principle then tells us that the change δT_c in T_c is a second-order effect. This is magnified in Zn because $\lambda - \mu^*$ is small and a second-order change inside the exponential of Eq. (1.1) can be significant. It is possible to avoid the assumption of small anisotropy by making a simplified model, such as the two-band model or the factorizable model.[110] Whitmore and Carbotte[116a] used this model for the electron–phonon attraction and found that nonzero (but very small) T_c's occur even if $\mu^* > \lambda$, provided that some anisotropy is present.

The expansions (16.8) and (16.11b) are now inserted into the basic equation (16.7). If this equation is then averaged over k, the result is the isotropic theory (11.10) plus a term involving $\delta K(k'; n, n')\delta(k'n')$ summed over k' and n'. This latter term is second order in the fluctuations, so we neglect it. Next, we subtract the isotropic equation (11.10) from the full equation (16.7). The result is (setting $\rho = 1$)

$$\delta(k, n) = \sum_{n'} w_{n'} \delta K(k; n, n')\Delta(n') + \sum_{k'n'} w_{k'} w_{n'} \delta K(k, k'; n, n')\delta(k'n'). \tag{16.16}$$

Surprisingly, the last term of Eq. (16.16) is not second order, because there is a piece of $\delta K(k, k'; n, n')$ which is zeroth order, namely,

$$-(\delta^{(2)}_{kk'} - 1)\delta_{nn'} \sum_{n''} V^{(+)}(n - n'') s_n s_{n''} . \tag{16.17}$$

This piece vanishes when either k or k' is averaged, and thus does not appear in Eq. (16.15). Using this piece of δK in the last term of Eq. (16.16), the result is $[1 - Z(i\omega_n)]\delta(k, n)$, where $Z(i\omega_n)$ is defined in Eq. (11.4). Finally, Eq. (16.16) becomes

[116a] M. D. Whitmore and J. P. Carbotte, *Phys. Rev. B* **23,** 5782 (1981).

$$\delta(k, n) = Z^{-1}(i\omega_n) \sum_{n'} \delta\lambda(k, n - n')[w_{n'}\Delta(i\omega_{n'}) - s_n s_{n'} w_n \Delta(i\omega_n)]. \quad (16.18)$$

The reason for the factor Z^{-1} is obvious from examining Eq. (16.2). Equation (16.18) provides a simple way to compute the anisotropic gap, given the isotropic solution $\Delta(i\omega_n)$ and the anisotropic spectral function $\alpha^2F(k, \Omega)$; it is exact to first order in anisotropy, and computationally much simpler than the full equations (16.7). The theory of Ref. 115 gave a more complicated but still tractable result and the differences are probably negligible.

The effect of impurities on the gap anisotropy $\delta(k, i\omega_n)$ occurs only in the factor $Z(i\omega_n)$ in the denominator, which can be written as $Z_0(i\omega_n) + \gamma_N/|2n + 1|$, where Z_0 is the renormalization function (11.4) for a pure superconductor. Thus nonmagnetic impurities reduce the gap anisotropy quite rapidly, similar to the way in which magnetic impurities weaken the isotropic pairing [see Eq. (15.1)]. However, the feedback effect that caused T_c to vanish for $n > n_c$ in the magnetic case does not operate in this case.

To calculate T_c of the anisotropic superconductor, we can improve the estimate of the largest eigenvalue ρ by using $\Delta + \delta$ as a trial function:

$$\delta\rho = \rho - \rho_0 = \frac{\langle\Delta + \delta|\hat{K}|\Delta + \delta\rangle}{\langle\Delta + \delta|\Delta + \delta\rangle} - \frac{\langle\Delta|\hat{K}|\Delta\rangle}{\langle\Delta|\Delta\rangle}, \quad (16.19)$$

where ρ_0 is used to denote the isotropic estimate. To lowest order, $\rho_0 = \langle\Delta|\hat{K}|\Delta\rangle/\langle\Delta|\Delta\rangle$ is 1. To second order, Eq. (16.19) can be written as

$$\delta\rho \simeq \frac{2\langle\delta|\hat{K}|\Delta\rangle + \langle\delta|\hat{K}|\delta\rangle - \langle\delta|\delta\rangle}{\langle\Delta|\Delta\rangle}. \quad (16.20)$$

This can further be simplified by defining a diagonal operator $\hat{Z}_0$ that is related to the isotropic renormalization (11.4). Equation (16.18) for $|\delta\rangle$ can then be written as

$$\hat{Z}_0|\delta\rangle = \delta\hat{K}|\Delta\rangle, \quad (16.21)$$

$$Z_0(k, k'; n, n') = \delta^{(2)}_{kk'}\delta_{nn'}Z(i\omega_n). \quad (16.22)$$

In the first term of Eq. (16.20), $\hat{K}$ can be replaced by $\delta\hat{K}$, since $\langle\delta|\hat{S}|\Delta\rangle$ is zero. The next term should be evaluated by using the isotropic approximation to $\hat{K}$; only the piece given in Eq. (16.17) contributes. Thus the first two terms of Eq. (16.20) are

$$\langle\delta|\hat{K}|\Delta\rangle = \langle\delta|\hat{Z}_0|\delta\rangle, \quad (16.23)$$

$$\langle\delta|\hat{K}|\delta\rangle \simeq \langle\delta|(\hat{1} - \hat{Z}_0)|\delta\rangle. \quad (16.24)$$

Thus Eq. (16.20) simplifies to the expression

$$\delta\rho = \frac{\langle\delta|\hat{Z}_0|\delta\rangle}{\langle\Delta|\Delta\rangle} = \frac{\sum_{kn} w_k w_n Z_0(i\omega_n)|\delta(k, n)|^2}{\sum_n w_n|\Delta(i\omega_n)|^2}. \tag{16.25}$$

This can be easily computed using Eq. (16.18) for $\delta(k, n)$.

Now we estimate $\delta T_c = T_c - T_0$, where T_0 is the isotropic estimate found from $\rho_0(T_0) = 1$ and T_c is the anisotropic value found from $\rho(T) = 1$, or

$$1 = \rho_0(T_c) + \delta\rho, \tag{16.26}$$

$$\rho_0(T_c) \cong \rho_0(T_0) + \delta T_c(d\rho_0/dT)|_{T_0}. \tag{16.27}$$

The first term in Eq. (16.27) equals 1. Thus the T_c enhancement is

$$\delta T_c = -\delta\rho/(d\rho_0/dT)|_{T_0}. \tag{16.28}$$

The denominator of Eq. (16.28) is readily computed from the isotropic theory of Section 11. Explicit formulas are cited in Ref. 115.

To get further insight, let us evaluate Eq. (16.28) using the square-well model, Eqs. (14.6), for $\delta\lambda(k, m)$. With no magnetic impurities, the isotropic solution is $\Delta(i\omega_n) = \Delta\theta(\omega_D - |\omega_n|)$. The renormalization $Z(i\omega_n)$ equals $1 + \lambda + \gamma_N/|2n + 1|$. The anisotropic part $\delta(k, m)$ of the gap becomes, from Eq. (16.18),

$$\frac{\delta(k, n)}{\Delta} \simeq \left(1 + \lambda + \frac{\gamma_N}{|2n + 1|}\right)^{-1}\left(\sum_{n'=0}^{N-1}\frac{1}{n' + 1/2} - 1\right)\delta\lambda_k, \tag{16.29}$$

where the sum can be evaluated as in Eq. (2.35). Thus the gap anisotropy is proportional to the λ_k anisotropy $\delta\lambda_k = \lambda_k - \lambda$. This equation was derived by Butler and Allen[39] for the special case $\gamma = 0$.

The most complete model solution of anisotropic Eliashberg theory so far is the calculation by Peter *et al.*[117] for Nb. Ashkenazi *et al.*[118] compared the exact numerical results for Nb with the square-well model (16.29). Overall agreement was fairly good, although some strong-coupling corrections were observed.

A square-well formula for the T_c enhancement is derived as follows. The numerator $\delta\rho$ of Eq. (16.28) follows from Eqs. (16.25) and (16.29):

$$\delta\rho = \frac{\langle\delta\lambda_k^2\rangle_{FS}}{1 + \lambda}\left[\log\left(\frac{1.13\omega_D}{T_c}\right) - 1\right]^2 f(\alpha'), \tag{16.30}$$

[117] M. Peter, J. Ashkenazi, and M. Dacorogna, *Helv. Phys. Acta* **50**, 267 (1977).

[118] J. Ashkenazi, M. Dacorogna, and P. B. Allen, *Solid State Commun.* **36**, 1051 (1980).

$$f(\alpha') = \sum_{n=0}^{N-1} \frac{1}{n + 1/2 + \alpha'} \bigg/ \sum_{n=0}^{N-1} \frac{1}{n + 1/2}, \tag{16.31}$$

$$\alpha' = \gamma_N/2(1 + \lambda) = 1/4\pi T_c \tau_N(1 + \lambda). \tag{16.32}$$

The function $f(\alpha')$ is the correction for impurities and goes to 1 when the impurity factor $\alpha' = 0$. Note that α' differs by ½ from the corresponding parameter α [Eq. (15.3)] used for paramagnetic impurities. The denominator of Eq. (16.28) is evaluated by observing that for a square well, $\rho_0 = \langle\Delta|S|\Delta\rangle/\langle\Delta|\Delta\rangle$ equals [from Eq. (11.10)]

$$\rho_0 = (\lambda - \mu^*) \ln\left(\frac{1.13\omega_D}{T}\right) - \lambda, \tag{16.33}$$

$$d\rho_0/dT = -(\lambda - \mu^*)/T. \tag{16.34}$$

Finally, combining Eqs. (16.28), (16.30), and (16.34) and using Eq. (14.9) to eliminate the logarithm, we get

$$\left(\frac{\delta T_c}{T_c}\right)_{\text{pure}} = \frac{(1 + \mu^*)^2}{(1 + \lambda)(\lambda - \mu^*)^3} \langle\delta\lambda_k^2\rangle_{FS}, \tag{16.35}$$

$$\left(\frac{\delta T_c}{T_c}\right)_{\text{dirty}} = \left(\frac{\delta T_c}{T_c}\right)_{\text{pure}} f(\alpha'). \tag{16.36}$$

These are essentially the well-known results of Markowitz and Kadanoff,[110] except generalized to a more realistic model of the interactions which includes strong-coupling corrections $1 + \lambda$, and does not assume that the interaction is factorizable, but does assume weak anisotropy. The sums in Eq. (16.31) for $f(\alpha)$ have to be performed differently from the treatment given in Section 15 for paramagnetic impurities, because there is no feedback effect causing N to increase as α increases. Thus the approximation in Eq. (15.5) is not always applicable here. The T_c-versus-n curves will have upward curvature, as in Fig. 11, rather than downward curvature as in Fig. 10. The sum (16.31) can be evaluated exactly using Eq. (15.2):

$$f(\alpha') = \left[\psi\left(\frac{\omega_D}{2\pi T_c} + 1 + \alpha'\right) - \psi(1/2 + \alpha')\right]\left[\psi\left(\frac{\omega_D}{2\pi T_c} + 1\right) - \psi(1/2)\right]^{-1}. \tag{16.37}$$

If α' is neglected compared with $\omega_D/2\pi T_c$ in the first digamma function, this agrees with Eq. (18) of Ref. 112, except for factors of $1 + \lambda$. In the nearly pure limit (small α') the formula is

$$f(\alpha') \sim 1 - \frac{\pi/8T_c\tau_N(1 + \lambda)}{\log(\omega_D/T_c)} \qquad (\alpha \ll 1), \tag{16.38}$$

while for large α (but with $1/\tau \ll \omega_D$)

$$f(\alpha') \sim 1 - \ln[1/2T_c\tau_N(1 + \lambda)]/\ln(\omega_D/T_c) \qquad (\alpha \gg 1). \qquad (16.39)$$

In these equations, the factor of 1.13 inside the log has been dropped. Probably ω_D should be replaced by $\omega_{\log}/1.2$, as in Eq. (1.1). Markowitz and Kadanoff found good agreement with experiment. The treatment by Allen[115] used a slightly different perturbation theory that seems equally rigorous and which gives a slightly different answer:

$$f_A(\alpha') = [1 + (\pi^2/2)\alpha'/\ln(\omega_D/T_c)]^{-1}. \qquad (16.40)$$

This agrees with Eq. (16.38) for small α', but disagrees somewhat with Eq. (16.39). Probably the difference is impossible to detect experimentally because of the difficulty of subtracting out the valence effect in the regime of large α'.

The role of low-frequency phonons in anisotropic superconductors was studied by Daams and Carbotte.[119] They found that the functional derivative $\delta T_c/\delta\alpha^2F(\Omega)$ becomes negative at low frequencies. In isotropic superconductors, $\delta T_c/\delta\alpha^2F(\Omega)$ is always positive, but goes to zero as ω^1, when ω tends to zero. In a model for anisotropic Pb, with $\langle\delta\lambda_k^2\rangle/\lambda^2 = 0.04$, they found $\delta T_c/\delta\alpha^2F(\Omega) \sim -\omega^{-1}$ as ω tends to zero. The interpretation is that small-ω phonons are essentially static defects which do not affect the isotropic T_0 but reduce the anisotropic enhancement δT_c. In their model, δT_c was about 3% of T_c, with low-frequency phonons causing a tiny reduction of 0.001%.

The parameter α' of Eq. (16.32) can be estimated from the residual resistance ratio $\mathrm{rrr} = \rho_{300}/\rho_0$. According to Boltzmann theory, $\mathrm{rrr} = (1/\tau_{ph} + 1/\tau_N)/(1/\tau_N)$ and $1/\tau_{ph} = 2\pi\lambda_{tr}T_{300}$. The difference between λ and λ_{tr} is small[120] and can be neglected. Thus

$$\alpha' = \frac{1}{2}\left(\frac{\lambda}{1+\lambda}\right)\left(\frac{T_{300}}{T_c}\right)\frac{1}{\mathrm{rrr} - 1}. \qquad (16.41)$$

This assumes the validity of Boltzmann theory, which seems to be accurate for transition elements,[120] but questionable for A15 compounds.[121]

17. "*p*-WAVE" SUPERCONDUCTORS

The BCS pairing scheme ($k\uparrow$, $-k\downarrow$) is a singlet spin state. Antisymmetrization requires a spatial part of the pair wave function which has even parity; for a spherically symmetric situation this means even angular mo-

[119] J. M. Daams and J. P. Carbotte, *Solid State Commun.* **33,** 585 (1980).

[120] F. J. Pinski, P. B. Allen, and W. H. Butler, *Phys. Rev. B* **23,** 5080 (1981).

[121] P. B. Allen, *in* "Superconductivity in d- and f-Band Metals" (H. Suhl and M. B. Maple, eds.), p. 291. Academic Press, New York, 1980.

mentum l, and in particular there is no reason to doubt that $l = 0$ in all known metals. The possibility of $l \neq 0$ pairing and, in particular, spin triplet $l = 1$ pairing has aroused interest since the beginning of BCS theory. The discovery[122] of a pairing transition in ^{3}He, now known to be $l = 1$, has aroused new interest in the possible exotic pairing schemes which could occur in metals.

Quite a few estimates have been made of possible transition temperatures into exotic states in metals. The level of reliability of these calculations is uniformly low. This section attempts to clarify the description of possible exotic states, why they are harder to calculate reliably than normal T_c's, and how they are affected by defects.

The basic equation (16.7) has a kernel $K(k, k'; n, n')$ which is invariant under the operations R of the point group of the crystal; that is,

$$K(Rk, Rk'; n, n') = K(k, k'; n, n') \tag{17.1}$$

for all rotations R, proper and improper, which leave the crystal invariant. The eigenvectors $\Delta(k, i\omega_n)$ can then be chosen to transform according to the irreducible representations of the point group; that is,

$$\Delta_{i\alpha}(Rk, i\omega_n) = \sum_{\beta} \Gamma^i(R)_{\alpha\beta}\Delta_{i\beta}(k, i\omega_n), \tag{17.2}$$

where $i\alpha$ labels the representation and the row and the matrix $\Gamma^i(R)$ is a transformation matrix for the representation i, which is irreducible in the usual sense of group representation theory. The highest symmetry crystalline phases are cubic, with 10 irreducible representations. Thus, in principle, in addition to the usual BCS pairing (Γ_1 or identity representation with $\Gamma^1(R) = 1$ for all R), there are nine exotic pairing schemes which are allowed. Probably the simplest alternative to Γ_1 is Γ_{15}, the vector representation analogous to $l = 1$ in spherical symmetry. We assume that the true meaning of "p-wave" pairing in metals is Γ_{15} pairing. In crystals with low symmetry there are many fewer irreducible representations (and none with degeneracy 3), so it seems that the phrase "p-wave" is probably meaningless or logically impossible in these cases.

The variational principle can be applied to seek the largest eigenvalue of $\hat{K}$ that corresponds to an eigenvector of Γ_{15} symmetry. Unlike the Γ_1 case, there is no uniquely reasonable zeroth-order choice for the k dependence of Δ:

$$\Delta(k, i\omega_n) \cong \Delta(i\omega_n) \qquad (\Gamma_1), \tag{17.3}$$

$$\Delta(k, i\omega_n) \cong \alpha v_{kx}\Delta_1(i\omega_n) + \beta k_x\Delta_2(i\omega_n) + \cdots ? \qquad (\Gamma_{15}). \tag{17.4}$$

In the Γ_1 case, Eq. (17.3) is always a reasonable first guess, whereas in Eq.

[122] D. D. Osheroff, R. C. Richardson, and D. M. Lee, *Phys. Rev. Lett.* **28,** 885 (1972).

(17.4), there are at least two equally reasonable vector functions of k, namely, $\mathbf{k}$ and $\mathbf{v}_k = \partial\epsilon_k/\partial\mathbf{k}$, which are completely dissimilar in a metal like Pd, but both reduce to spherical harmonics Y_{lm} in the spherical case. Furthermore, if there is more than one sheet of Fermi surface, the coefficients α and β will vary from sheet to sheet, having no reason at all to stay similar in magnitude or sign, unlike the Γ_1 case. Thus while a no-parameter variational estimate (17.3) yields an accurate approximation T_0 in the Γ_1 case, one should have at least $2s - 1$ parameters (where s is the number of sheets) in the Γ_{15} case. So far this has not been done. Another reason for greater computational difficulty in exotic phases is that there is less reason to trust square-well models. The non-s-wave spectral functions analogous to $\alpha^2F(\Omega)$ can oscillate in sign.

In ^{3}He it appears that the triplet-paired state is enhanced by spin–fluctuation interactions (or "paramagnon" effects, to be discussed in Section 20). It is presumed that paramagnon effects may also be helpful in promoting exotic phases in metals such as Pd. A discussion of paramagnon interactions in $l = 0$ and $l = 1$ superconductivity is given by Nakajima.[123] In this section we shall discuss only the electron–phonon case.

Butler and Allen[39] found that the exact anisotropic Eqs. (10.15) and (10.16) can be written as

$$\Delta_J(i\omega_n) = \sum_{J',n'} w_{n'}K_{JJ'}(n, n')\Delta_{J'}(i\omega_{n'}), \tag{17.5}$$

$$\begin{aligned} K_{JJ'} = {} & \lambda_{JJ'}(\omega_n - \omega_{n'}) - \mu^*_{JJ'} \\ & - \delta_{nn'} \sum_{n''} \Lambda_{JJ'}(\omega_n - \omega_{n''})s_n s_{n''} + \delta_{nn'}(\gamma_{JJ'} - \Gamma_{JJ'}). \end{aligned} \tag{17.6}$$

These equations are valid for *all* irreducible representations. They are simply Eqs. (10.12) and (10.14) with Z eliminated to give a single equation for Δ. The simplest derivation is to start with Eq. (16.7) in k space and to recast into J space using the orthogonality property (3.27) of Fermi surface harmonics $F_J(k)$. In an obvious notation, where Eq. (3.31) can be written as

$$\alpha^2F(J, J'; \Omega) = \langle J|\alpha^2F(k, k'; \Omega)|J'\rangle, \tag{17.7}$$

the as yet undefined parameters Λ, γ, and Γ are

$$\Lambda_{JJ'}(\omega_\nu) = \langle JJ'|\lambda(k, k'; \omega_\nu)|0\rangle, \tag{17.8}$$

$$\gamma_{JJ'} = \langle J|(1/\tau_N - 1/\tau_P)|J'\rangle/2\pi T_c\,, \tag{17.9}$$

$$\Gamma_{JJ'} = \langle JJ'|(1/\tau_N + 1/\tau_P)|0\rangle/2\pi T_c\,, \tag{17.10}$$

where $\langle k|0\rangle = F_0(k) = 1$. Thus in the Γ_1 case, using only $J = J' = 0$ in

[123] S. Nakajima, *Prog. Theor. Phys.* **50,** 1101 (1973).

lowest order, $\gamma_{00} - \Gamma_{00} = -1/\pi T_c \tau_P$, and nonmagnetic impurities cancel from Eq. (17.6) (Anderson's theorem).

In a square-well model[39,124] the solution of Eq. (17.5) for a pure superconductor is

$$T_c = 1.13\omega_D \exp(-1/\lambda_{\text{eff}}), \tag{17.11}$$

$$\lambda_{\text{eff}} = \text{max eigenvalue } [(\hat{1} + \hat{\Lambda})^{-1/2}(\hat{\lambda} - \hat{\mu}^*)(\hat{1} + \hat{\Lambda})^{1/2}]. \tag{17.12}$$

This generalizes the usual result (14.9). The matrices are all block diagonalized by using functions F_J, which transform as basis functions for irreducible representations. The "*p*-wave" transition temperature uses Eq. (17.11) with λ_{eff}, the largest eigenvalue of the Γ_{15} submatrix in Eq. (17.12). The coupling constants seem likely to be quite small in all but the Γ_1 submatrix.

For an impure non-Γ_1 superconductor, an equation analogous to Eq. (15.2) holds. The pair-breaking parameter α'' is

$$\alpha'' = (\Gamma_{jj} - \gamma_{jj})/2(1 + \Lambda_{jj}), \tag{17.13}$$

where $|j\rangle$ is the eigenvector of Eq. (17.12). In principle, the k dependence of $|j\rangle$ varies with impurity concentration, but this will not affect Eq. (17.13) much. For non-Γ_1 channels, $\gamma_{jj} \ll \Gamma_{jj}$ and $\Gamma_{jj} \simeq \Gamma_{00} = (1/\tau)/2\pi T_c$, where $1/\tau$ counts *both* magnetic and nonmagnetic scattering rates, and $\Lambda_{jj} \simeq \Lambda_{00} = \lambda$. Thus we have, for any symmetry except Γ_1,

$$\alpha'' = 1/4\pi T_c \tau(1 + \lambda). \tag{17.14}$$

The reason for the equations $\Gamma_{jj} \simeq \Gamma_{00}$, and so on, is that for any normalized function $F_j(k)$, $F_j^2(k) = 1 +$ terms with $j \neq 0$, and the $\langle 0|(\quad)|0\rangle$ matrix elements always dominate. Note the factor of ½ in Eq. (17.14) relative to Eq. (15.3), a result which occurred also in Eq. (16.32). Then, by reference to Eq. (15.7), the critical scattering rate for the destruction of non-Γ_1 superconductivity is

$$1/\tau_c = 2(1 + \lambda)T_c/1.13. \tag{17.15}$$

Using the argument that led to Eq. (16.41), the critical resistance ratio rrr_c below which superconductivity is destroyed is

$$\text{rrr}_c = 1 + 1.13\pi\left(\frac{\lambda}{1 + \lambda}\right)\frac{T_{300}}{T_c}. \tag{17.16}$$

Thus for the case of Pd with $\lambda \simeq 0.4$, there is not much point in searching for *p*-wave superconductivity below 0.01°K unless rrr exceeds 30,000.

[124] F. J. Pinski, P. B. Allen, and W. H. Butler, *in* "Superconductivity in d- and f-Band Metals" (H. Suhl and M. B. Maple, eds.), p. 215. Academic Press, New York, 1980.

V. Complications and Speculations

18. Anharmonic Effects[125]

The smallness of the parameter u/a (lattice displacement over lattice constant) at $T = T_c$ guarantees that anharmonic effects are small for most superconductors. However, if very light atoms are important (as in PdH) or if harmonic restoring forces are very small (near a second-order phase transition, as in Nb_3Sn or V_3Si), then anharmonic effects may be significant. The "reverse isotope effect"[126,127] in PdH (i.e., higher T_c in PdD) is most easily interpreted by the assumption that anharmonic renormalization of phonon frequencies is larger for PdH than for PdD, making $M\langle\omega^2\rangle$ larger and thus λ smaller.[128] Alternatively,[129] zero-point motional averaging, which reduces electron–phonon coupling through Debye–Waller factors, is larger in PdH. Experiment[130,131] seems to support the former.

There are well-known correlations between high-T superconductivity and peculiarities of lattice dynamics, including structural instabilities. One interpretation is that both are separate consequences of large electron–phonon coupling. However, Testardi has argued[132] that the anomalous lattice dynamics plays an important role in raising T_c.

In spite of the possible importance of anharmonic effects, there has been little theoretical work to incorporate anharmonicity into the theory of superconductivity. To some extent, anharmonicity is already built into the theory by the assumption that the phonon spectrum ω_Q is taken from experiment, and thus already includes anharmonic renormalization. However, there are other ways in which anharmonicity enters. Karakozov and Maksimov[129] (KM) made a nice formulation of this problem: They enlarged the starting Hamiltonian, Eqs. (3.18) and (3.19), to include anharmonic effects of an arbitrary type in the lattice Hamiltonian $\mathscr{H}_L$, and higher than linear terms in the coupling term $\mathscr{H}_{ep}$. Their coupling term was

[125] This and related subjects are reviewed by P. B. Allen, *in* "Dynamical Properties of Solids" (G. K. Horton and A. A. Maradudin, eds.), Vol. 3, Chapter 2. North-Holland Publ., Amsterdam, 1980.

[126] T. Skoskiewicz, *Phys. Status Solidi A* **11,** K123 (1972).

[127] R. J. Miller and C. B. Satterthwaite, *Phys. Rev. Lett.* **34,** 144 (1975).

[128] B. N. Ganguly, *Z. Phys.* **265,** 433 (1973).

[129] A. E. Karakozov and E. G. Maksimov, *Zh. Eksp. Teor. Fiz.* **74,** 681 (1978); *Sov. Phys.—JETP (Engl. Transl.)* **47,** 358 (1978).

[130] J. M. Rowe, J. J. Rush, H. G. Smith, M. Mostoller, and H. E. Flotow, *Phys. Rev. Lett.* **33,** 1297 (1974).

[131] A. K. Rahman, K. Skold, C. Pellizari, and S. K. Sinha, *Phys. Rev. B* **14,** 3630 (1976).

[132] L. R. Testardi, *Phys. Rev. B* **5,** 4342 (1972).

$$\mathscr{H}'_{\text{ep}} = \sum_{il} [V_0(\mathbf{r}_i - \mathbf{R}_l) - \langle V_0(\mathbf{r}_i - \mathbf{R}_l)\rangle_{\text{L}}]$$

$$= \sum_i \int d\mathbf{R}\, V_0(\mathbf{r}_i - \mathbf{R})[\rho_{\text{L}}(\mathbf{R}) - \langle \rho_{\text{L}}(\mathbf{R})\rangle_{\text{L}}], \tag{18.1}$$

$$\rho_{\text{L}}(\mathbf{R}) = \sum_l \delta(\mathbf{R} - \mathbf{R}_l), \tag{18.2}$$

where $\mathbf{r}_i$ and $\mathbf{R}_l$ are electron and nuclear coordinates, respectively, V_0 is the unscreened electron–ion interaction, $\rho_{\text{L}}(R)$ is the nuclear density operator, and $\langle\ \ \rangle_{\text{L}}$ denotes a thermal average over the lattice coordinates. The original Hamiltonian $\mathscr{H}_{\text{ep}}$ [Eq. (3.19)] kept only linear terms in $u_l = R_l - \langle R_l\rangle_{\text{L}}$.

The result of the KM analysis is that Eliashberg theory remains unaltered, except that the spectral function $\alpha^2F(\Omega)$ is replaced by a more complicated object $\alpha^2F'(\Omega)$. The form of this interaction can be derived from linear response theory. The effective time-dependent interaction $V_{\text{eff}}(x, x'; t - t')$ is the potential felt by an electron at position x and time t because of the fact that there was an electron at position x' for an instant t'. The presence of an electron for an instant at (x', t') causes a perturbation on the lattice at points (R', τ') given by

$$\mathscr{H}_{\text{pert}} = V(R' - x')\delta\rho_{\text{L}}(R')\delta(\tau' - t'), \tag{18.3}$$

$$\delta\rho_{\text{L}}(R) = \rho_{\text{L}}(R) - \langle\rho_{\text{L}}(R)\rangle_{\text{L}}, \tag{18.4}$$

where V is now the screened potential $\epsilon^{-1}V_0$ at R′ caused by the electron at x'. We are assuming that the electronic screening is instantaneous by comparison with ionic screening. Next, the lattice responds to the perturbation (18.3) by developing a lattice density distortion $\delta\rho(R\tau)$ at point $R\tau$:

$$\langle\delta\rho(R\tau)\rangle = \int dR'\, \chi(R, R'; \tau - t')V(R' - x'), \tag{18.5}$$

where χ is the lattice susceptibility or density response function, given by the Kubo-type formula

$$\chi(R, R'; t - t') = i\theta(t - t')\langle[\delta\rho_{\text{L}}(R, t), \delta\rho_{\text{L}}(R', t')]\rangle. \tag{18.6}$$

Finally the interaction $V_{\text{eff}}(x, x'; t - t')$ is the potential felt by the electron at xt because of the lattice density distortion (18.5):

$$V_{\text{eff}}(x, x'; t - t') = \int dR\, dR'\, V(x - R)\chi(R, R'; t - t')V(R' - x'). \tag{18.7}$$

Again the screened potential V appears rather than V_0, because the potential due to Eq. (18.5) is screened by the other electrons. The interaction which appears in Eliashberg theory is actually the Fourier transform

$V_{\text{eff}}(x, x'; \omega)$ evaluated at $\omega = i\omega_\nu$, which can be evaluated as follows:

$$V_{\text{eff}}(x, x'; i\omega_\nu) = \int dR\, dR'\, V(x - R)S(R, R'; i\omega_\nu)V(R' - x'), \tag{18.8}$$

$$S(R, R'; i\omega_\nu) = \frac{1}{2}\int_{-\beta}^{\beta} d\tau\, e^{i\omega_\nu \tau} S(R, R'; \tau), \tag{18.9}$$

$$S(R, R'; \tau) = -\langle T_\tau \delta\rho_{\text{L}}(R, \tau)\delta\rho_{\text{L}}(R', 0)\rangle. \tag{18.10}$$

In k space, the effective interaction is given by the Bloch wave matrix elements for scattering from k to k':

$$V_{\text{eff}}(k, k'; i\omega_\nu) = \int dx\, dx'\, \psi_k(x)\psi_{k'}^*(x)V_{\text{eff}}(x, x'; i\omega_\nu)\psi_{k'}(x')\psi_k^*(x'). \tag{18.11}$$

Let us evaluate this to the lowest nonvanishing order in the lattice displacements u_l. The nuclear density operator, Eq. (18.2), can be Taylor expanded, using $\mathbf{R}_l = \mathbf{l} + \mathbf{u}_l$:

$$\rho_{\text{L}}(R) = \sum_{\mathbf{l}} \delta(\mathbf{R} - \mathbf{l}) - u_{l\alpha}\nabla_{R\alpha}\delta(\mathbf{R} - \mathbf{l}) + \cdots . \tag{18.12}$$

The first term contributes nothing to $\delta\rho_{\text{L}}$ [Eq. (18.4)], and the second term equals $\delta\rho_{\text{L}}$ to first order, because $\langle u_{l\alpha}\rangle_{\text{L}} = 0$. Inserting this into Eq. (18.10) and using Eqs. (18.8) and (18.9), the effective interaction to lowest order is

$$V_{\text{eff}}(k, k'; i\omega_\nu) \cong \langle k|\nabla_\alpha V|k'\rangle D_{\alpha\beta}(k - k'; i\omega_\nu)\langle k'|\nabla_\beta V|k\rangle. \tag{18.13}$$

The self-energy Σ_{ep} in Eq. (3.20) can thus be written as

$$\Sigma_{\text{ep}}(k, i\omega_n) = -\frac{1}{\beta}\sum_{k',\nu} V_{\text{eff}}(k, k'; i\omega_\nu)G(k', i\omega_n - i\omega_\nu). \tag{18.14}$$

The result of KM is that the full interaction (18.8) and (18.11) should be used in place of Eq. (18.13) in the self-energy (18.14).

The new expression for $\alpha^2F(\Omega)$ is derived by first making a spectral representation for $S(k, k'; i\omega_\nu)$. By analogy with Eqs. (3.11) and (3.13),

$$S(R, R'; i\omega_\nu) = \frac{1}{\pi}\int_0^\infty d\Omega\, \bar{S}(R, R'; \Omega)\frac{2\Omega}{\omega_\nu^2 + \Omega^2}, \tag{18.15}$$

$$\bar{S}(R, R'; \Omega) = -\frac{1}{\pi}\,\text{Im}[S(R, R'; \Omega + i\delta)]. \tag{18.16}$$

Then the equation that replaces Eq. (3.22) for $\alpha^2F(k, k', \Omega)$ is

$$\alpha^2F'(k, k', \Omega)$$
$$= N(0)\int dR\, dR'\langle k|V(x - R)|k'\rangle\bar{S}(R, R'; \Omega)\langle k'|V(x' - R')|k\rangle. \tag{18.17}$$

This is equivalent to Eq. (28) of Ref. 129. As Karakozov and Maksimov point out, $\bar{S}$ has Fourier coefficients $S(Q + G, Q + G'; \Omega)$, and the diagonal part of this ($G = G'$) is the neutron inelastic cross section for scattering by phonons in Born approximation. If the phonons are governed by a completely harmonic Hamiltonian, then Eq. (18.17) can in principle be exactly evaluated, giving a theory of superconductivity that includes exactly all the multiphonon interactions as well as the usual one-phonon exchange. Two-phonon exchange processes have been discussed by Kumar[133] and Ngai[134] as possible mechanisms for enhancing superconductivity. It is necessary to be careful when using the electron–two phonon coupling $\frac{1}{2}u_\alpha u_\beta \nabla_\alpha \nabla_\beta V$. Several authors have shown that second-order interband processes involving $u_\alpha \nabla_\alpha V$ must also be included in order to have a theory which is translationally invariant[37,135,136] and representation independent.[137] There are also Debye–Waller factors which reduce the coupling strength and partially cancel the two-phonon processes.[34,129]

If the phonons are not harmonic, Eq. (18.17) still provides a formally exact theory, but there is no straightforward way of evaluating it. A qualitative discussion of some aspects is given in Ref. 129.

19. Energy Dependence of Density of States

In recent years there have been many attempts to incorporate the effects of the rapidly varying electronic density of states $N(\epsilon)$ near the Fermi level ϵ_F into Eliashberg theory.[138–150] The motivation is that some A15 super-

[133] N. Kumar, *Phys. Rev. B* **9,** 4993 (1974).

[134] K. L. Ngai, *Phys. Rev. Lett.* **32,** 215 (1974).

[135] R. Zeyher, *in* "Light Scattering in Solids" (M. Balkanski, R. C. C. Leite, and S. P. S. Porto, eds.), p. 87. Flammarion, Paris, 1976.

[136] P. B. Allen and V. Heine, *J. Phys. C* **9,** 2305 (1976).

[137] J. Ashkenazi, M. Dacorogna, and M. Peter, *Solid State Commun.* **29,** 181 (1979).

[138] P. Horsch and H. Rietschel, *Z. Phys. B* **27,** 153 (1977).

[139] S. J. Nettel and H. Thomas, *Solid State Commun.* **21,** 683 (1977).

[140] S. G. Lie and J. P. Carbotte, *Solid State Commun.* **26,** 511 (1978).

[141] W. E. Pickett, *Phys. Rev. B* **21,** 3897 (1980).

[142] S. G. Lie and J. P. Carbotte, *Solid State Commun.* **35,** 127 (1980).

[143] S. G. Lie, B. Mitrović, and J. P. Carbotte, *Conf. Ser.—Inst. Phys.* **55,** 559 (1980).

[144] P. Müller, G. Ischenko, H. Adrian, J. Bieger, M. Lehmann, and E. L. Haase, *in* "Superconductivity in d- and f-Band Metals" (H. Suhl and M. B. Maple, eds.), p. 369. Academic Press, New York, 1980.

[145] W. E. Pickett and B. M. Klein, *Solid State Commun.* **38,** 95 (1981).

[146] B. Mitrović and J. P. Carbotte, *Solid State Commun.* **40,** 249 (1981).

[147] M. D. Whitmore and J. P. Carbotte, *Physica B + C (Amsterdam)* **107,** 707 (1981).

[148] G. Kieselmann and H. Rietschel, *J. Low Temp. Phys.* **46,** 27 (1982).

[149] E. Schachinger, B. Mitrović, and J. P. Carbotte, *J. Phys. F* **12,** 1771 (1982).

[150] W. E. Pickett, *Phys. Rev. B* (submitted for publication).

conducting materials with high T_c, such as V_3Si, V_3Ga, and Nb_3Sn (T_c = 14–18°K), have long been suspected of having unusually sharp peaks in $N(\epsilon)$,[11] with structure on the scale of $\omega_D \sim 30$ meV. Several recent band structure calculations have found sharp peaks.[151,152] Unfortunately, there is not enough agreement between various calculations on the scale of ω_D to make the results completely convincing.[153] Nevertheless, a general argument has been given by Ho *et al.*[151] which explains why one can expect sharp peaks in $N(\epsilon)$ with six or more transition element atoms per unit cell.

Since the superconducting pairing correlations extend by $\pm\omega_D$ above and below the Fermi surface, it is obviously important to account for any variation in $N(\epsilon)$ on the scale of ω_D in order to obtain accurate numerical values for T_c. The effects of nonconstant $N(\epsilon)$ on T_c were first treated by Labbé *et al.*[154] within BCS theory. First of all, sharp peaks in $N(\epsilon)$ should permit large values of $N(\epsilon_F)$, which is favorable for superconductivity. Normally, the finite width ω_D of the phonon spectrum limits the attractive range of electron–phonon interaction. However, the extraordinarily narrow peak in $N(\epsilon)$ invoked by Labbé *et al.*[154] (they used the Labbé–Friedel model $N(\epsilon) = N(0)(\epsilon + \epsilon_m)^{-1/2}\epsilon_m{}^{1/2}$, with $\epsilon_m = 1.8$ meV and $E_F = 0$) replaces ω_D in limiting the attractive range of interaction. In this way they have offered an explanation for the vanishing isotope effect in some A15 compounds, which is unfortunately not supported by recent band theory.

Before we write down the Eliashberg equations generalized to include nonconstant $N(\epsilon)$, it should be stressed that the large residual resistivities of most samples of high-T_c A15 materials, between 2 and 60 $\mu\Omega$ cm,[155,156] should make any structure in $N(\epsilon)$ on the scale of T_c ($\sim$1–2 meV) disappear. This is based on the simple order of magnitude estimate of the scale, $\Delta\epsilon$, on which any fine structure in $N(\epsilon)$ is washed out by scattering for the given value of residual resistivity ρ_0:

$$\Delta\epsilon\ (\text{meV}) \simeq \frac{1}{\tau} = \frac{\Omega_p^2}{4\pi}\rho_0 = \frac{\Omega_p^2\ (\text{eV})}{7.5}\rho_0\ (\mu\Omega\ \text{cm}). \tag{19.1}$$

For a typical value of Drude plasma energy in high-T_c A15 materials, Ω_p = 3–4 eV, $\rho_0 = 1\ \mu\Omega$ cm causes a smearing of the structure in $N(\epsilon)$ on

[151] K. M. Ho, M. L. Cohen, and W. E. Pickett, *Phys. Rev. Lett.* **41,** 815 (1978).

[152] B. M. Klein, D. A. Papaconstantopoulos, and L. L. Boyer, *in* "Superconductivity in d- and f-Band Metals" (H. Suhl and M. B. Maple, eds.), p. 455. Academic Press, New York, 1980.

[153] J. Ruvalds and C. M. Soukoulis, *Phys. Rev. Lett.* **43,** 1263 (1979).

[154] J. Labbé, S. Barišić, and J. Friedel, *Phys. Rev. Lett.* **19,** 1039 (1967).

[155] A. J. Arko, D. H. Lowndes, A. T. Van Kessel, H. W. Myron, F. H. Mueller, F. A. Muller, L. W. Roeland, J. Wolfrat, and G. W. Webb, *J. Phys., Colloq. (Orsay, Fr.)* **C6,** Suppl. to Vol. 39, No. 8, p. 1385 (1978).

[156] A. K. Ghosh and M. Strongin, *in* "Superconductivity in d- and f-Band Metals" (H. Suhl and M. B. Maple, eds.), p. 305. Academic Press, New York, 1980.

the scale $\sim$1–2 meV. Thus the details of $N(\epsilon)$ are likely to be relevant only in the regime $\rho_0 \lesssim$ 10–20 $\mu\Omega$ cm, when $\Delta\epsilon \lesssim \omega_D$ ($\sim$30–50 meV).

In all treatments so far of the influence of nonconstant $N(\epsilon)$ on superconducting properties, the interband effects have been ignored. This may not be justified for A15 materials with a large number of narrow bands near the Fermi level. The Eliashberg theory including all the complexities of the electronic band structure has been worked out in 1967 by Garland.[157] However, at the present stage, any concrete numerical solution of Eliashberg equations has to ignore the interband effects, owing to the complexity of the relevant equations and the lack of knowledge of the appropriate matrix elements for A15 materials.

If the interband effects are ignored, then Eqs. (8.2)–(8.6) are general enough to incorporate the anisotropy effects and the energy dependence. The energy dependence of $N(\epsilon)$ is explicitly separated in these equations from the remaining (ϵ, ϵ') dependence of the functions $\alpha^2F(\epsilon, \epsilon', \Omega)$, $V(\epsilon, \epsilon')$, etc. The definition (3.25) and (3.31) of $\alpha^2F(\epsilon, \epsilon', \Omega)$ shows that (ϵ, ϵ') variation arises from changes in the averaged matrix element $|g_{kk'}|^2$ and the phonon frequency $\omega_{k-k'}$ as the wave vectors (k, k') are averaged over the separate surfaces (ϵ, ϵ'). This variation is probably slow compared to the variation of $N(\epsilon)$. If the (ϵ, ϵ') variation is neglected and only $N(\epsilon')$ is kept, Eqs. (8.2)–(8.6) show that the various self-energy parts are independent of ϵ, which simplifies the analysis. Possible energy dependence of various matrix elements on the scale of ω_D around E_F would imply analogous ϵ dependence of various self-energy parts, which would complicate the numerical solution of Eliashberg equations. Mitrović and Carbotte[158] have shown that for the separable model

$$\alpha^2F(\epsilon\epsilon', \Omega) = [1 + a(\epsilon)]\alpha^2(\Omega)F(\Omega)[1 + a(\epsilon')], \tag{19.2}$$

analogous to the usual model[110] for anisotropy, one can formulate and numerically solve the T_c equation. Here, we shall assume that anisotropy and the (ϵ, ϵ') dependence of matrix elements can be ignored. Whitmore and Carbotte[147] have studied the effects of interplay between anisotropy in electron–electron interactions and a sharp peak in $N(\epsilon)$ on T_c within the BCS limit. They concluded that for a peak in $N(\epsilon)$ of the half-width $\sim \omega_D$ and for a typical value of the anisotropy parameter $\langle a^2\rangle = 0.03$, the depression of T_c by normal impurity scattering happens because of simultaneous washing out of gap anisotropy and smearing of the peak in $N(\epsilon)$. The two effects cannot be distinguished from one another in the dependence of T_c on the amount of disorder for a given value of $\langle a^2\rangle$ and a half-width of the peak in $N(\epsilon)$.

[157] J. W. Garland, *Phys. Rev.* **153,** 460 (1967).

[158] B. Mitrović and J. P. Carbotte, to be published.

At T_c the Eliashberg equations generalized to include nonconstant $N(\epsilon)$ are

$$\phi(i\omega_n) = \pi T_c \sum_{m=\infty}^{+\infty} \left[\lambda(n-m) - \mu^*(\omega_c)\theta(\omega_c - |\omega_m|) + \delta_{nm}\left(\frac{1}{2\tau_{iN}}\right)\Big/(\pi T_c)\right]$$
$$\times \frac{1}{\pi}\int_{-\infty}^{+\infty} dE \frac{N(E)}{N(\mu_b)} \frac{\phi(i\omega_m)}{[\omega_m Z(i\omega_m)]^2 + [E - \mu + \chi(i\omega_m)]^2}, \tag{19.3}$$

$$\omega_n Z(i\omega_n) = \omega_n + \pi T_c \sum_{m=-\infty}^{+\infty} \left[\lambda(n-m) + \delta_{nm}\left(\frac{1}{2\tau_{iN}}\right)\Big/(\pi T_c)\right]$$
$$\times \frac{1}{\pi}\int_{-\infty}^{+\infty} dE \frac{N(E)}{N(\mu_b)} \frac{\omega_m Z(i\omega_m)}{[\omega_m Z(i\omega_m)]^2 + [E - \mu + \chi(i\omega_m)]^2}, \tag{19.4}$$

$$\chi(i\omega_n) = -\pi T_c \sum_{m=-\infty}^{+\infty} \left[\lambda(n-m) + \delta_{nm}\left(\frac{1}{2\tau_{iN}}\right)\Big/(\pi T_c)\right]$$
$$\times \frac{1}{\pi}\int_{-\infty}^{+\infty} dE \frac{N(E)}{N(\mu_b)} \frac{E - \mu + \chi(i\omega_m)}{\omega_m Z(i\omega_m) + [E - \mu + \chi(i\omega_m)]^2}. \tag{19.5}$$

Here μ_b is the "bare" band structure chemical potential at $T = 0$ and we interpret the quantities $\lambda(n - m)$, $\mu^*(\omega_c)$, and $1/(2\tau_{iN})$ to contain a factor $N(\mu_b)$ in their definitions [see Eqs. (3.22), (9.5), and (6.12)] in place of $N(0)$. We reserve $N(0)$ to denote $N(\mu)$, where $\mu = \mu_b + \delta\mu$ is the true interacting chemical potential at $T = T_c$. In writing Eqs. (19.3)–(19.5) we have switched, for the sake of clarity, from the notation in which electronic energies ϵ are measured with respect to μ to the one in which the electronic state of lowest energy is $E = 0$ [$\epsilon(E) = E - \mu$]. Equations (19.3)–(19.5) have to be supplemented by the equation that expresses the conservation of the number of electrons (per unit volume):

$$2\int_0^{\mu_b} dE\, N(E) = 2T_c \sum_{m=-\infty}^{+\infty} \int_0^{+\infty} dE\, N(E) G_{11}(E, i\omega_m) e^{i\omega_m 0^+}$$
$$= -2\pi T_c N(\mu_b) \sum_{m=-\infty}^{+\infty} e^{i\omega_m 0^+} \frac{1}{\pi}\int_{-\infty}^{+\infty} dE \frac{N(E)}{N(\mu_b)}$$
$$\times \frac{i\omega_m Z(i\omega_m) + [E - \mu + \chi(i\omega_m)]}{[\omega_m Z(i\omega_m)]^2 + [E - \mu + \chi(i\omega_m)]^2}, \tag{19.6}$$

where one should keep in mind that $N(E) = 0$ for $E < 0$. This equation determines $\mu(T)$.[23]

Equations (19.3)–(19.6) allow for scattering by nonmagnetic impurities, Eq. (6.12), and we have assumed that the impurity potential is short ranged and that it scatters isotropically, that is, that $\langle k|V_N|k'\rangle$ is a constant, V_N.

It should be stressed that the effects of Coulomb interaction do not appear in Eqs. (19.4) and (19.5) for diagonal components of the self-energy, owing to the assumption that they have been included in calculating the band structure. Also, the static screening approximation is assumed.

Equations (19.3)–(19.6) have to be solved self-consistently in order to find T_c. This was done by Pickett and Klein[145] for $1/(2\tau_{iN}) = 0$ and by Pickett[150] for $1/(2\tau_{iN}) \neq 0$. We now argue that χ and $\Delta\mu$ tend to cancel and can be ignored in Eqs. (19.3)–(19.6) for $1/(2\tau_{iN}) = 0$.

In the normal state at $T = 0$ and for $1/(2\tau_{iN}) = 0$, the real part of the analytic continuation of χ to just above the real-frequency axis at $\omega = 0$, $\mathrm{Re}[\chi(\omega = 0 + i0^+)]$, gives the shift $\delta\mu$ of the chemical potential due to the electron–phonon interaction.[159] A nonvanishing χ arises because of asymmetry in $N(\epsilon)$ near the Fermi level.[160] Numerical calculations[161] show that for a peak in $N(\epsilon)$ comparable to the one found in band calculations for Nb_3Sn,[152] $\chi(\omega + i0^+)$ varies slowly with ω and can be replaced by $\delta\mu$ in the range where $\mathrm{Re}[\chi(\omega)]$ makes an important contribution to the integrals (19.3)–(19.6), that is, the range $\omega\, \mathrm{Re}[Z(\omega + i0^+)] \leq \mathrm{Re}[\chi(\omega + i0^+)]$. Results are almost identical to the ones obtained by Mitrović and Carbotte[146] for the superconducting state at $T = 0$. Thus in the equation

$$-\frac{1}{\pi}\,\mathrm{Im}\left(\int_{-\infty}^{+\infty} dE\, N(E) G(E, \omega + i0^+)\right)$$
$$= -\frac{1}{\pi}\,\mathrm{Im}\left(\int_{-\infty}^{+\infty} dE\, N(E)\, \frac{1}{\omega Z(\omega + i0^+) - (E - \mu_b) + \delta\mu - \chi(\omega + i0^+)}\right)$$

one can assume that there is a complete cancellation between $\delta\mu$ and Re χ and Im χ can be ignored compared to ω Im Z, with the result

$$-\frac{1}{\pi}\,\mathrm{Im}\left(\int_{-\infty}^{+\infty} dE\, N(E) G(E, \omega + i0^+)\right)$$
$$= -\frac{1}{\pi}\,\mathrm{Im}\left(\int_{-\infty}^{+\infty} d\epsilon\, N(\epsilon + \mu_b)\, \frac{1}{\omega Z(\omega + i0^+) - \epsilon}\right).$$

It is expected that the same conclusions will hold at superconducting temperatures for materials in which the Fermi level is away from the bottom of the band by many multiples of T_c, that is, $\sim 10\omega_D$, which is the case for A15 materials. Then Eqs. (19.3)–(19.6) can be replaced by the simpler set of equations

[159] J. M. Luttinger and J. C. Ward, *Phys. Rev.* **118,** 1417 (1960).

[160] This can be understood from Schrieffer's perturbative argument. J. R. Schrieffer, "Theory of Superconductivity," pp. 133–135. Benjamin, New York, 1964.

[161] B. Mitrović, Ph.D. Thesis, McMaster University, Hamilton, Ontario (1981) (unpublished).

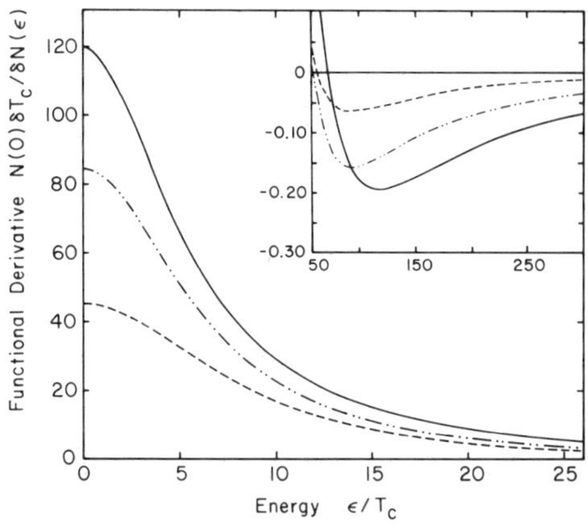

FIG. 12. Functional derivative of T_c by $N(\epsilon)$ calculated for Ta (———), Nb (—· · —), and Nb_3Sn (---) using spectra α^2F derived from tunneling and making the assumption $N(\epsilon) = N(0)$. Reprinted with permission from S. G. Lie and J. P. Carbotte, *Solid State Commun.* **26,** 511 (1978).

$$\phi(i\omega_n) = \pi T_c \sum_{m=-\infty}^{\infty} [\lambda(n-m) - \mu^*(\omega_c)\theta(\omega_c - |\omega_m|)] \times \frac{\phi(i\omega_m)}{|\omega_m Z(i\omega_m)|}\left(\frac{1}{\pi}\int_{-\infty}^{+\infty} d\epsilon \frac{N(\epsilon)}{N(0)} \frac{|\omega_m Z(i\omega_m)|}{[\omega_m Z(i\omega_m)]^2 + \epsilon^2}\right), \quad (19.7)$$

$$\omega_n Z(i\omega_n) = \omega_n + \pi T_c \sum_{m=-\infty}^{+\infty} \lambda(n-m)s_m \times \left(\frac{1}{\pi}\int_{-\infty}^{+\infty} d\epsilon \frac{N(\epsilon)}{N(0)} \frac{|\omega_m Z(i\omega_m)|}{[\omega_m Z(i\omega_m)]^2 + \epsilon^2}\right), \quad (19.8)$$

where now the energies ϵ are measured with respect to the bare band structure chemical potential μ_b and no shift of the chemical potential is necessary.

Lie and Carbotte[140] have calculated the functional derivative $\delta T_c/\delta N(\epsilon)$ within the framework of Eqs. (19.7) and (19.8). For a small change $\Delta N(\epsilon)$ in $N(\epsilon)$ the corresponding change in critical temperature T_c is given by

$$\Delta T_c = \int_{-\infty}^{+\infty} d\epsilon \frac{\delta T_c}{\delta N(\epsilon)} \Delta N(\epsilon). \quad (19.9)$$

Thus $\delta T_c/\delta N(\epsilon)$ indicates how important the values of $N(\epsilon)$ are at various energies in determining T_c. Lie and Carbotte have found that the values of $N(\epsilon)$ only within 5–10 times T_c around the Fermi level $\epsilon = 0$ have an appreciable effect on T_c; $\delta T_c/\delta N(\epsilon)$ is approximately of Lorentzian form centered at $\epsilon = 0$ and becomes negative only at larger energies $\epsilon \gtrsim 50T_c$ (see Fig. 12). The special sensitivity of T_c to the values of $N(\epsilon)$ within $|\epsilon| \leq (5\text{–}10)T_c$ can be qualitatively understood by examining Eq. (19.7). Since the expression in the large parentheses is of the order of unity [it is exactly

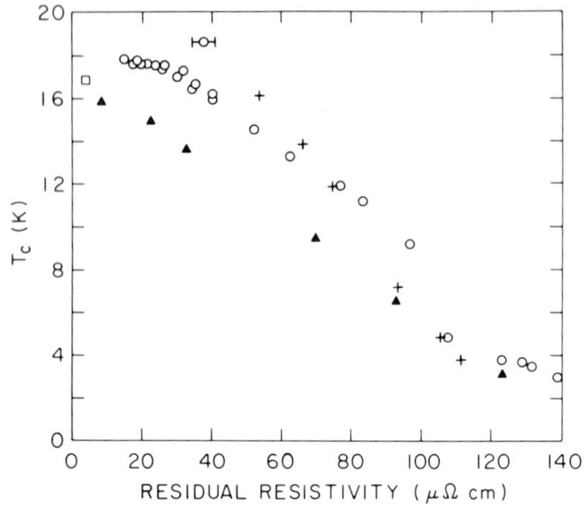

FIG. 13. T_c as a function of residual resistivity in radiation-damaged thin film samples of Nb_3Sn (○), Nb_3Al (+), and V_3Si (▲). Also shown is a value for bulk Nb_3Al (●) and bulk single-crystal V_3Si (□). Both electron and α-particle radiation is shown. The behavior of T_c shows a striking regularity. Reprinted with permission from A. K. Ghosh and M. Strongin, *in* "Superconductivity in d- and f-Band Metals" (H. Suhl and M. B. Maple, eds.), p. 305. Academic Press, New York, 1980.

1 for $N(\epsilon) = N(0)$], the most important terms in the sum over m are, as usual, the ones with low values of m, owing to the $1/\omega_m$ factor. For a given m the most important values of $N(\epsilon)$ in contributing to the integral

$$\int_{-\infty}^{+\infty} d\epsilon \, \frac{N(\epsilon)}{N(0)} \frac{1}{\pi} \frac{|\omega_m Z(i\omega_m)|}{[\omega_m Z(i\omega_m)]^2 + \epsilon^2}$$

are determined by the half-width $|\omega_m Z(i\omega_m)|$ of the Lorentzian factor in the integrand. For $m = 0$, $|\omega_m Z(i\omega_m)| \lesssim \pi T_c(1 + \lambda)$, that is, a few times T_c. The general form of $\delta T_c/\delta N(\epsilon)$ implies that if $N(\epsilon)$ is constant for $|\epsilon| \leq \omega_D$ ($\sim 20T_c$–$30T_c$ for high-T_c A15 materials), it is not crucial to include all the details of $N(\epsilon)$ into the Eliashberg theory to obtain accurate values of T_c.

Figure 13 shows experimental values of T_c as a function of residual resistivity ρ_0 for radiation-damaged A15 samples. Various explanations have been offered for the observed degradation of T_c.[153,162,163] One explanation focuses on the dependence of λ on ρ_0, which is primarily governed by the smearing of $N(\epsilon)$ with disorder. The data on T_c versus ρ_0 are usually unfolded[156] to obtain λ versus ρ_0 by using McMillan's[14] or Allen–Dynes[76] formula for T_c. However, the question arises as to how adequate these formulas are for the case of rapidly varying $N(\epsilon)$. The problem has been

[162] D. E. Farrell and B. S. Chandrasekhar, *Phys. Rev. Lett.* **38,** 788 (1977).

[163] M. Gurvitch, A. K. Ghosh, B. L. Gyorffy, H. Lutz, O. F. Kammerer, J. S. Rosner, and M. Strongin, *Phys. Rev. Lett.* **41,** 1616 (1978).

treated by Nettel and Thomas,[139] who performed an analysis of analytically continued Eqs. (19.7) and (19.8), along the line of McMillan's work.[14] They argued that the standard McMillan's equation can still be approximately applied, provided that $N(0)$ is replaced by the average

$$\bar{N} \equiv \left(\int_{k_B T_c}^{k_B \theta_D / 1.45} \frac{d\omega \, \{N[\mathrm{Re}\, Z(\omega)\, \omega] + N[-\mathrm{Re}\, Z(\omega)\, \omega]\}}{\omega \,\mathrm{Re}\, Z(\omega)} \right) \left[\ln\left(\frac{\theta_D}{1.45 T_c} \right) \right]^{-1} \tag{19.10}$$

and assuming $\lambda \propto \bar{N}$. This neglects damping effects, but may be reasonable if peaks are not too narrow. Simpler averaging procedures have been used[164] and may give similar results. If the Fermi level E_F falls on the maximum of the peak in $N(\epsilon)$, then, as can be seen from $\delta T_c / \delta N(\epsilon)$ in Fig. 12, these averaging procedures would underestimate T_c, whereas if E_F falls on the slope in $N(\epsilon)$, they should give a reasonable result.[145] However, for residual resistivities $\gtrsim$ 20–30 $\mu\Omega$ cm the structure in $N(\epsilon)$ on the scale of ω_D is largely washed out and, as indicated by the general form of $\delta T_c / \delta N(E)$ or by Eq. (19.10), one should not worry about modifications of Eliashberg equations or approximate T_c equations due to varying $N(\epsilon)$. It should be noted that possible sharp peaks in $N(\epsilon)$ near the Fermi level can affect the superconducting tunneling experiments or, more specifically, the values of inverted $\alpha^2 F$.[146,148] There the effects of nonconstant $N(\epsilon)$ should be more pronounced, since one is measuring the frequency dependence of $|\Delta(\omega)|$, which is in turn "modulated" by $N(\omega)$.

20. Paramagnons

Recent "first-principles" calculations for d-band elements and compounds have tended somewhat to overestimate T_c.[165–167] This may indicate a fault[4] in the scheme of calculation, or it may be taken as a sign[165] that the repulsive interaction is larger than the ordinary estimate of $\mu^* \sim 0.1$–0.13. The culprit most often blamed is spin fluctuations or "paramagnons."

In metals like Pd, where the spin susceptibility is strongly exchange enhanced,[8] there can exist at low temperatures medium-range spin order persisting over moderately long periods of time. This corresponds to an effective attraction of parallel-spin electrons, or a repulsion of antiparallel spins, which would tend to suppress singlet pairing and enhance triplet pairing.[123,168] We discuss only the singlet case.

In the study of magnetic properties of metals it is common to employ

[164] L. R. Testardi and L. F. Mattheiss, *Phys. Rev. Lett.* **41,** 1612 (1978).

[165] H. Rietschel and H. Winter, *Phys. Rev. Lett.* **43,** 1256 (1979).

[166] H. Rietschel, H. Winter, and W. Reichardt, *Phys. Rev. B* **22,** 4284 (1980).

[167] D. Glötzel, D. Rainer, and H. R. Schober, *Z. Phys. B* **35,** 317 (1979).

[168] D. Fay and J. Appel, *Phys. Rev. B* **22,** 3173 (1980).

the Hubbard Hamiltonian[50] in which (d-band) electrons repel each other via a contact potential $I\delta(r_1 - r_2)$. Thus only electrons in opposite spin bands interact. The random phase approximation (RPA) gives an enhanced long-wavelength static spin susceptibility[169]

$$\chi_{zz}(0, 0) = \chi_0/(1 - \bar{I}), \tag{20.1}$$

where χ_0 is the noninteracting Pauli susceptibility and $\bar{I} = N(0)I$. Here χ_{zz} is the longitudinal susceptibility. More relevant for paramagnons is the traverse susceptibility, $\chi^{-+}(q, \omega)$.[50] RPA calculations of $-(1/\pi)\,\mathrm{Im}\,\chi^{-+}(q, \omega)$ within the Hubbard model show that for fixed q this function is peaked at some frequency $\omega_{\mathrm{sf}}(q)$ that decreases as $\bar{I}$ increases.[50] Thus although the spin fluctuations are heavily damped excitations—$-(1/\pi)\,\mathrm{Im}\,\chi^{-+}$ does not have the form $\delta[\omega - \omega_{\mathrm{sf}}(q)]$—their excitation energy can be low for large enough $N(0)$ or I. The virtual scattering of electrons near the Fermi surface by paramagnons [which are loosely defined as interacting particle–hole pairs of spin 1 and characteristic energy $\sim \omega_{\mathrm{sf}}(q)$] can therefore lead to considerable renormalization.[170,171]

A modification of Eliashberg theory which incorporates the effects of electron–paramagnon interaction was given by Berk and Schrieffer.[8,172] The equations at $T = T_c$ in isotropic approximation are

$$Z(i\omega_n)\Delta(i\omega_n) = \pi T_c \sum_m [\lambda(n - m) - \lambda_{\mathrm{sf}}(n - m) - \mu^*(\omega_c)\theta(\omega_c - |\omega_m|)] \times \frac{\Delta(i\omega_m)}{|\omega_m|} I_m, \tag{20.2}$$

$$Z(i\omega_n) = 1 + \frac{\pi T_c}{|\omega_n|} \sum_m s_n s_m [\lambda(n - m) + \lambda_{\mathrm{sf}}(n - m)] I_m, \tag{20.3}$$

$$I_m = \frac{1}{\pi} \int_{-\infty}^{\infty} d\epsilon \, \frac{N(\epsilon)}{N(0)} \frac{|\omega_m Z(i\omega_m)|}{\omega_m^2 Z(i\omega_m)^2 + \epsilon^2}. \tag{20.4}$$

The cutoff ω_c of $\mu^*(\omega_c)$ must be chosen to be greater than $[\omega_{\mathrm{sf}}(q)]_{\mathrm{max}}$. For constant $N(\epsilon)$, the correction factor I_m is unity. This approximation has always been made so far. By analogy with $\lambda(n - m)$, defined in Eq. (11.6), $\lambda_{\mathrm{sf}}(n - m)$ is defined in terms of an electron–paramagnon spectral function $P(\Omega)$ as

$$\lambda_{\mathrm{sf}}(n - m) = \int_0^{+\infty} d\Omega \, P(\Omega) \frac{2\Omega}{\Omega^2 + (\omega_n - \omega_m)^2}. \tag{20.5}$$

$P(\Omega)$ is related to the spectral density of the spin-fluctuation propagator

[169] P. A. Wolff, *Phys. Rev.* **120,** 814 (1960).

[170] S. Doniach and S. Engelsberg, *Phys. Rev. Lett.* **17,** 750 (1966).

[171] W. F. Brinkman and S. Engelsberg, *Phys. Rev.* **169,** 417 (1968).

[172] N. F. Berk and J. R. Schrieffer, *Phys. Rev. Lett.* **17,** 433 (1966).

χ_{sf}[8,172] by

$$P(\Omega) = N(0) \sum_{k,k'} \left(-\frac{1}{\pi} \operatorname{Im} \chi_{sf}(k, k'; \Omega + i0^+) \right) \delta(\epsilon_k)\delta(\epsilon_{k'}) \Big/ \sum_{k,k'} \delta(\epsilon_k)\delta(\epsilon_{k'}). \quad (20.6)$$

This is closely analogous to Eqs. (3.22) and (5.12).

It should be emphasized that Eqs. (20.2)–(20.6) are simply an *ad hoc* effort to build a model which contains the physical effects intuitively expected to accompany an exchange-enhanced susceptibility. The true underlying Hamiltonian contains no Hubbard term, only $e^2/|r_1 - r_2|$ repulsion. In Section 5, a subset of Feynman graphs was *arbitrarily* selected, because they are readily summed and interpreted in terms of dielectric screening. Now we are admitting that the experimental enhancement of χ shows that further Feynman graphs are needed. It is possible to find another subset of graphs that introduces terms having the structure of Eqs. (20.2)–(20.5) and which can be interpreted in terms of magnetic susceptibility. However, it is again *not* possible to find any small parameter or other argument to justify omitting all the other Feynman graphs. Furthermore, even if one believed that the Hubbard Hamiltonian contained the appropriate terms omitted in Section 5, there would still remain the difficulty that the RPA treatment of the Hubbard term cannot be justified. This is equivalent to the absence of a Migdal's theorem for paramagnons[173] and will be discussed further below. Ultimately, one is forced either to accept on faith Eqs. (20.2)–(20.6) as a plausible model to be fitted to something, and nothing more, or else to reject all of it (including all details in Section 5) as uncontrolled speculation. We decline to advocate a position on this.

In the original work of Berk and Schrieffer[172] χ_{sf} is given as the sum of ladder diagrams for particle–hole scattering. This neglects the spin zero component of spin fluctuations,[171,174] which in the limit of a contact interaction is given by the sum of bubble diagrams with an even or odd number of bubbles, depending on whether they appear in the effective pairing interaction between opposite-spin electrons or in the contribution to the diagonal self-energy component.[123,168,171] This omission is usually patched up by introducing a factor of $^3/_2$ into the definition (20.6) of $P(\Omega)$, which is correct within RPA only in the limit of a contact interaction.[171,175]

Model calculations[8,165,166] of $P(\Omega)$ based on Schrieffer's[175] prescription indicate that $P(\Omega)$ is a peaked function with peak frequency $\Omega_{sf} \sim (0.3–0.5)\epsilon_f$, ϵ_f being the Fermi energy, and a maximum paramagnon frequency $\Omega_{sf}^{max} \sim 8\epsilon_F$. Although paramagnons appear in Eqs. (20.2) and (20.3) in a way similar to that in which phonons do, the large values of parameters

[173] J. A. Hertz, K. Levin, and M. T. Beal-Monod, *Solid State Commun.* **18,** 803 (1976).

[174] D. Fay and J. Appel, *Phys. Rev. B* **16,** 2325 (1977).

[175] J. R. Schrieffer, *J. Appl. Phys.* **39,** 642 (1968).

Ω_{sf}/ϵ_f and $\Omega_{sf}^{max}/\epsilon_F$ indicate an essential difficulty with Eqs. (20.2) and (20.3), namely, the absence of Migdal's theorem for paramagnons.[173] In the derivation of Eqs. (20.2) and (20.3) it is implicitly assumed that some sort of Migdal theorem applies. The differences between phonons and paramagnons were first discussed by Pethick.[176] Hertz, Levin, and Beal-Monod[173] have explicitly shown that the vertex corrections for paramagnons are of the same order as the bare electron–paramagnon vertex, essentially because $\Omega_{sf}/\epsilon_F \sim 1$. However, in the subsequent formulation of strong-coupling theory for paramagnon-induced triplet pairing in ^{3}He, Levin and Valls[177] have assumed that some higher-order vertex corrections can be lumped into the value of the coupling constant $\bar{I}$, which is a fitting parameter. One may hope that the higher-order vertex corrections can be included in $P(\Omega)$, so that Eqs. (20.2) and (20.3) remain a sensible model. The effective renormalization parameter λ_{sf} due to spin fluctuations is defined as

$$\lambda_{sf} \simeq \sum_m \lambda_{sf}(\omega_0 - \omega_m) s_m I_m \,. \tag{20.7}$$

If the Fermi level happens to fall within a peak in $N(\epsilon)$ of width $\sim\Omega_{sf}$, then the approximation $N(\epsilon) \approx N(0)$ gives $I_m = 1$, which overestimates λ_{sf}. In Pd for instance, the peak width is $\sim$0.7 eV.[178] This possibility was hinted at in the original work of Berk and Schrieffer.[172] Note that the approximation $N(\epsilon) \approx N(0)$ is not too serious in the Δ equation, Eq. (20.2), owing to the $|\omega_m|^{-1}$ factor. Thus for a peak in $N(\epsilon)$ with a half-width of less than a few times ω_D and $\lesssim\Omega_{sf}$, the correction I_m may be set equal to 1 in Eqs. (20.2) and (20.3) when multiplying $\lambda(n - m)$, while the full energy dependence in $N(\epsilon)$ has to be retained for the paramagnon contribution to Z.

Daams *et al.*[179] have considered Eqs. (20.2) and (20.3), with constant $N(\epsilon)$, in the approximation when $P(\Omega)$ does not lead to any significant Ω dependence on the scale of a few times the maximum phonon frequency. Then the spin-fluctuation contributions in Eqs. (20.2) and (20.3) can be described by the square-well model

$$\lambda_{sf}(n - m) \to \lambda_{sf}\,\theta(\Omega_{sf} - |\omega_n|)\theta(\Omega_{sf} - |\omega_m|), \tag{20.8}$$

$$\pi T_c |\omega_n|^{-1} \sum_m \lambda_{sf}(n - m) s_n s_m \to \lambda_{sf}\,\theta(\Omega_{sf} - |\omega_n|), \tag{20.9}$$

with $\mu^*(\omega_c)$ rescaled to $\mu^*(\Omega_{sf})$. After dividing both Eqs. (20.2) and (20.3) by $1 + \lambda_{sf}$ and denoting the paramagnon-renormalized renormalization

[176] C. J. Pethick, *Lect. Theor. Phys.* **11B**, 245 (1968).

[177] K. Levin and O. T. Valls, *Phys. Rev. B* **17**, 191 (1978).

[178] F. M. Mueller, A. J. Freeman, J. O. Dimmock, and A. M. Furdyna, *Phys. Rev. B* **1**, 4617 (1970).

[179] J. M. Daams, B. Mitrović, and J. P. Carbotte, *Phys. Rev. Lett.* **46**, 65 (1981).

function $Z/(1+\lambda_{sf})$ by Z^*, we obtain

$$Z^*(i\omega_n)\Delta(i\omega_n) = \pi T_c \sum_m^{|\omega_m|<\Omega_{sf}} \left(\frac{\lambda(n-m) - \lambda_{sf} - \mu^*(\Omega_{sf})}{1+\lambda_{sf}}\right) \frac{\Delta(i\omega_m)}{|\omega_m|}, \tag{20.10}$$

$$Z^*(i\omega_n) = 1 + \frac{\pi T_c}{|\omega_n|} \sum_m \frac{\lambda(n-m)}{1+\lambda_{sf}} s_n s_m . \tag{20.11a}$$

These are valid for $|\omega_n| < \Omega_{sf}$. From these equations it can be seen that when $\Omega_{sf} \gg \omega_D$, the main effect of paramagnons is to increase the effective repulsion between opposite-spin electrons and renormalize all interactions and self-energies. Equations (20.10) and (20.11a) were tested for a model where $P(\Omega)$ is a δ function $P(\Omega) = \frac{1}{2}\lambda_{sf}\Omega_{sf}\delta(\Omega - \Omega_{sf})$. It was found that these equations describe quite well all equilibrium thermodynamic properties, including superconductive tunneling.[179,180]

It should be kept in mind that the cutoff in Eq. (20.9) is Ω_{sf}. This is important in the case when one wants to translate effects of paramagnons as described by Eqs. (20.8) and (20.9) into various approximate T_c formulas.[181] To illustrate this we take the square-well model for $\lambda(n-m)$ analogous to Eqs. (20.7) and (20.8). First, the effective Coulomb pseudopotential $\mu^*_{eff}(\Omega_{sf}) = [\mu^*(\Omega_{sf}) + \lambda_{sf}]/(1+\lambda_{sf})$ should be rescaled to the phonon cutoff ω_D. It is $\mu^* = \mu^*(\omega_D)$ which enters McMillan's equation. The result is

$$\mu^*_{eff}(\omega_D) = \frac{\mu^*_{eff}(\Omega_{sf})}{1 + \mu^*_{eff}(\Omega_{sf})\ln(\Omega_{sf}/\omega_D)} . \tag{20.11b}$$

Equations (20.9) and (20.10) reduce to

$$Z(i\omega_n)\Delta(i\omega_n) = \pi T_c[\lambda^* - \mu^*_{eff}(\omega_D)] \sum_m^{|\omega_m|<\omega_D} \frac{\Delta(i\omega_m)}{|\omega_m|}, \tag{20.12}$$

$$Z(i\omega_n) - 1 = \lambda^* \equiv \lambda(1+\lambda_{sf})^{-1}. \tag{20.13}$$

The solution of these equations (accurate for weak coupling) is

$$T_c = 1.13\omega_D \exp\{-(1+\lambda^*)/[\lambda^* - \mu^*_{eff}(\omega_D)]\}. \tag{20.14a}$$

Only for the case of very small λ_{sf} and $\mu^*_{eff}(\Omega_{sf})$, that is, both of the order 0.1, does Eq. (20.14) give T_c comparable to the value obtained from

$$T_c = 1.13\theta_D \exp\{-(1+\lambda+\lambda_{sf})/[\lambda - \mu^*(\Omega_{sf}) - \lambda_{sf}]\}, \tag{20.14b}$$

which is an equation which has been used[182] (with a different prefactor) in the analysis of data.

[180] B. Mitrović and J. P. Carbotte, *Solid State Commun.* **41**, 695 (1982).

[181] C. R. Leavens, private communication.

[182] T. P. Orlando and M. R. Beasley, *Phys. Rev. Lett.* **46**, 1598 (1981).

The validity of Eqs. (20.2)–(20.6) is unclear, because these equations rest on a model. However, Eqs. (20.10) and (20.11a), which are derived from them for $\Omega_{sf} \gg \omega_D$, incorporate effects of spin fluctuation in a way one intuitively expects. Only one thing is clear. Palladium has strong enough electron–phonon coupling[183] to expect superconductivity with $T_c \gtrsim 0.1°K$. The absence of superconductivity is clearly correlated with enhanced spin susceptibility.

21. Is There a Maximum T_c?

As of January 1982, there has been a maximum T_c of ~23°K for the last 8 years.[184] This represents a normal fluctuation in the steady trend of the 3°K increase of T_c per decade[185] that has occurred since 1911. However, the investment of manpower and money in the last decade has been large and the results disappointing. Nevertheless, it is clearly dangerous to assert[186] that T_c is saturating at a maximum. Two different sensible arguments were advanced in the past[15,187] to set a limit for T_c, and each was later shown to be wrong.[76,188] Meanwhile the maximum T_c jumped 3°K.

The first suggested limit to T_c was McMillan's "$\lambda = 2$ limit," which claimed, first, that a higher T_c was mainly achieved by softening phonons and, second, that the structure of the T_c equation made this process saturate at $\lambda = 2$. Both aspects seemed correct at the time, but now appear to be wrong. In Refs. 76 and 79 it is argued that η or $(\lambda\langle\omega^2\rangle)^{1/2}$ rather than λ governs T_c of high-T_c materials, and this is not enhanced by softening phonons. In Ref. 76 it was shown that McMillan's T_c formula is too pessimistic when λ is large.

The second limit to T_c was a suggestion by Cohen and Anderson,[187] that the requirement $\epsilon(q) > 0$ would demand a repulsive net interaction in a homogeneous system. This was used to generate a limit for T_c. In Ref. 188 it is shown that $\epsilon(q) > 0$ is not compulsory, except at $q = 0$, and is in fact violated in many real systems. Thus there is no limit to T_c from this source.

The actual limiting factor in practice is usually lattice stability, for example, a second-order phase transition driven by a "soft"-phonon mode. The same electron–phonon interaction responsible for high T_c causes the phonon softening. The low-T phase has reduced $N(\epsilon_F)$ and smaller T_c

[183] F. J. Pinski and W. H. Butler, *Phys. Rev. B* **19**, 6010 (1979).

[184] J. R. Gavaler, *Appl. Phys. Lett.* **23**, 480 (1973).

[185] P. B. Allen, *in* "Dynamical Properties of Solids" (G. K. Horton and A. A. Maradudin, eds.), Fig. 1. North-Holland Publ., Amsterdam, 1980.

[186] B. T. Matthias, *Comments Solid State Phys.* **3**, 93 (1970); *Phys. Today* August, p. 23 (1971).

[187] M. L. Cohen and P. W. Anderson, *AIP Conf. Prof.* **4**, 17 (1972).

[188] O. V. Dolgov, D. A. Kirzhnits, and E. G. Maksimov, *Rev. Mod. Phys.* **53**, 81 (1981).

(although this is not easily demonstrated experimentally.) Less dramatic, but probably more common, are first-order instabilities which prevent the most promising compounds from forming in the crystal structure which would be favorable to a high T_c. Often the system phase separates—for example, A_3B refuses to form, yielding grains of pure A and grains of A_2B_5. It is easy to make qualitative arguments but difficult to form predictive theories of these effects.

Another practical limit to T_c may be magnetism. It has been found that ferromagnetic order will destroy superconductivity, but so far, with Y_4Co_3 possibly excepted,[189,190] superconductivity has not been found to destroy magnetic order. Usually this competition occurs in materials with $T_c \lesssim 10$ °K, but as more exotic narrow-band and lower-dimensional materials are investigated, the competition between superconductivity and magnetism may play a role in limiting T_c.

Appendix A. The Digamma Function

The digamma function $\psi(z)$ and the polygamma functions $\psi^{(n)}(z)$ are derivatives of the logarithm of the gamma function[191]:

$$\psi(z) = d[\ln \Gamma(z)]/dz, \tag{A.1}$$

$$\psi^{(n)}(z) = d^n\psi(z)/dz^n. \tag{A.2}$$

Two fundamental properties of the gamma function are the recurrence and reflection formulas

$$\Gamma(z + 1) = z\Gamma(z), \tag{A.3}$$

$$\Gamma(z)\Gamma(1 - z) = \pi \csc(\pi z). \tag{A.4}$$

From these follow the recurrence and reflection formulas of the digamma function:

$$\psi(z + 1) = \psi(z) + 1/z, \tag{A.5}$$

$$\psi(1 - z) = \psi(z) + \pi \cot(\pi z). \tag{A.6}$$

Similar formulas obviously hold for the polygamma function.

The usefulness of the digamma function for summing thermal Green's function expressions over the Matsubara frequencies can be appreciated from the following formula:

[189] A. Kolodziejczyk, B. V. B. Sarkissian, and B. R. Coles, *J. Phys. F* **10,** L333 (1980).

[190] E. Gratz, J. O. Strom-Olsen, and M. J. Zuckermann, *Solid State Commun.* **40,** 833 (1981).

[191] P. J. Davis, *in* "Handbook of Mathematical Functions" (M. Abramowitz and I. A. Stegun, eds.), p. 253. Dover, New York, 1965.

$$\psi(n+1+z) - \psi(z+1) = \sum_{m=1}^{n} \frac{1}{m+z} \tag{A.7}$$

This identity follows directly from Eq. (A.5) by iteration. This formula, with $z = -1/2$, is used in Section 2 to get the BCS T_c equation. Another important sum formula is found by taking the limit of Eq. (A.7) as n goes to infinity. This first requires an asymptotic formula for $\psi(z)$, which can be found from the asymptotic formula for $\ln \Gamma(z)$ (basically Stirling's formula):

$$\psi(z) \sim \ln z - \frac{1}{2z} - \sum_{n=1}^{\infty} \frac{B_{2n}}{2nz^{2n}}, \tag{A.8}$$

where B_{2n} are the Bernoulli numbers. In order to handle the logarithm of $z + n + 1$ for large n, Euler's constant γ can be added on:

$$\gamma = \lim_{n\to\infty} \left(\sum_{m=1}^{n} \frac{1}{m} - \log n \right) = 0.577\ldots. \tag{A.9}$$

Then, taking the $n \to \infty$ limit of Eq. (A.7) with the help of Eqs. (A.8) and (A.9), the result is

$$\psi(1+z) = -\gamma + \sum_{m=1}^{\infty} \frac{z}{m(m+z)}. \tag{A.10}$$

In the Russian literature, the symbol γ and the term *Euler's constant* are often used to denote e^{γ}.

By judicious manipulations of these formulas, many of the sums needed in Matsubara space can be performed. Some results which will be needed later are

$$N(\Omega) + 1/2 = 1/2 \coth\left(\frac{\Omega}{2T}\right) = \frac{T}{\Omega} + \frac{1}{\pi} \operatorname{Im} \psi\left(1 + i\frac{\Omega}{2\pi T}\right), \tag{A.11}$$

$$1/2 - f(\epsilon) = 1/2 \tanh\left(\frac{\epsilon}{2T}\right) = \frac{1}{\pi} \operatorname{Im} \psi\left(\frac{1}{2} + i\frac{\epsilon}{2\pi T}\right), \tag{A.12}$$

$$\operatorname{Im}[\psi(1+ix) - \psi(n+1+ix)] = \sum_{m=1}^{n} \frac{x}{m^2 + x^2}, \tag{A.13}$$

$$\operatorname{Im}[\psi(1/2+ix) - \psi(n+1/2+ix)] = \sum_{m=1}^{n} \frac{x}{(m-1/2)^2 + x^2}. \tag{A.14}$$

Equations (A.11) and (A.12) follow directly from Eqs. (A.5) and (A.6) with special choices of z, while Eqs. (A.13) and (A.14) are easily derived from Eq. (A.7). Equation (2.28) follows from Eq. (A.12) and the $n \to \infty$ limit of Eq. (A.14). In Section 2 we need the value of $\psi(1/2)$. From Eq. (A.10) this can be written as

$$\psi(1/2) = -\gamma - 2 \sum_{m=1}^{\infty} \left(\frac{1}{2m-1} - \frac{1}{2m}\right) = -\gamma - 2(1 - 1/2 + 1/3 - 1/4 + \cdots).$$

The last is a famous series, yielding the result

$$\psi(1/2) = -\gamma - 2 \ln 2. \tag{A.15}$$

Appendix B. Derivation of Eq. (3.51)

Rather than attempt to integrate directly on Eq. (3.44), it seems easier to go back a few steps and calculate instead $R(i\omega_n, \Omega)$, using Eq. (3.40) for I:

$$R(i\omega_n, \Omega) = T \sum_{i\omega_\nu} \frac{2\Omega}{\omega_\nu^2 + \Omega^2} \int_{-\infty}^{\infty} \frac{d\omega'}{i\omega_n - i\omega_\nu - \omega'} \tag{B.1}$$

$$= T \sum_{i\omega_\nu} \frac{2\Omega}{\omega_\nu^2 + \Omega^2} \lim_{L\to\infty} \log\left(\frac{L + i\omega_n - i\omega_\nu}{-L + i\omega_n - i\omega_\nu}\right) \tag{B.2}$$

$$= i\pi T \sum_{i\omega_\nu} \frac{2\Omega}{\omega_\nu^2 + \Omega^2} \operatorname{sign}(\omega_\nu - \omega_n). \tag{B.3}$$

The result (B.3) follows, because the phase difference between the numerator and the denominator of the argument of the logarithm is either $+\pi$ or $-\pi$, depending on whether $i\omega_\nu - i\omega_n$ is in the upper or lower half-plane. The same result, Eq. (B.3), emerges in a somewhat more natural way in Eq. (10.6) by taking the normal limit of $Z(i\omega_n)$ in the superconducting state. Careful consideration of the terms in the sum (B.3) shows that terms with positive and negative values of ω_ν cancel, unless $|\omega_\nu| < |\omega_n|$; in the latter case the sign of the result depends on whether or not $\omega_n > 0$:

$$R(i\omega_n, \Omega) = -i\pi T \operatorname{sign} \omega_n \sum_{i\omega_\nu}^{|\omega_\nu|<|\omega_n|} \frac{2\Omega}{\omega_\nu^2 + \Omega^2} \tag{B.4}$$

$$= -2i \operatorname{sign} \omega_n \left(\frac{1}{2x} + \sum_{\nu=1}^{n} \frac{x}{\nu^2 + x^2}\right). \tag{B.5}$$

In the last equation, x is $\Omega/2\pi T$. The sum in Eq. (B.5) appears also in Eq. (A.13), from which we get

$$R(i\omega_n, \Omega)$$

$$= -2i \operatorname{sign} \omega_n \left[\frac{\pi T}{\Omega} + \operatorname{Im} \psi\left(1 + i\frac{\Omega}{2\pi T}\right) - \operatorname{Im} \psi\left(n + 1 + i\frac{\Omega}{2\pi T}\right)\right]. \tag{B.6}$$

Next, Eq. (A.11) is used, and n is converted into $(\omega_n/2\pi T) - 1/2$:

$$R(i\omega_n, \Omega) = -2\pi i \operatorname{sign} \omega_n \times \left\{ N(\Omega) + \tfrac{1}{2} - \frac{1}{2\pi i}\left[\psi\left(\frac{1}{2} + i\frac{\Omega - i\omega_n}{2\pi T}\right) - \psi\left(\frac{1}{2} - i\frac{\Omega + i\omega_n}{2\pi T}\right)\right]\right\}. \quad \text{(B.7)}$$

Finally, analytic continuation is accomplished by replacing $i\omega_n$ by z, which will vary in the upper half-plane and later become $\omega + i\delta$. The factor sign ω_n becomes $+1$ in the process, since we ignore what happens in the lower half-plane ($\omega_n < 0$) in making the continuation:

$$R(z, \Omega) = -2\pi i[N(\Omega) + \tfrac{1}{2}] + \psi\left(\frac{1}{2} + i\frac{\Omega - z}{2\pi T}\right) - \psi\left(\frac{1}{2} - i\frac{\Omega + z}{2\pi T}\right). \quad \text{(B.8)}$$

This function is the unique physical analytic continuation, provided that it is bounded for all Im $z > 0$.[24] The digamma function is unbounded when its argument is a negative integer or zero, but this occurs in Eq. (B.8) only for Im $z < 0$. The separate functions ψ in Eq. (B.8) behave as ln z for large $|z|$, but the difference between the two terms remains bounded.

Acknowledgments

Support from National Science Foundation Grant No. DMR79-00837 is gratefully acknowledged.

Spectroscopic and Morphological Structure of Tetrahedral Oxide Glasses

J. C. PHILLIPS

Bell Laboratories, Murray Hill, New Jersey

I. Introduction

Among inorganic glasses the tetrahedral oxide glasses g-SiO_2, g-GeO_2, and the multiple-oxide glasses based upon them are especially significant for historical, technological, and scientific reasons. Tetrahedral building blocks are the geometrical feature that makes these glasses so important, whereas the Si–O and Ge–O bonds are the significant chemical features. Both features are exceptionally stable and little affected by additives.

Many attempts have been made to determine the structure of these glasses

ISBN 0-12-607737-1

on a molecular scale, but in the absence of Bragg scattering the results have often proved discouraging, in spite of the accuracy and completeness of much of the experimental data. In the case of SiO_2, for example, there are many known crystalline phases with complex unit cells which typically contain from 2 to 16 formula units with almost the same short-range order (bond lengths and bond angles).[1] Not only radial diffraction but also spectral data are essentially one dimensional and are ill suited to the task of differentiating three-dimensional fragments of such complex structures.

Many authors have postulated that the pronounced glass-forming tendency of SiO_2 might be explained by assuming that microclusters of one or another of the phases are present in the liquid and that when the supercooled liquid is quenched to form a glass, a misoriented mixture of these fused microphases results that is sufficient to inhibit crystallization.[2] The paracrystallites, as they are sometimes called, are supposed to be embedded in a matrix of higher defect concentration whose structure is unspecified. This rather complex picture has often been simplified by both crystallographers and mathematicians into a more morphologically homogeneous model[3] which is often called a continuous random network (CRN). Models of the CRN type are convenient and easily generated and have formed the basis for almost all recent quantitative attempts to interpret microscopic data on the molecular structure of these glasses.

In an earlier series of papers[4] on chalcogenide glasses such as As_2Se_3 and $GeSe_2$ I rejected this entire train of reasoning and suggested a fundamentally different morphological approach. My view is that the fundamental origin of the glass-forming tendency in inorganic materials such as SiO_2 or As_2Se_3 is the same as in organic materials such as glycerine or glucose, which consist of large molecules. Similarly, in the chalcogenide glasses clusters are formed based on the principle of partially broken chemical order. As a result of broken chemical order, the clusters are of two types, donor and acceptor, with the latter type of clusters stabilized at their boundaries by chalcogen dimers. Incidentally, the reader may note that As_2Se_3 has only one known crystal structure and yet it is an excellent glass former. Thus its crystallization is not frustrated by the existence of multiple phases, but rather by topological factors which I have characterized precisely in algebraic terms and which also correctly predict the optimal glass-forming composition of Ge_xSe_{1-x} alloys.[4]

[1] W. L. Bragg and G. F. Claringbull, "Crystal Structure of Minerals," Chapter 6. Cornell Univ. Press, Ithaca, 1965.

[2] E. A. Porai-Koshitz, *Glastech. Ber.* **32,** 450 (1959); B. Eckstein, *Mater. Res. Bull.* **3,** 199 (1968); J. E. Stanworth, *Glass Technol.* **17,** 194 (1976); K. S. Evstropyev and E. A. Porai-Koshits, *J. Non-Cryst. Solids* **11,** 170 (1972).

[3] W. H. Zachariasen, *J. Am. Chem. Soc.* **54,** 3841 (1932).

[4] J. C. Phillips, *J. Non-Cryst. Solids* **34,** 153 (1979); *Phys. Status Solidi* B **101,** 473 (1980).

In oxide glasses the ionic interactions are very strong and chemical order is almost perfect. However, the prevalence of the glass-forming tendency in other oxides (B_2O_3, P_2O_5, etc.) strongly suggests that glass formation occurs in oxides not for topological reasons, but rather for a specific chemical reason, namely, the ability of oxygen to form double rather than single bonds at little expense in enthalpy. Nonbridging oxygen atoms (such as singly coordinated oxygen atoms, which I denote by O^*), as they are generally called by glass scientists, are often thought to be present in homogeneous CRN models as point defects (denoted by O_p^*), where they serve the function of reducing network strain energy. However, although such strain-energy reduction facilitates the avoidance of crystallization, kinetically it is not so important a mechanism as the formation of clusters with noncoalescing interfaces. Indeed, one can see immediately that if the cluster surfaces are saturated by O_s^*, crystallization will be almost completely suppressed. Moreover, if the cluster surface is covered by O_s^*, then the cost in energy of forming steps on the cluster surface may be very much smaller than is commonly the case in non-glass-forming materials. This means that cluster interfaces may be highly adaptive and may fit together very snugly with little void volume. In turn this means that one can dispense with the defective matrix hypothesis of vitroid theory: The cluster surfaces are true surfaces in the usual sense of crystalline grain boundaries.

One sees, then, that the critical topological feature which can be used to distinguish morphologically homogeneous CRN models from weakly polymerized cluster models is expected phenomenologically to be the arrangement of O^* as O_p^* or O_s^*, respectively. Fortunately in SiO_2 and GeO_2 there is a large difference in atomic masses, with $m_O \ll m_{Si}$ and m_{Ge}. This means that the vibrational modes associated with O_p^* and O_s^* are strongly localized in space and may be split off in frequency as narrow sidebands above the bulk continuum bands of similar character. These bands are readily identified, but they have often been misinterpreted in the past, because O^* has been regarded only as a point defect O_p^* whose concentration should vary rapidly with preparative method and chemical composition (concentration of network additive in silicates and germanates). In fact, in some cases the strength and position of the narrow O_s^* localized mode is one of the most stable features of the spectrum, whereas the scattering strengths of the bulk continuum bands, which depend on the morphology of the cluster interior, are highly variable. This is just what one would expect from a cluster model, but it is the exact opposite of what is expected from conventional morphologically homogeneous CRN models.

The plan of this article is as follows. In Section II I summarize the general features of the vibrational spectrum of a tetrahedral oxide glass containing $O_{p,s}^*$ atoms, without specifically restricting the discussion to SiO_2 or GeO_2 or to neutron, infrared, or Raman spectra. This has been done before in other ways by many workers, but the aim of this section is to present in the simplest possible way the essential elements of the spectra freed from the irrelevant and unnecessary computational details[5] that arise in specific force field calculations based on specific ball-and-stick models. I next give a review of the experimental spectra for g-SiO_2 obtained by neutron scattering (Section III), infrared absorption (Section IV), and Raman scattering (Section V), placing special emphasis on the bond-bending O* mode in the 400–500 cm^{-1} range. For the reader's convenience the data are reproduced and enlarged for this sensitive and significant spectral region.

In Section VI I review the O* spectra of SiO_2 obtained in CRN models and demonstrate the failure of these models to explain the data for SiO_2. In Sections VII and VIII I present specific morphological cluster models for g-SiO_2 and g-GeO_2 and show that they are capable of explaining the O_s^* spectra. Remarkably quantitative results are obtained without detailed calculation for the dimensions of the clusters, their surface textures, and their interfacial spacing. In succeeding sections these models are tested against a very wide range of data and are found to be remarkably consistent with many observed phenomenological trends.

The presentation of this paper is dictated by two considerations. First, I desire to display the concepts in a balanced way that brings out the most informative features of the data; second, I recognize that if scientific truth were determined democratically, the world would be flat, the sun would revolve around the earth, and the CRN model of inorganic glasses would be validated at present by an overwhelming majority. Because I believe that the CRN model is fundamentally incorrect, I have gathered together all the data that bear on its shortcomings. I have also made generous use of the careful and very extensive calculations of vibrational spectra of CRN models of g-SiO_2 and g-GeO_2 that have been carried out by Bell and co-workers.[6]

Systematic criticism of these CRN models is possible only because of these authors' thorough and thoughtful calculations. At times when I am criticizing the CRN concept it might appear that I am criticizing their calculations, but just the opposite is the case; my criticisms can be taken seriously only to the extent that the CRN calculations of Bell and co-workers are very accurate, and had they not carried out these calculations I would have been obliged to do so myself. Space will not permit a restatement here of all their results, so I urge the reader who wishes to appreciate my discussion to study their papers carefully. Finally, in considering the relation-

[5] G. Herzberg, "Molecular Spectra and Molecular Structure," Vol. II, p. 100. Van Nostrand, Princeton, New Jersey, 1945; K. Nakamoto, "Infrared and Raman Spectra of Inorganic and Coordination Compounds," p. 134. Wiley, New York, 1963.

[6] R. J. Bell, P. Dean, and D. C. Hibbins-Butler, *J. Phys. C* **3**, 2111 (1970); **4**, 1214 (1971).

ship of this model to the work of Bell and co-workers, I am reminded of Newton's famous remark to Hooke, that if he had seen further, it was by standing on the shoulders of giants.

II. Schematic Spectrum of a Tetrahedral Oxide Glass

The normal vibrational modes of AX_4 tetrahedral molecules are well known[5] and will not be reproduced here. They consist of a nondegenerate scalar A_1 (symmetric breathing) mode, a doubly degenerate E tensor mode, and two triply degenerate vector bending and stretching $F_2^{1,2}$ modes. All modes are Raman active, but only the $F_2^{1,2}$ vector modes are infrared active.

Various efforts have been made in the context of CRN models to justify the separation of continuum modes in the glass and to establish their connection to the normal modes of AX_4 or A_2X free molecules.[6,7] One can combine the symmetries A_1, E, and $F_2^{1,2}$ of AX_4 and the parity symmetry of the quasilinear AXA molecule to predict the positions and widths of the main vibrational bands of g-SiO_2 from the eigenfrequencies of SiF_4 molecules, and similarly with g-GeO_2 and GeF_4. This is done in Appendix A without the use of elaborate algebra or giant ball-and-spoke models. The central result[8–11] is that the $F_2^{1,2}$ tetrahedral modes retain their identity in the glass, but the A_1 and E modes are strongly mixed by the combination of AX_4 and A_2X symmetries.

In addition to the continuum modes, one must include the localized O_p^* and O_s^* modes. These consist of two sets of two infrared-active and Raman-active narrow bands L_{1p} and L_{2p} and L_{1s} and L_{2s} split off above the continuum bands because of the small value of m_O. The bond-stretching localized mode L_2 is nondegenerate and splits off from the upper $F_2{}^2$ band, while the bond-bending localized mode L_1 is twofold degenerate and splits off from the lower $F_2{}^1$ band. The magnitudes of the splittings can be calculated for any localized mode by Green's function methods, but we will be interested primarily in the shifts of the L_1 frequency from neutron to infrared to Raman data. In Section VI it will be shown that these shifts are

[7] P. N. Sen and M. F. Thorpe, *Phys. Rev. B: Solid State* [3] **15,** 4030 (1977).

[8] G. Herzberg, "Molecular Spectra and Molecular Structure," Vol. II, pp. 66 and 181. Van Nostrand, Princeton, New Jersey, 1945. K. Nakamoto, "Infrared and Raman Spectra of Inorganic and Coordination Compounds," p. 22. Wiley, New York, 1963.

[9] P. M. Bridenbaugh, G. P. Espinosa, J. E. Griffiths, J. C. Phillips, and J. P. Remeika, *Phys. Rev. B: Condens.* Matter [3] **20,** 4140 (1979).

[10] N. Kumagai, J. Shirafuji, and Y. Inuishi, *J. Phys. Soc. Jpn.* **42,** 1261 (1977).

[11] K. Nakamoto, "Infrared and Raman Spectra of Inorganic and Coordination Compounds," p. 136. Wiley, New York, 1963.

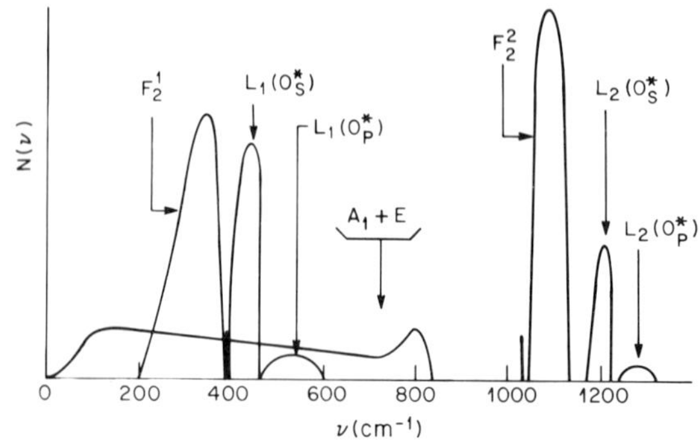

FIG. 1. Schematic diagram of bulk continuum and localized nonbridging oxygen O* contributions to the vibrational spectral density $N(\nu)$ of g-SiO_2. The continuum contributions are labeled according to tetrahedral symmetry modes A_1, E, and $F_2^{1,2}$, as described in Appendix A. The O* contributions are separated according to surface and point O_s^* and O_p^* modes, respectively. The $F_2^{1,2}$ frequencies of the SiF_4 molecule at 390 and 1030 cm^{-1} are also marked, because, as discussed on Appendix A, these frequencies are expected to fall close to the $F_2^{1,2}$ continuum and $L_{1,2}(O_s^*)$ modes of g-SiO_2.

nearly zero for O_p^* atoms regarded as isolated defects in a CRN model but that they follow a systematic pattern in a cluster model that is based on the dispersion of the surface O_s^* vibrational bands.

The importance of the L_1 and L_2 modes has been discussed by Bell *et al.*[6] in terms of the end O* atoms in their CRN model, but, as these authors noted, their localized modes are not surface modes in the commonly accepted sense, because their O* atoms just happen to be situated at the surface as a consequence of the method of model construction. Had these O* atoms been located in the interior of the model, they would be expected to have higher frequencies corresponding to the increased density in the interior.

In the morphological models considered here large surface areas (corresponding to low-index crystal faces) are covered by O_s^* atoms which are nearly equivalent and are bonded to the cluster interior as siliconyl units, namely, $O_s^*{=}Si{-}(O_{1/2})_2$. Associated with these surface arrays are L_1 and L_2 vibrational bands analogous to the surface vibrational bands found with crystals.[12] Quite generally, however, one expects that

$$\nu[L_{1,2}(O_p^*)] > \nu[L_{1,2}(O_s^*)], \tag{2.1}$$

because of the higher density in the vicinity of O_p^* atoms compared to O_s^* atoms.

In Fig. 1 I placed $\nu[L_{1,2}(O_s^*)]$ at considerably higher values than did Bell

[12] R. F. Wallis, *Phys. Rev.* **116,** 302 (1959).

and co-workers. Corresponding to these authors' values of $\nu[L_{1,2}(O_s^*)]$ of 300 and 850 cm^{-1}, I utilized values of 420 and 1150 cm^{-1}, respectively. For their O* stretching and bending force constants Bell *et al.*[6] employed the *same* parameters as in the interior of their model. These are *single-bond* values, but, as indicated above, the appropriate force constants for O* must take into account the local chemical bonding configuration, which is *double bonded*, that is, $O^*{=}Si{-}(O_{1/2})_2$.

In general, double-bonding force constants are much larger than single-bonding ones, for example, $\nu(N{=}O)$ is 1450 cm^{-1}, as compared to 1075 cm^{-1} for $\nu(N{-}O)$; that is, the double-bonded force constants may be nearly twice the single-bonded ones.[13] The exact ratio of bond-bending and bond-stretching force constants for $O^*{=}Si$ compared to $O{-}Si$ is not known, but my choice of $\nu[L_{1,2}(O_s^*)] \simeq 1.35\nu(Si{-}O)$ is in much better agreement with the experimental spectra of g-SiO_2 and with chemical experience in molecules.

The schematic spectrum shown in Fig. 1 contains both broad and narrow bands. Generally speaking, in a noncrystalline solid described by, for example, a CRN, one would expect to find only broad bands. If one blindly accepts the CRN hypothesis, then the narrow bands must be ascribed to "defects" such as O_p^*. The problem with this interpretation is that one would then expect that by increasing the concentration of point "defects," for example, by neutron bombardment, all the narrow bands would grow in strength at about the same rate. In fact, in the virgin glass some of the narrow bands are very strong and some are weak, and only the weak bands grow in strength as the glass is damaged; the strong narrow bands are unchanged, as are the broad strong bands. This raises grave doubts about the validity of the point "defect" description of both strong and weak narrow bands.

In the present model the internal surfaces are an intrinsic feature of the glass structure and they give rise to the strong narrow bands. It is important to appreciate that the surface bands[12] are narrow for several reasons. The direct $O_s^*{-}O_s^*$ interaction is only a second-neighbor interaction, compared to the $Si{-}O$ first-neighbor interaction, which determines the width of the bulk or cluster interior bands; the strengths of the interactions differ by about a factor of 10. Second, the surface molecules (such as siliconyl) gen-

[13] K. Nakamoto, "Infrared and Raman Spectra of Inorganic and Coordination Compounds," p. 224. Wiley, New York, 1963. Another useful comparison is the $C{-}O$ stretch frequency (1100 cm^{-1}) cited by Nakamoto on p. 437, compared to the 75% enhanced $C{=}O$ frequency in $F_2C{=}O$ of 1944 cm^{-1} as measured by A. H. Nielsen, T. G. Burke, P. J. H. Woltz, and E. A. Jones, *J. Chem. Phys.* **20,** 596 (1952). Because of the strong multiple-bonding tendencies of C compared to Si, I prefer the 35% estimated increase in frequency for $Si{=}O$ compared to $Si{-}O$ mentioned in the text.

erally have a symmetry (e.g., quasitrigonal) that is different from the tetrahedral symmetry of the bulk molecules. This means that even when the narrow surface band overlaps the broad bulk band, the resonant mixing will be reduced by the symmetry difference with respect to the surface plane and planes normal to the surface plane.

At various times efforts have been made to explain the narrow bands in terms of "defect" molecular structures embedded in a fully three-dimensional CRN. Since neither of the considerations mentioned above applies to these "defects," it seems unlikely that they could give rise to strong narrow features in any of the vibrational spectra.

III. Neutron Scattering Spectra: g-SiO_2

Assuming that one has surmounted the hurdles associated with the differences in force constants between Si—O and Si=O bonds, one can begin by analyzing the density of vibrational states as determined by neutron scattering.[14] The measured scattering density $G(\omega)$ is proportional to the true density of states $Z(\omega)$ weighted by the factor $f_N(\omega)$, which averages the vibrational amplitudes $U^2(\alpha)$, the neutron scattering lengths b_α^2, and the masses M_α^{-1} over the atoms participating in a given normal mode. The ratio $(b_O/b_{Si})^2$ is about 1.7, and so is $(M_O/M_{Si})^{-1}$. Thus a mode localized on an O atom is weighted about 1.5 times as heavily as a mode with equal amplitudes on Si and 2O; moreover, in the model I give, the increase in the double-bonding force constants compared to the single-bonding ones compensates for the fact that the single-bonded oxygen is twice bonded. Thus the connection between the measured and true localized densities of states is

$$\frac{Z(O^*)}{Z(O)} = \frac{G(O^*)}{1.5G(O)}. \tag{3.1}$$

I have smoothed $G(\omega)$ as measured by Leadbetter and Stringfellow,[14] subtracted their estimated multiphonon background, and separated the contribution of the localized modes, which appears to peak near 410 cm^{-1} and which I associate with $L_1(O_s^*)$ in Fig. 1. The results are shown in Fig. 2. Unfortunately, two different monochromators were used to obtain the data in the interval 350–420 cm^{-1}, so there is some ambiguity in the data in the critical range. However, careful inspection of the original data in the overlapping region reveals very similar structures, with both sets of data indicating a satellite peak near 410 cm^{-1}. Nevertheless, this is the most important region of the spectrum for the neutron data and I look forward

[14] A. J. Leadbetter and M. W. Stringfellow, *Neutron Inelastic Scattering, Proc. Grenoble Conf., 1972* pp. 501–514 (1974); A. J. Leadbetter, private communication.

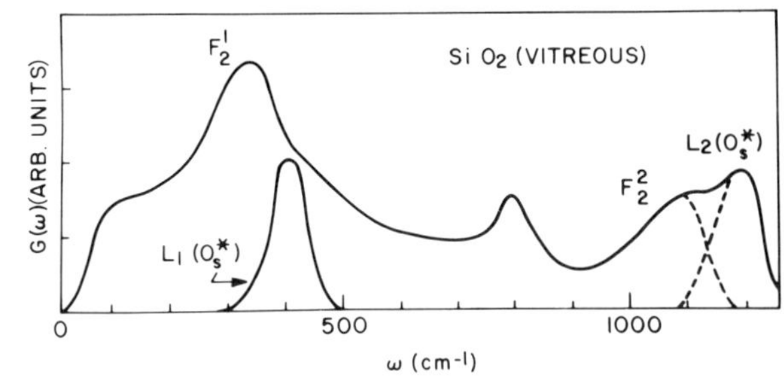

FIG. 2. Smoothed neutron scattering density of states $G(\omega)$ for g-SiO_2 as measured by Leadbetter and Stringfellow.[14] The density associated with the bending nonbridging internal surface oxygen atoms $L_1(O_s^*)$ is shown as a separate contribution.

with great interest to the results that will be obtained when the data are remeasured.[14]

According to the model discussed in Appendix A and illustrated in Fig. 1, of the nine vibrational degrees of freedom per formula unit, six are contained in the continuum spectrum between 0 and 900 cm^{-1}. On the other hand, there are two $L_1(O_s^*)$ degrees of freedom per surface $(O_{1/2})_2$—Si—O_s^* molecule. Therefore if x_s is the fraction of surface molecules per cluster, then the integrated area A_1 in Fig. 2 associated with the $L_1(O_s^*)$ mode, compared to the area associated with the continuum modes A_0 up to 900 cm^{-1}, should be given by

$$A_1/A_0 = y/(1 - y), \tag{3.2}$$

with

$$y = (1.5)2x_s/6, \tag{3.3}$$

where the factor of 1.5 was estimated previously as the enhancement in scattering strength of the $L_1(O_s^*)$ mode compared to an average mode. I have determined graphically that $A_1/A_0 = 0.11$ and thus $y = 0.10$ and

$$x_s = 0.20. \tag{3.4}$$

In Section VII I will compare this value of x_s with the value obtained from the cluster dimensions as measured by transmission electron microscopy (TEM) and will show that the two values are in good agreement.

In Fig. 2 the remaining features that are shown separately in Fig. 1 are marked and presented together as part of the continuum scattering. A localized bond-stretching O_s^* mode, $L_2(O_s^*)$, seems to be present near 1200 cm^{-1}, but the scatter is large in this region.

At this point it is instructive to compare the neutron spectrum of SiO_2 with the vibrational spectra calculated from the random network model

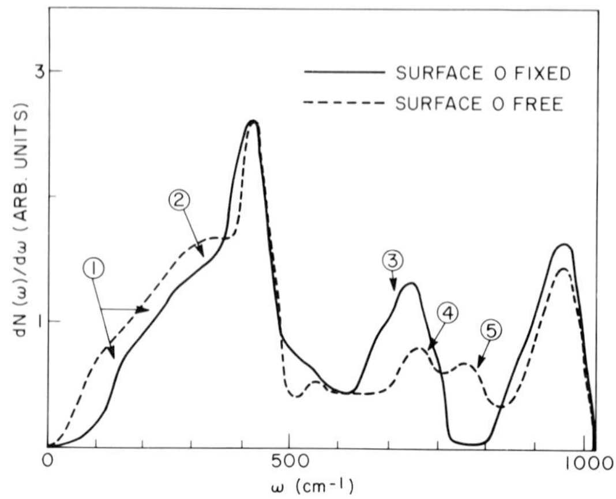

FIG. 3. The spectral densities $dN(\omega)/d(\omega)$ calculated by Bell and Dean from their model for fixed and free surface O atoms. Attention is drawn to (1) the downward shift or softening of the acoustic modes with free surface atoms, (2) the weak localized resonance of the surface O atom bending mode, (3) the upper cutoff peak of the $A_1 + E$ tetrahedral continuum for fixed surface O atoms, (4) the same for free surface O atoms, and (5) the localized modes of $A_1 + E$ surface molecules $(O_{1/2})_3—Si—O$ for free surface O atoms.

by Bell and co-workers.[6,15] Again I am interested in isolating the localized mode contribution to the spectrum from the continuum contribution. The calculations of Bell *et al.* were based on an atomic model[16] containing 188 SiO_4 units with 100 singly coordinated nonbridging oxygen atoms present on the model surface as $(O_{1/2})_3—Si—O$ units. The surface locally resembles an oxidized surface of a (111) face of cristobalite and thus is fundamentally different from the (100) model presented here. In fact, the surface units employed in the model of Bell *et al.* violate the stoichiometry of the glass and the model contains an extra 50 O atoms, which from a chemical point of view should be described as O^- anions.

The size of the cluster used by Bell and Dean was chosen to be as large as possible (consistent with their computational capabilities at that time). The fraction of surface molecules x_s is $^{100}/_{188} = 0.54$, which is much larger than the value derived in Eq. (3.4). Moreover, each surface molecule contributes one nonbridging O_s^* atom, so that the fraction of nonbridging oxygen atoms is $z_s = {}^{100}/_{426} = 0.24$. In the cristobalite model presented in this article this fraction should be $z_s = x_s/2 = 0.10$ according to Eq. (3.4).

According to the random network hypothesis, the violation of stoichiometry of the sample surface is an irrelevant artifact of the ball-and-stick model. Calculations of the vibrational spectra were therefore carried out

[15] R. J. Bell, *Methods Comput. Phys.* **15,** 216 (1976).
[16] R. J. Bell and P. Dean, *Philos. Mag.* [8] **25,** 1381 (1972).

with two boundary conditions, fixed and free nonbridging oxygen atoms. Presumably, in an infinite CRN model the true spectrum lies somewhere between these two spectra. From the present point of view, however, the two calculations provide the local modes associated with the $(O_{1/2})_3{-}Si{-}O$ surface molecules. Their difference spectra (Fig. 3) show that in the free-end model the acoustic mode spectral density is shifted downward by about 50 cm^{-1}. This effect is interesting in its own right and is a manifestation in the Bell–Dean (BD) model of surface softening, which accounts for the additional T^3 term in the low-temperature specific heat.[17,18] In addition, there are two very weak resonances of localized states near 270 (labeled N, or NB, by Bell) and 500 cm^{-1} (not labeled). Their integrated areas contain only about one-fifth as many states as the localized mode near 410 cm^{-1} in the neutron data, yet if all the bending surface modes were localized, the model peak would be four times larger than the neutron peak.

The reason for the 20-fold reduction in the strength of the localized mode in the BD model is that by using single-bond force constants for their non-bridging oxygen atoms, Bell and Dean have immersed these modes in the acoustic continuum, giving rise to strong resonance mixing which is described as autoionization in the analogous electronic case. Because the localized modes were regarded as artifacts of the model, Bell and Dean attached little importance to the feature.

One other significant difference in the spectra is that the single spectral peak near 750 cm^{-1} for fixed ends splits into two peaks, a continuum one at 750 cm^{-1} and a localized one at 850 cm^{-1}, for free ends. Experimentally only one Raman-active (but weakly infrared-active) peak is observed near 800 cm^{-1}. In the $(O_{1/2})_2{-}Si{=}O$ (100) cristobalite model localized surface modes are associated only with the $F_2^{1,2}$ continuum bands. However, in the BD (111) model the extra back-bonded oxygen atom in the $(O_{1/2})_3{-}Si{-}O$ unit may generate a spurious extra surface mode. I strongly suspect that the absence of this mode from the experimental spectrum is good evidence against their surface molecular geometry. Note that in Fig. 1 the 800-cm^{-1} peak is associated with the cutoff of the $A_1 + E$ tetrahedral band. In my model the surface molecules are trigonal, not tetrahedral, so one would not expect to find a localized surface molecule mode associated with the cutoff of the tetrahedral continuum.

IV. Infrared Spectra: g-SiO_2

It has long been the custom, when comparing theoretical and experimental spectra, to describe the agreement as good. Early workers (e.g., Gas-

[17] J. C. Phillips, *J. Non-Cryst. Solids* **43,** 37 (1981), see especially p. 49.

[18] M. T. Loponen, R. C. Dynes, V. Narayanamurti, and J. P. Garno, *Phys. Rev. Lett.* **45,** 457 (1980).

kell[19]) showed that peaks in the vibrational spectra of g-SiO_2 could be fitted within 50 (100) cm^{-1} with a model based on corner-sharing tetrahedra using three (two) force constants. In view of the fact that the shifts in peak positions, for example, from β-cristobalite to g-SiO_2, are less than 50 cm^{-1}, and that even in disilicates shifts of less than 50 cm^{-1} are also typical, this 5–10% level of agreement is what one would expect. However, if one wishes to use the spectra to elucidate the structure of the glass, a much higher level of accuracy (at least of the order of 10 cm^{-1}, or 1%) is required.

Many studies of the infrared spectra of silica and silicates have been carried out, but it appears that only Gaskell and Johnson[20] have made the proper Kramers–Kronig-transformed reflectivity experiments, which establish the peak positions and line shapes with the precision needed for structural analysis. As they pointed out, and as is evident from chemical trends in the Raman spectra of silicates, the bond-bending spectra below 550 cm^{-1} contain the most information of structural significance. The data of Gaskell and Johnson for $\epsilon_2(\omega)$ in the 200–525 cm^{-1} range are shown in Fig. 4.

At first it may seem that there is nothing special about the experimental spectrum. It is true, as stressed by Gaskell and Johnson, that the 460-cm^{-1} peak is asymmetric. By scrutinizing the spectrum more carefully still, one notices that the asymmetry arises from a low-frequency tail that dies off very rapidly with $\omega_{max} - \omega$, where $\omega_{max} = 460$ cm^{-1} is the peak frequency. For reasons that will be developed in Section VI, I suspect that this decay is actually exponential. In addition, the full width at half maximum (FWHM) of the peak is only 39 cm^{-1}, which is very small for a "random" network.

To justify these comments I have plotted in Fig. 4 the infrared absorption spectrum for g-SiO_2 calculated from the random network model by Bell and Hibbins-Butler.[21] The agreement between their peak frequency (470 cm^{-1}) and experiment (460 cm^{-1}) is indeed very much superior to that obtained by other workers, which is why I used their results as a benchmark. Nevertheless, their FWHM is 180 cm^{-1}, which is much too large, and their peak dies off much too slowly at low frequencies, being an order of magnitude too large at $\omega = 200$ cm^{-1}. This is true for both of the peaks shown in Fig. 4, which were calculated with free surface O atoms. (The results are much worse with fixed O atoms, so that a large part of the peak strength must be associated with these surface atoms. The two curves in Fig. 4 correspond to different choices of effective charges, and the overall scale of

[19] P. H. Gaskell, *Phys. Chem. Glasses* **8,** 69 (1967); see also J. Etchepare, *Spectrochim. Acta, Part A* **26A,** 2147 (1970).

[20] P. H. Gaskell and D. W. Johnson, *J. Non-Cryst. Solids* **20,** 153, 171 (1976).

[21] R. J. Bell and D. C. Hibbins-Butler, *J. Phys. C* **9,** 1171 (1976).

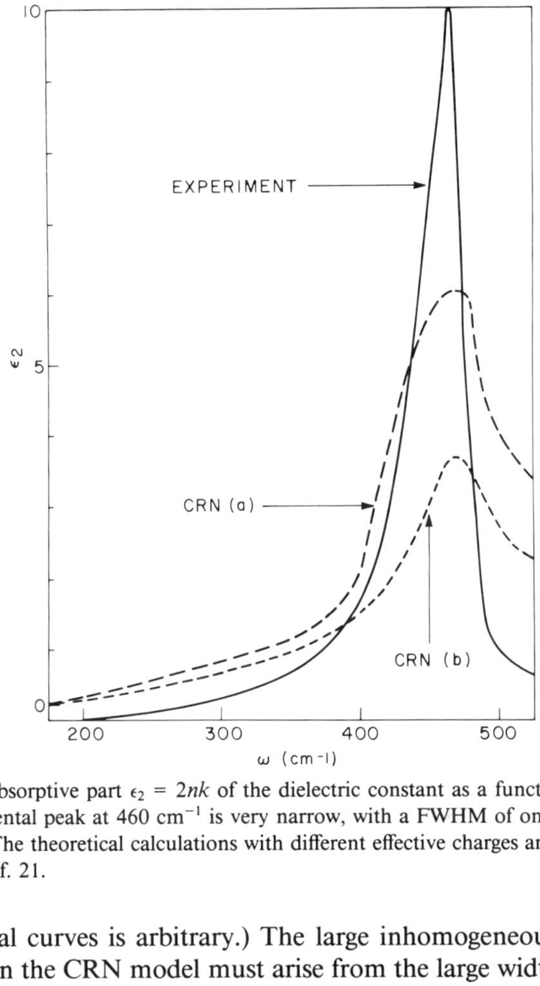

FIG. 4. The absorptive part $\epsilon_2 = 2nk$ of the dielectric constant as a function of frequency ω. The experimental peak at 460 cm^{-1} is very narrow, with a FWHM of only 39 cm^{-1} (data from Ref. 20). The theoretical calculations with different effective charges are based on CRN models from Ref. 21.

the theoretical curves is arbitrary.) The large inhomogeneous broadening of this peak in the CRN model must arise from the large width of the ring-statistical distribution, as I will discuss later in Section IX.

Another way of interpreting the CRN model calculation is to separate the spectrum into a bulk or continuum background and a peak localized on the surface atoms, as in Fig. 1. In this case the peak is superposed on a continuum background that is about one-third as high as the peak itself for the CRN (a) curve in Fig. 4. With this interpretation the FWHM of the peak is about 70 cm^{-1}, only about twice as large as experiment. However, the high background level calculated with the CRN model is nearly completely absent (<5% of the peak height) in the experimental data.

Gaskell and Johnson analyzed[20] their peak widths and attempted to relate them to the measured widths of the first- and second-neighbor peaks as measured by X-ray diffraction.[22] They assumed that the widths of these peaks could be translated into frequency widths by assuming that $d \ln \omega / d \ln R$ was of the order -1.5 and that the effects of the three different widths added incoherently. With these assumptions the authors obtained reasonably narrow peaks and therefore concluded that the intense absorption bands tell one no more about the structure than that "the local symmetry and geometry are disordered by amounts similar to those measured by X-ray diffraction." This train of reasoning was contradicted by the later explicit results of the CRN model[21] that yielded a FWHM nearly five times larger than experiment, although the model itself had been *constructed* to yield a radial distribution function in "good" agreement with experiment (but see Appendix B for a discussion of just how good the agreement really is). In any case it appears that the weakness in the Gaskell–Johnson argument is the assumption that the effects related to bond length and angle disorder must add incoherently. I will return to this point in Section IX.

V. Raman Spectra: g-SiO_2

For many years the weakness of Raman scattering limited its usefulness as a tool for studying molecular structure, but the widespread availability of lasers has reversed this situation and now Raman spectra are more abundant not only than neutron spectra, but even more so than infrared reflectance spectra of comparable structural sensitivity. Unfortunately, the interpretation of Raman spectra is complicated by the fact that even the simplest phenomenological model, the bond polarizability model, introduces three coupling parameters that connect the vibration density of states to the observed Raman scattering spectrum.[23] Thus when the atomic structure is well understood, Raman scattering can be a powerful tool for substantiating the model, as has recently been demonstrated for the chalcogenide $GeSe_2$ glasses, where the tetrahedra are nearly decoupled.[9] However, in general one cannot expect to be able to invert Raman data on glasses to establish the structure, unless a good initial model is already available.

To illustrate this point one can compare the experimental data[24] on g-SiO_2 (shown for the reader's convenience with HH polarization in Fig. 5) with the theoretical curves obtained from the CRN model.[23] In this case, in addition to free and fixed ends, there are three different theoretical spectra corresponding to three different coupling constants. In general, none of these coupling constants should be zero, so the actual spectrum should be

[22] R. L. Mozzi and B. E. Warren, *J. Appl. Crystallogr.* **2**, 164 (1969).

[23] R. J. Bell and D. C. Hibbins-Butler, *J. Phys. C* **9**, 2955 (1976).

[24] B. O. Mysen, D. Virgo, and C. M. Scarfe, *Am. Mineral.* **65**, 690 (1980).

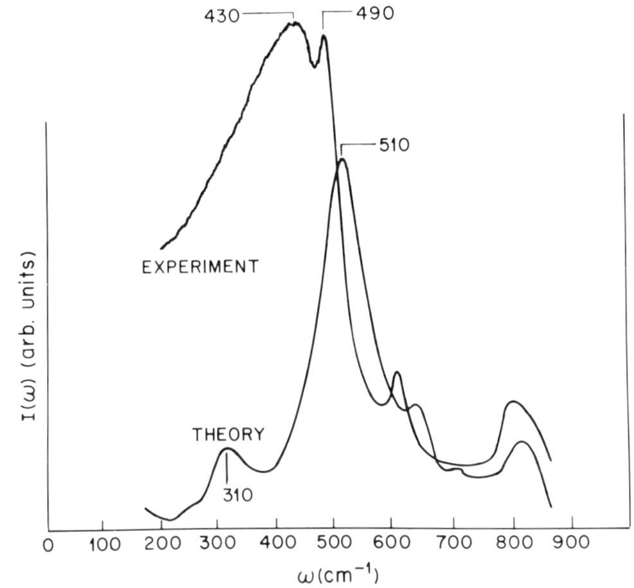

FIG. 5. Polarized (HH) Raman spectrum of g-SiO_2 (reproduced here for the reader's convenience from Ref. 24). Also shown is the polarized CRN model theoretical spectrum from Ref. 23.

obtained as a linear superposition of the three different spectra (with positive weights). Actually no choice of coupling constants consistent with the bond polarizability model is possible, since the spectra are strongly polarized[25] between 200 and 600 cm^{-1} and almost completely depolarized from 600 to 1300 cm^{-1}. This remarkable result cannot be explained by a simple mechanical and dielectric model, but I will present a tentative explanation for it in Sections X and XII.

Suppose one takes the phenomenological point of view, that the polarized spectrum (HH or VV) is more significant simply because it contains much more structure, and regard the depolarized scattering (HV) as a weaker effect of secondary interest. Although almost complete depolarization has been observed[26] in the Raman spectrum of a-Si, this may be a result of the overconstrained nature of the amorphous network.[4] By the same token, Raman scattering from the nearly completely relaxed glassy network may be very strongly polarized in certain regions of the spectrum where relax-

[25] F. L. Galeener and G. Lucovsky, *Phys. Rev. Lett.* **37,** 1474 (1976); F. L. Galeener, *Phys. Rev. B* **B19,** 4292 (1979); S. K. Sharma, D. Virgo, and I. Kushiro, *J. Non-Cryst. Solids* **33,** 235 (1979).

[26] R. Alben, J. E. Smith, Jr., M. H. Brodsky, and D. Weaire, *Phys. Rev. Lett.* **30,** 1141 (1973).

ation is most effective (Section X), which includes the region near 500 cm^{-1} of greatest structural significance.

These arguments lead one to conclude that among the six curves calculated for the CRN model,[23] one should compare with experiment in Fig. 5 the theoretical curve for free ends (as in the infrared case, to emphasize the surface O atoms) with the polarized spectrum ($G^2 = 0$). In fact, this is the curve that gives the best agreement,[23] but as can be seen from Fig. 5, the agreement is only "partial."[23]

The main peak at 510 cm^{-1} agrees fairly well with the experimental peaks at 430 and 490 cm^{-1}, but it does not explain why the former is so broad and the latter so narrow. The theory also gives subsidiary peaks at 100, 310, and 650 cm^{-1}, which are not present in the experimental data. The problem here primarily arises once again from misplacement of the surface oxygen resonant frequency near 310 cm^{-1}, just below the $dN/d\omega$ density of continuum states peak near 350 cm^{-1} (see Fig. 2). Actually the surface oxygen atoms produce the narrow experimental peak near 490 cm^{-1}, while the continuum modes produce the broad Raman peak near 430 cm^{-1}, near the peak in $dN/d\omega$, as discussed in Section VI.

In this section I have been forced to resort to a complex analysis in order to make a meaningful comparison between theory and experiment, but I do not claim that the analysis is convincing in its own right. I will jump ahead, however, to discuss a feature of the spectrum shown in Fig. 5 that varies dramatically in silicate glasses such as $(Na_2O)_x(SiO_2)_{1-x}$, which will be discussed in more depth in Section XII. This is the dependence on x of $\gamma = S(430)/S(490)$, that is, the relative integrated scattering strengths of the broad and narrow bending modes corresponding to bulk and surface O atom motion. It turns out that $\gamma(x)$ is nearly proportional to $(0.2 - x)^{1/2}$ up to $x = 0.2$; that is, the strength of the narrow surface mode is almost independent of x while the broad bulk mode disappears rapidly, and almost completely quenched at $x = 0.2$. This result is completely inexplicable within the context of a CRN model, and it has not previously been discussed, although the result has been in the literature[27] for more than 10 years.

VI. Dispersion of Internal Surface Modes

In the preceding sections I have shown that the infrared reflectance and neutron and Raman scattering spectra of g-SiO_2 can be explained on the assumption that these spectra consist of two contributions, a bulk or continuum contribution and a localized contribution. The latter is divided into two parts again, that associated with extended internal surfaces and that associated with point defects (isolated broken bonds). The internal surface

[27] M. Haas, *Phys. Chem. Solids* **31**, 415 (1970).

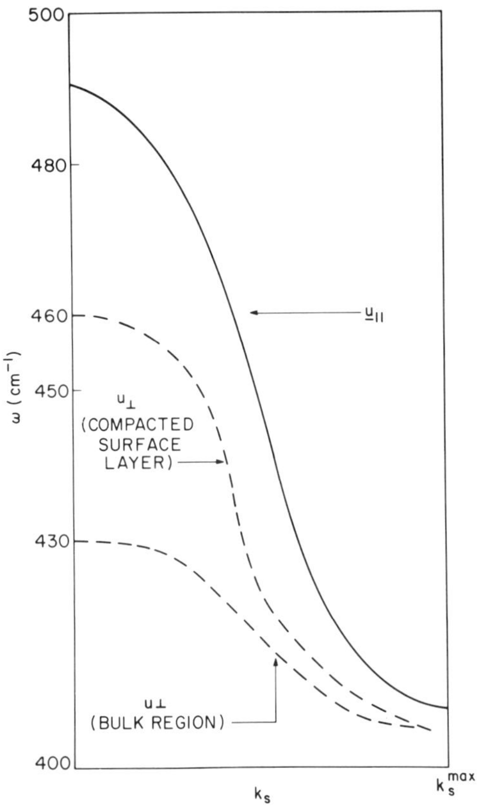

FIG. 6. Dispersion of localized surface vibrational modes $\omega(\mathbf{k}_s)$ as a function of surface wave vector $\mathbf{k}_s$. Also shown is an estimate of the effect of compaction due to polishing (cold-working) the surface layer (depth $\sim 10^3$ Å) on the transverse optic modes measured by infrared absorption in this layer.

contribution to the three types of vibrational spectra is much larger than that of the isolated point defects, and the internal surface itself can be regarded as an intrinsic feature of the glassy structure.[4] (It is *not* merely an artifact produced by the necessarily finite extent of a ball-and-stick cluster model of a CRN.) Moreover, because of the large electronegativity difference between Si and O, almost all of the surface molecules must be neutral and describable through the configuration $(O_{1/2})_2{-}Si{=}O_s^*$.

In Section VII I take up the question of cluster dimensions and morphology. In this section I consider certain general features of the vibrational modes localized on the nonbridging surface O_s^* atoms that are necessarily present, regardless of the internal cluster dimensions and morphology. These features are summarized in Fig. 6, which shows the dispersion of the surface vibrational modes as a function of surface wave number k_s.

Ordinarily in solids one describes crystalline dispersion spectra $\omega(\mathbf{k})$ in terms of a three-dimensional Bloch wave vector **k** associated with the periodicity of the lattice; however, for qualitative purposes, similar isotropic dispersion curves $\omega(k)$ can be constructed for any dense interacting system of nearly uniformly spaced and nearly identical units, for example, the phonon and roton modes in liquid He. In the case of surface lattice vibrations, these curves will have two branches α associated with in-plane vibrational displacements **u** parallel or perpendicular to the two-dimensional surface wave vector $\mathbf{k}_s$. These correspond to longitudinal and transverse optic vibration modes in crystals.

Again, by analogy with the crystalline result, but without demanding crystalline periodicity (since the wavelength of light is much larger than the intermolecular spacing, while the neutron scattering length is much smaller), one can establish the general features of $\omega_s{}^{\alpha}(k_s)$ from the infrared reflectance and neutron and Raman scattering data, namely, $\omega_s{}^{\parallel}(0) = 490$ cm^{-1} (Raman) and $\omega_s{}^{\perp}(0) = 460\ \text{cm}^{-1}$ (infrared), while $\omega_s^{\parallel,\perp}(k_s^{\max}) \approx 405$ cm^{-1}. Here $k_s^{\max}$ is a cutoff for the surface wave vector, which is of order $2\pi/b$, where b is the average surface intermolecular spacing. The peak in the neutron density of localized states, $G_1(\omega)$, is expected to come near $\omega = \omega(k_s^{\max})$, because the number of states is proportional to k_s and in a model of nearest-neighbor interactions only, the dispersion is given by

$$\omega(k_s) = \omega_0 - a\cos(k_s b/2). \tag{6.1}$$

The foregoing description is generally valid, provided that the surface modes are genuinely localized, either because their frequencies lie outside the continuum bulk bands completely or (for practical purposes in a glass) because the density of bulk states that they overlap is very small. In the case of the CRN model of Bell and co-workers,[6,15,16,21] their unphysical choice of surface molecule $(O_{1/2})_3{-}Si{-}O$ apparently had the unfortunate consequence of placing the $L_1(O_s^*)$ bare surface bending resonant frequency close to 400 cm^{-1}, which is close to the peak in the continuum density of states. This caused the free-end localized mode resonance to split into two parts, one near 270 cm^{-1}, which the authors identified, and one near 550 cm^{-1}, which they overlooked. (The latter may actually be localized on the Si atoms in the surface molecules.) This accidental behavior produces many anomalies in the calculated spectra of Bell and co-workers that are especially evident, for example, in the enormous variability with the different boundary conditions and polarizability weighting factors of the authors' Raman spectra.[23] Behavior of this kind often occurs in the calculation of surface electronic energy bands, which is why great care must be exercised when one attempts to carry out the calculations using a simplified approach such as the tight-binding method.[28]

[28] K. C. Pandey and J. C. Phillips, *Phys. Rev. B: Solid State* [3] **13**, 750 (1976).

The dichotomy here that separates vibrational modes into bulk continuum and internal surface modes has led to the description of the surface modes as shown in Fig. 6. The parallel–perpendicular splitting at $k_s = 0$ is derivable from macroscopic reasoning[29] that can be generalized[30] to a material containing several infrared-active optic modes. The relative weight of each mode depends on the product of the local polarizability and effective charge, neither of which is known. One can also use the transverse infrared optical spectrum $\epsilon_2(\omega)$ to predict positions of longitudinal optical resonances in $\mathrm{Im}(\epsilon_1 + \iota\epsilon_2)^{-1}$ according to classical plasmon theory.[31] These resonances occur at the places where $\epsilon_1(\omega) = 0$ and $d\epsilon_1/d\omega > 0$.

According to my present model, the successful interpretation of most features of the spectrum of g-GeO_2 in terms of the TO and LO modes of some unspecified microscopic entity[31] actually referred to modes localized on internal surfaces. However, for the surface modes $L_1(O_s^*) = (460, 490)$ cm^{-1} of g-SiO_2 shown in Fig. 6, they predict an $\omega_{LO}^s - \omega_{TO}^s$ splitting of about 60 cm^{-1}, which differs by a factor of 2 from the observed splitting of 30 cm^{-1}. Their infrared spectrum for g-SiO_2 shows, moreover, substantial absorption near 380 cm^{-1}, which is absent from the data of Gaskell and Johnson.[20] The extra absorption probably arises from OH groups chemisorbed in the macroscopically porous surface of their sample.[32]

To see whether the hydroxyl absorption is significant, I applied standard dispersion theory[30] to the data of Gaskell and Johnson and corrected for the asymmetry of their 460-cm^{-1} oscillator. The result again was a LO–TO splitting close to 60 cm^{-1} for the $L_1(O_s^*)$ mode, rather than 30 cm^{-1} as required by experiment. One possible way of resolving this difficulty might appear to lie in the following observation. The proper condition for the internal surface plasmon resonance is [31]

$$\epsilon_{b1}(\omega_s) + \epsilon_{s1}(\omega_s) = 0, \qquad (6.2)$$

where $\tilde{\epsilon}_b$ and $\tilde{\epsilon}_s$ are the complex dielectric functions of the cluster interiors and surfaces, respectively. Assuming that the weighted volumes are f_b and f_s, respectively, then the infrared experiment measures the weighted average $\langle\tilde{\epsilon}\rangle = f_b\tilde{\epsilon}_b + f_s\tilde{\epsilon}_s$. Thus application of the macroscopic rule for calculating

[29] M. Born and K. Huang, "Dynamical Theory of Crystal Lattices," Sect. 8. Oxford Univ. Press (Clarendon), London and New York, 1954.

[30] A. S. Barker, Jr., *Phys. Rev.* [2] **136A,** 1290 (1964).

[31] F. L. Galeener and G. Lucovsky, *Phys. Rev. Lett.* **37,** 1474 (1976). Because of the discrepancies they encountered for g-SiO_2, Galeener and Lucovsky expressed misgivings about the condition for longitudinal plasmon resonances, namely, that it is given by the maximum in $\mathrm{Im}\ \epsilon^{-1} = \epsilon_2/(\epsilon_1^2 + \epsilon_2^2)$. For a homogeneous medium, this condition is macroscopically exact. For a discussion of interfacial plasmons see F. A. Stern and R. A. Ferrell, *Phys. Rev.* **120,** 130 (1960).

[32] R. B. Laughlin, J. D. Joannopoulous, C. A. Murray, K. J. Hartnett, and T. J. Greytak, *Phys. Rev. Lett.* **40,** 461 (1978).

longitudinal resonances

$$f_b\epsilon_{b1}(\omega_0) + f_s\epsilon_{s1}(\omega_0) = 0 \tag{6.3}$$

will lead to a value of $\omega_0 \neq \omega_s$. From Sections III and VII it is known that $f_b \sim 0.8$ and $f_s \sim 0.2$. If one identifies the 460-cm^{-1} peak as a resonance in $\epsilon_{2s}(\omega)$, then one can show that $\omega_s > \omega_0$. Thus use of the proper condition (6.2) in place of (6.3) does not reduce the discrepancy between the observed difference $\omega_{LO}^s - \omega_{TO}^s = 30$ cm^{-1} and that close to 60 cm^{-1} predicted by the infrared data. The discrepancy also cannot be removed if one associates the 460-cm^{-1} infrared peak with $\epsilon_{2b}(\omega)$, since the 430-cm^{-1} peak in the Raman scattering has been assigned to the bulk modes.

It is possible that the actual origin of the discrepancy is extrinsic. Suppose that the interfacial regions of the outermost surface region of depth of order 10^3 Å of the sample have been substantially compacted (b' reduced, see Section VII), for example, by cold-working during polishing, and that the infrared frequency of $\omega = 460$ cm^{-1} measured by Gaskell and Johnson is characteristic of these compacted interfacial regions. At the greater depths (of order 10^4 Å) probed by Raman scattering ω_t may be as low as 430 cm^{-1}; that is, as shown in Fig. 6, the TO mode may be nearly that in bulk. If the compaction effect is approximately linear in density versus depth over a range of the order of 10^3 Å, then, because the light is absorbed exponentially, an exponential tail would be found on the low-frequency side of the infrared peak. This is actually true of the data of Gaskell and Johnson shown in Fig. 4, which is plausible although not conclusive evidence for the compaction mechanism.

This somewhat *ad hoc* mechanism can be subjected to a decisive experimental test.[33] After the sample has been polished and its infrared and Raman spectra measured, it can be heavily damaged by neutron bombardment; the primary effect of this bombardment is to introduce a high level of point defects, especially nonbridging oxygen atoms. This leads to an order-of-magnitude enhancement of the Raman scattering strength of the 606-cm^{-1} peak, as shown in Fig. 7, which I assign to $L_1(O_p^*)$ in Fig. 1.

[33] R. H. Stolen, J. T. Krause, and C. R. Kurkjian, *Discuss. Faraday Soc.* **50,** 103 (1970). These authors gave a cautious and reasonable discussion of "defect" lines, but they did not consider the internal surface mechanism presented here, which was also not considered in defect discussions by other authors: J. B. Bates, R. W. Hendricks, and L. B. Shaffer, *J. Chem. Phys.* **61,** 4163 (1974); G. N. Greaves, *J. Non-Cryst. Solids* **32,** 295 (1979); G. Lucovsky, *Philos. Mag.* **39,** 513 (1979); E. J. Friebele, D. L. Griscom, M. Stapelbrocck, and R. A. Weeks, *Phys. Rev. Lett.* **42,** 1346 (1979); A. R. Silin and P. J. Bray, *Bull. Am. Phys. Soc.* [2] **26,** 218 (1981); F. L. Galeener, *J. Non-Cryst. Solids* **40,** 527 (1980). Several of these authors suggested that the 606-cm^{-1} line is a point defect associated either with a dangling O atom or an O—O peroxy bridge; both interpretations would be categorized as O_p^* in my notation.

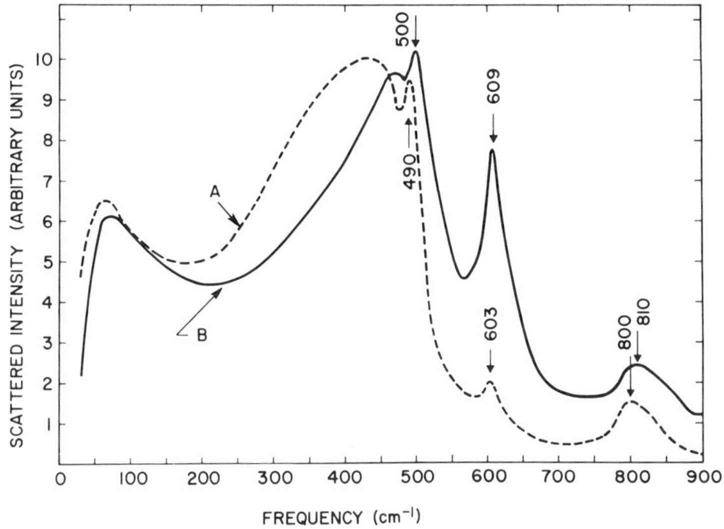

FIG. 7. The polarized Raman spectrum for (a) fused quartz and (b) neutron-bombarded (10^{20} cm^{-2}, energy > 10 keV) fused quartz, with respective densities 2.20 and 2.25 g/cm^3. The data from Ref. 33 are reproduced here for the reader's convenience.

The uniform level of damage associated with neutron bombardment also would tend to mitigate the difference in interfacial spacing in the cold-worked surface layer probed by infrared absorption compared to that of the deeper layers probed by Raman scattering. This would tend to increase $\omega_{LO}^{s} - \omega_{TO}^{s}$ from the value of 30 cm^{-1} observed in unirradiated g-SiO_2. In fact,[33] this is the case; in the neutron-bombarded sample the infrared value of ω_{TO}^{s} is decreased by about 10 cm^{-1} and the Raman value of ω_{LO}^{s} is increased by about 10 cm^{-1}, so that the $\omega_{LO} - \omega_{TO}$ difference increases to 50 cm^{-1}, which is much closer (although, of course, some difference due to cold-working of the surface still remains) to the Kramers–Kronig analysis.[31] In spite of the subtlety of this mechanism, I regard these shifts as decisive confirmation of my proposal. Thus polarized surface waves satisfactorily account for the existence, pairing, and splitting of localized modes in g-SiO_2 and g-GeO_2, both qualitatively and quantitatively.

At this point some readers may feel that surface compaction is merely a *post hoc* explanation of the difference between the $\omega_{TO} = 430$ cm^{-1} required by the Kramers–Kronig analysis and the $\omega_{TO} = 460$ cm^{-1} measured by infrared reflection. Thus it is of special significance that the (ω_{LO}, ω_{TO}) pair frequencies in g-SiO_2 can be measured directly in a single experiment that probes only the sample bulk and not the cold-worked surface region measured by infrared absorption. This experiment is hyper-Raman (two-photon) scattering,[34] a very weak and hence extremely surface-insensitive effect. Because two photons can be scattered with phonon emission or absorption, both TO and LO modes are Raman active, and (ω_{TO}, ω_{LO}) phonon pairs are resolved in the unpolarized spectrum of g-SiO_2 for scattering angle $\theta = 90°$ at (470, 530) cm^{-1} and (1065, 1250) cm^{-1}. The magnitudes of the splittings, especially the $L_1(O_s^*)$ splitting of 60 cm^{-1}, are in excellent agreement with those predicted by the Kramers–Kronig analysis,[31] and I suppose that the splitting is measured more accurately in this difficult experiment than are either of the peak frequencies alone.

An interesting feature of the hyper-Raman scattering spectrum[34] is that the broad strong peak seen at 430 cm^{-1} in ordinary Raman scattering is completely absent. This is the only peak that I have assigned to bulk or cluster interior modes. It is possible that over the large lengths $\gtrsim 2.10^3$ cm that contribute to two-photon scattering[31] only the internal surface vibrational modes remain coherent, so that Raman scattering from bulk or cluster interior modes is largely suppressed. If this is the case, then hyper-Raman scattering should prove to be an extremely sensitive and powerful tool for studying chemical trends in internal surface morphologies in silicates and germanates (see Secs. XII and XIII).

Having removed this serious quantitative discrepancy for $L_1(O_s^*)$ for g-SiO_2, we can examine the implications of the other (ω_{TO}, ω_{LO}) pairings for g-SiO_2. The most important conclusion is that the $F_2{}^2$ pairing at 1065 and 1200 cm^{-1} is predicted correctly. This means that the 1065-cm^{-1} infrared peak is due predominantly to absorption by surface stretching modes and not by bulk stretching modes. Since there is also no infrared absorption corresponding to the Raman peak at 430 cm^{-1}, we can conclude that the electric dipole oscillator strength for the bulk or interior $F_2^{1,2}$ modes is very much smaller than that for the surface modes. If we denote this oscillator strength by $(Z_{b,s}^*)^2$, then the condition

$$(Z_s^*)^2 \gg (Z_b^*)^2 \tag{6.4}$$

must hold.

There are several reasons why Eq. (6.4) may be valid. The vibrational amplitudes of the onefold-coordinated O_s^* atoms may be much larger then those of the O interior atoms that are twofold coordinated. Also, a strong self-consistent screening of internal electric field fluctuations may occur because of the self-consistent interactions among the internal oxygen atoms. I develop this idea at some length in Section IX.

Within the context of the present model the existence of (ω_{LO},ω_{TO}) paired

[34] V. N. Denisov, B. N. Mavrin, V. B. Podobedov, and K. E. Sterin, *Sov. Phys.—Solid State (Engl. Transl.)* **20,** 2016 (1978); *JETP Lett. (Engl. Transl.)* **32,** 316 (1980); A. S. Barker and R. Loudon, *Rev. Mod. Phys.* **44,** 18 (1972).

narrow bands is easily understood in terms of surface modes. The necessary conditions for the existence of these modes are only that two narrow bands be degenerate in the long-wavelength limit in the absence of Coulomb forces; and that modes in these bands have a wave character that is distinguishably longitudinal or transverse. Many scientists have found, on the other hand, that it is difficult to imagine that such paired modes can exist in a glass which is modeled as a CRN of disordered molecular units. It should therefore be mentioned here that Denisov *et al.*[34] have carried out a decisive experiment with hyper-Raman scattering near $\theta = 0$ that shows the same kind of polariton effects in g-SiO_2 as are found in simple crystals such as GaP.[34] For the polariton (photon–phonon mixed state) effect to exist one demands only that $R/\lambda \ll 1$ (where R is the average radius of curvature of the surface and λ is the wavelength of the light). The remaining conditions, that narrow vibrational bands exist with long-wavelength λ character and $w_{TO} \rightarrow$ const ($\neq 0$) as $\lambda \rightarrow \infty$, are automatically satisfied by the surface vibrational modes described here. I emphasize once again that it is essential that these internal surfaces have a definite paracrystalline surface texture in order that the bands be narrow compared to the ω_{LO}–ω_{TO} splitting. With random surface textures of the Bell and Dean type this is not expected to be the case.

VII. Cluster Dimensions and Morphology (SiO_2)

In the preceding sections I have shown that the localized mode contributions to infrared absorption and neutron and Raman scattering spectra not only contain some very narrow peaks, but also that these peaks shift with the probing interaction in a way that is inexplicable in CRN models, which predict that the O_p^* shifts must be comparable to their peak widths. On the other hand, the behavior of the localized mode spectra is readily understood if one postulates that the glass contains a large internal surface area covered by nearly equivalent double-bonded O_s^* atoms, with the local quasitrigonal configuration $O_s^*{=}Si{-}(O_{1/2})_2$.

The failure of the CRN model to produce sufficiently narrow localized peaks is a strong indication that the substrate network to which the quasitrigonal configuration is bonded is much more homogeneous than one would expect from a CRN model with a wide distribution of ring statistics.[16] This naturally suggests a cluster model with a paracrystalline surface texture. Paracrystalline models have often been discussed in the older literature on the grounds that they provide the most physical way to picture the glass as a frozen liquid.[2] However, for some time many other workers believed that diffraction data (including small-angle scattering) could be used to exclude cluster models. Actually, in Section XVIII I discuss abundant evidence that

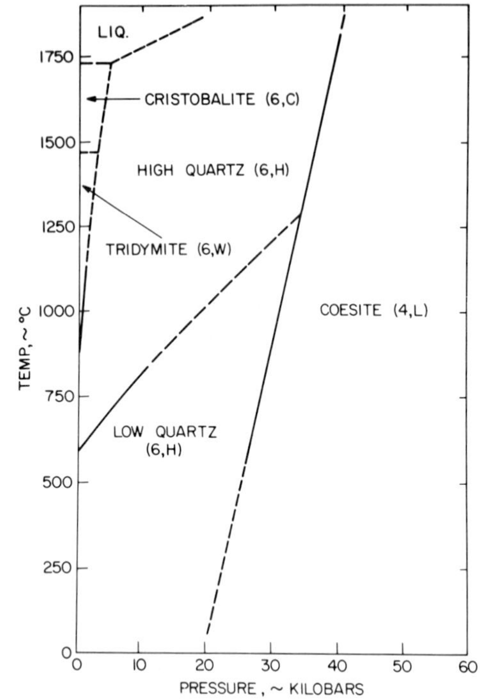

FIG. 8. The phase diagram for SiO_2 from Ref. 35. The Si coordination numbers are indicated for each phase by (CN).

shows that the diffraction data, including the most recent ultra-high-resolution electron micrographs, actually support the cluster model. For completeness I discuss the older diffraction ideas in Appendix B and show why they are no longer relevant as "proofs" against the cluster model.

Granted the intuitive attractiveness of the cluster model, as well as the strong experimental evidence from all sides that supports it, what is the most appropriate cluster morphology? There are many arguments that suggest that for g-SiO_2, β-cristobalite is the proper choice.[35] I have gathered these arguments together in the following paragraphs. An equilibrium phase diagram[35] for SiO_2 is shown in Fig. 8.

The structure of β-cristobalite is that of the cubic Si crystal with oxygen atoms statistically distributed, one oxygen atom in one of six equivalent

[35] J. T. Randall, H. P. Rookesby, and B. S. Cooper, *Nature (London)* **125,** 458 (1930); J. F. G. Hicks, *Science* **155,** 459 (1967); F. R. Boyd and J. L. England, *JGR, J. Geophys. Res.* **65,** 752 (1960); F. E. Wagstaff, *J. Am. Ceram. Soc.* **52,** 650 (1969); A. G. Boganov and V. S. Rudenko, *Sov. J. Glass Phys. Chem. (Engl.Transl.)* **2,** 31 (1976).

sites clustered around the midpoint of the line connecting two Si nearest neighbors.[36] This cluster site structure is nearly isotropic and can generate at its surface many nearly equivalent trigonal silinyl configurations O_s^* $=Si-(O_{1/2})_2$. It is not uniaxial, like quartz, which makes it much more suitable for forming spheroidal clusters with nearly equivalent O_s^* surface atoms. According to Fig. 7, cristobalite and tridymite are the stable forms of silica above the glass transition temperature (~1200°C), and the glass density[37] (2.20) is only slightly less than the cristobalite density (2.32), but much less than the density of quartz (2.65). A 5% difference in density is reasonable, but a 20% difference in density is not. Finally the ring statistics of β-cristobalite, with two formula units per unit cell, are very simple and are dominated entirely by six Si—O units. The much more compact structure of quartz also includes rings with six Si—O units, but these always lie nearly in a plane normal to the *c* axis. The formation of dense clusters with a strong spatial preference would seem unlikely when a more nearly isotropic alternative phase is available.

It has sometimes been suggested that tridymite (density, 2.26) may also compete with cristobalite as a candidate for cluster morphology. Tridymite bears the same relation to cristobalite as Si in the wurtzite structure does to Si in the diamond structure. Because of a smaller density difference, tridymite clusters would be twice as large as cristobalite clusters, according to the calculation given below, so that I prefer the cristobalite morphology.

The older thermochemical literature contains many attempts to demonstrate that clusters (usually with a cristobalite morphology) are the fundamental structural units of g-SiO_2. However, these models, although physically quite plausible, have failed to gain general acceptance because of several quantitative inadequacies. The first of these is their inability to estimate the cluster size, a shortcoming that I remedy later. The second is their failure to identify the chemical bonding mechanism that can generate a stoichiometric (chemically neutral) cluster surface; that is, in these theories I have not been able to find any mention of the unique role played by the double-bonding proclivities of oxygen and, to a lesser degree, Si. (I have already shown in Section II how failure to appreciate the strength of the Si=O bond has produced incorrect vibrational spectra of localized O* modes.) This recurrent blind spot has greatly hindered the development of satisfactory quantitative cluster models of g-SiO_2.

Although the thermochemical arguments in favor of clusters with a cris-

[36] A. F. Wright and A. J. Leadbetter, *Philos. Mag.* [8] **31,** 1391 (1975). For further discussion of the microtwinning of cristobalite, see S. Hansen, L. Fälth, and S. Anderson, *J. Solid State Chem.* **39,** 137 (1981).

[37] W. A. Weyl and E. C. Marboe, "The Constitution of Glasses," p. 464. Wiley-Interscience, New York, 1964.

tobalite morphology were too qualitative to gain decisive acceptance, an example of this work should be mentioned here. Solution- (in HF) calorimetric studies of 100-μm grains of g-SiO_2, quartz, and cristobalite showed that the speeds of solution were generally in the ratio 9:1:3, respectively; that is, g-SiO_2 behaves much more like cristobalite than like quartz. The same measurements[38] yielded estimates of enthalpy differences between cristobalite and g-SiO_2 of 0.5 kcal/mol, compared to 2.3 kcal/mol for quartz and g-SiO_2. These results are consistent with the density differences.

Density differences can also be used to estimate cluster size. Because the Si sublattice in β-cristobalite has the same symmetry as c-Si, one can idealize this sublattice near a cluster–cluster interface simply by removing a (100) plane of Si atoms from c-Si. To simulate noncoalescence, the two semi-infinite Si crystals thus obtained should be displaced by a small amount parallel to their interface. They will also, after the Si (100) plane has been removed and the O atoms rebonded to give β-cristobalite with one double-bonded O atom per surface Si, relax toward each other, reducing the density deficit that would otherwise be overestimated. Assuming that the density deficit arises entirely from this source, and not, for example, from O_p^* atoms, what is the smallest cluster size that is compatible with the 5% density deficit between g-SiO_2 and β-cristobalite?

The answer to this question depends on how much accommodation or adaptive relaxation takes place at the interface. This could appear to be a complex problem that requires determination of the intercluster van der Waals potential. However, an analog structure exists that can be used to estimate the relaxation directly. Layer GeS_2 is isovalent to SiO_2, but it is constructed by combining tetrahedra to form S–Ge–S layers.[39] One can compare the nearest S–S *intra*layer spacing b with the nearest S–S *inter*layer spacing b' to estimate the interlayer relaxation (see Figs. 9 and 10). This shows that $b' = b/3$; that is, ⅔ of the original density deficit is recovered by relaxation. In the case of g-SiO_2, secondary relaxation also takes place via shortening of Si=O compared to Si—O; according to Pauling's rules this reduction is about 10%. On the other hand, there is a much stronger Coulomb repulsion between O^- in SiO_2 than between S in GeS_2. This effect may well overcompensate the double-bonding contraction. My final estimate for the reduced spacing between the β-cristobalite layers is $b' = 0.6b$.

The cubic lattice constant for β-cristobalite is 7.2 Å and this cube contains four (100) lattice planes. Removing one (100) SiO_2 trilayer and relaxing to an intercluster spacing S that is

[38] C. Hummel and H. E. Schwiete, *Glastech. Ber.* **32,** 327 (1959); *Chem Abstr.* **54,** 6288e (1960).

[39] G. Dittmar and H. Schafer, *Acta Crystallogr., Sect. B* **B32,** 1188, 2726 (1976).

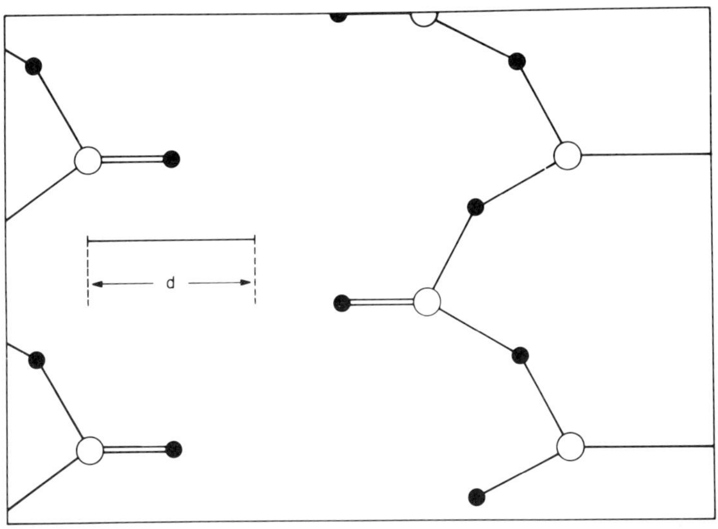

FIG. 9. Sketch of cristobalite (100) interface before relaxation, formed from cristobalite by removing a molecular plane of thickness d = 3.7 Å.

$$S = 2s = 0.6(7.2/4)\ \text{Å},$$

$$s = 0.55\ \text{Å}, \tag{7.1}$$

leads to a fractional density difference δ between a sphere of radius R and one filled only to $R - s$ that is approximately

$$\delta = \delta\rho/\rho = 0.05 = 3s/R, \tag{7.2}$$

from which

$$R = 3s/\delta = 33\ \text{Å}. \tag{7.3}$$

The value of x_s, the fraction of surface molecules, implied by Eqs. (7.1)–(7.3), is

$$x_s(\text{g-SiO}_2) = 0.17, \tag{7.4}$$

which is in excellent agreement with the value $x_s = 0.20$ that was previously obtained from the density of vibrational states, Eq. (3.4).

The cluster or paracrystalline structures of silica, germania, soda-silicate ($Na_2O \cdot 2SiO_2$) and a commercial silicate glass have been studied[40] directly in beautiful TEM experiments that may be the most fundamental experiments ever performed on glass. Zarzycki's experiments are in a class by themselves, because the samples were fine-drawn glass fibers that were frac-

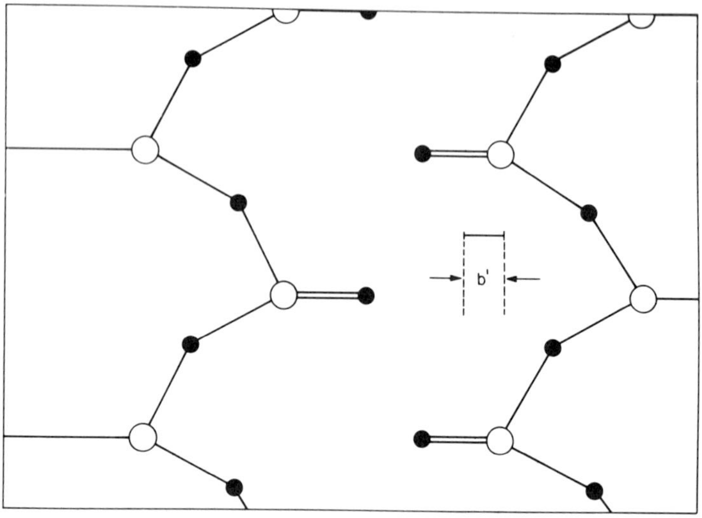

FIG. 10. Sketch of cristobalite (100) interface after relaxation of b to b'. In g-SiO_2 I estimate $b' = 0.6b$, so that the actual relaxation shown has been exaggerated for emphasis.

tured and micrographed entirely in an Ar atmosphere.[40] The domain radii for g-SiO_2 were measured to be 25–35 Å, in very good agreement with Eq. (7.3). The radii for the silicate glasses are larger, of order 50 Å.

One of the common criticisms of Zarzycki's experiments is that the observed structure is an artifact associated with chemical contamination of the surface. In fact, this criticism is a double red herring, because Zarzycki showed that the degradations of the structure upon exposure to air[40] (dry or humid) are exactly what would be expected from a cluster model. He found that the intercluster interfacial boundaries are highly reactive (in the order $K_2O \cdot 2SiO_2$, $Na_2O \cdot 2SiO_2$, SiO_2, commercial glass) and that they disappear upon contamination to air (presumably with H_2O). Another criticism that might be considered is that fracture during sample preparation and/or heating by an electron beam has induced phase separation. Clearly this particular objection cannot explain the results observed for g-SiO_2 and g-GeO_2. Further discussion of Zarzycki's silicate and GeO_2 data will be given in Section XVIII, but at this point I will comment that the chemical trends exhibited in Zarzycki's data correspond very closely to what I would have expected on intrinsic grounds. Although experiments of this kind are delicate and can easily be spoiled, I believe that Zarzycki's experiments are correct throughout and extremely informative. Readers who demand direct

[40] J. Zarzycki and R. Mezard, *Phys. Chem. Glasses* **3,** 163 (1962); J. Zarzycki, *C. R. Hebd. Seances Acad. Sci., Ser. B* **271,** 242 (1970).

evidence for the existence of clusters in oxide glasses will find everything they could expect (and I found much more) in these papers.[40] More recent ultra-high-resolution experiments that reveal the internal cluster morphology of silicate glasses will also be discussed in Section XVIII.

VIII. Morphology of Clusters in g-GeO_2

Whereas there are many crystalline polymorphs of SiO_2, crystalline GeO_2 is found generally in only two forms, quartz with CN(Ge) = 4 and rutile with CN(Ge) = 6. The O bond angle is about 10° smaller in the quartz form of GeO_2 than SiO_2, and the increased O–O interactions are presumably responsible for the nonoccurrence of GeO_2 in the cristobalite, tridymite, coesite, etc. structures.[41]

The densities of GeO_2 (quartz) and GeO_2 (rutile) are 4.23 and 6.24, respectively, compared to a density of 3.63 for g-GeO_2. If, by analogy with g-SiO_2, one postulates the existence of paracrystalline clusters in g-GeO_2, these must have the hexagonal quartz topology, not the cubic crystobalite morphology. Moreover, for GeO_2 the normalized density deficit $\delta = \delta\rho/\rho$ is nearly 0.14, compared to 0.05 for SiO_2.

The fractional density deficit δ is determined by two factors, the cluster size R and the intercluster interface width s. According to Zarzycki's electron micrograph studies[40] R is about 30 Å in both SiO_2 and GeO_2. Therefore s in GeO_2 should be nearly three times larger than in SiO_2. According to the discussion of the previous section, this implies that the relaxed average intercluster spacing in g-GeO_2 should be given by $b' = 1.8b$, that is, one and a half to two molecular diameters.

It appears that the most likely explanation for interfacial width expansion is that the cluster morphology in g-GeO_2 is much less adaptive than in g-SiO_2. This is partly because the packing of clusters with a uniaxial, essentially cylindrical morphology is much more complex than that of clusters with a cubic morphology. Moreover, the density of quartz is itself about 15% greater than cristobalite; that is, the structure of quartz is already collapsed and it is much less compressible and deformable by shear than cristobalite. It is also likely that the energy of step formation is much larger for the compacted quartz structure than it is for the more open cristobalite structure. Of course, since the paracrystallites are in general misaligned, there

[41] K. J. Seifert, H. Nowotny, and E. Hauser [*Monatsh. Chem.* **102,** 1006 (1971)] reported the synthesis of a "noncubic crystobalite GeO_2 powder" with a density of 3.93 g/cm^3, about 7% greater than g-GeO_2 from a NH_4–germanate zeolite. This report has not been confirmed by other workers, and the purity of the (possibly metastable) samples appears to be open to doubt.

must be a high surface step density to produce the kind of close interfacial accommodation illustrated in Figs. 9 and 10.

There are two experimental checks on this model that says that the intercluster spacing in g-GeO_2 is about three times larger than in g-SiO_2. First, the larger intercluster spacing explains the TEM observation[40] that "the inhomogeneity of the [micellar] structure [is] *much more marked* in the case of vitreous germania" compared to vitreous silica. Second, one can examine the relative solubilities of inert gases (such as He, Ne, and Ar) in g-GeO_2 compared to g-SiO_2[42] as a function of the hard-core radius R_c of the gas atom. What trends does one expect to see?

According to the model for g-SiO_2, its (100)-like interfacial width is small ($b' = 0.6b$) and nearly constant. For the quartz paracrystallites in g-GeO_2 the interfaces generally have different textures and the interfacial widths are expected to be distributed over a wide range, with average interfacial spacing $b' = 1.5b$. The solubilities of the noble gases He, Ne, and Ar in g-SiO_2 and g-GeO_2 are believed[42] to measure the distribution of cavity sizes in the glasses, so that one may expect these solubilities to reflect the distribution of interfacial widths.

Suppose one defines the effective radius R_c of a noble gas atom as the value of its interatomic potential at which the hard-core repulsive interaction[42] reaches the level of 2×10^{-3} atomic units, or about 0.055 eV. This corresponds to a temperature of about 600°K, which lies in the range in which the solubilities were measured.[41] I have plotted these solubilities against $2R_c$, the so-defined diameter of the gas atom in angstroms, in Fig. 11.

The general features of Fig. 11 bear out my expectations. According to my estimate for g-SiO_2, $b' = 0.6b = 2.2$ Å, which is slightly larger than $2R_c(\text{He}) = 1.8$ Å, and smaller than $2R_c(\text{Ne}) = 2.6$ Å or $2R_c(\text{Ar}) = 3.3$ Å. Thus only He shows a large solubility in g-SiO_2, and there is an abrupt drop to the Ne and Ar solubilities, which are nearly an order of magnitude

[42] W. W. Brandt, B. Rauch, and J. J. Wagner, *Z. Naturforsch., A* **27A,** 617 (1972); R. L. Matcha and R. K. Nesbet, *Phys. Rev.* **160,** 72 (1967). Differently weighted values for He solubilities have been obtained by J. E. Shelby [*J. Appl. Phys.* **43,** 3068 (1972)]. Measurements of the pressure dependence of these solubilities [J. E. Shelby, *J. Appl. Phys.* **47,** 135 (1976)] have shown that "the solubility of inert gases in oxide glasses is a simple physical process, similar to [internal] surface adsorption, with no strong chemical interaction with the glass network." Although Brandt's more extensive experiments involved a powder technique that weights the porosity differently from Shelby's bulk (3 mm thick) permeation (percolation) technique, the two experiments agree to within 50% (J. E. Shelby, private communication), where they overlap. More recent experiments on helium migration in binary additive silicates have shown abrupt changes in behavior that are attributed to cluster morphology [J. E. Shelby, *J. Non-Cryst. Solids* **45,** 411 (1981)]. These changes are quite compatible with the present picture of intercluster diffusion channels (analogous to polycrystalline grain boundaries), which make the dominant contribution to helium migration.

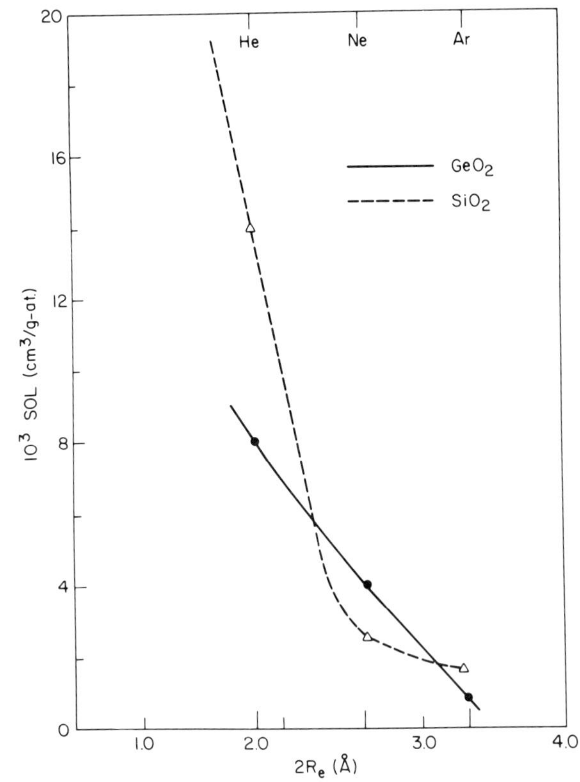

FIG. 11. Chemical trends in noble gas solubilities in g-SiO_2 and g-GeO_2 plotted against noble gas radii as defined in the text. (Solubility data are from Ref. 42.)

smaller. On the other hand, the solubilities of He and Ne in g-GeO_2 differ only by a factor of 2, while Ar is an order of magnitude smaller. Thus the critical size of channels in g-GeO_2 is of the order of 3 Å in g-GeO_2, compared to 2 Å in g-SiO_2. This is not a factor of 3 difference, but one should keep in mind that the g-GeO_2 widths are very broadly distributed, which from a percolative point of view generates blocking effects and prevents use of noble gas solubilities as a direct measure of channel width. Nevertheless, the trend is qualitatively correct.

The effect of intense neutron bombardment on the Raman spectrum of g-GeO_2 has been studied by Galeener.[33] Several similarities to neutron-bombarded g-SiO_2 (see Fig. 7) are observed. The shifts in bond-bending and bond-stretching frequencies (which increase and decrease, respectively) are indicative of damage-induced compaction, which is a well-known effect for g-SiO_2. The weak mode at 520 cm^{-1} in the HH Raman spectrum of the virgin glass (see Fig. 14) is observed to grow in scattering strength, analogously to the 606-cm^{-1} mode in g-SiO_2 (Fig. 7). It is therefore ascribed to a point defect, probably $(O_{1/2})_3Ge\cdot\cdot\cdot O{-}Ge(O_{1/2})_3$, in agreement with analogous assignments[33] of the 606-cm^{-1} peak in g-SiO_2. As I have mentioned frequently, my own preference for O_p^* is $O{=}Ge(O_{1/2})_2$.

Perhaps the most remarkable feature of Galeener's data is one that easily eludes attention; this is that whereas intense damage increases the strength of the 606-cm^{-1} peak in g-SiO_2 by more than an order of magnitude, the same level of damage no more than doubles the strength of the 520-cm^{-1} peak in g-GeO_2. Because the valence, size, and electronegativities of Si and Ge are nearly identical, it is quite difficult to explain this difference of nearly a factor of 10 in the neutron-bombardment-induced O_p^* population enhancement solely in terms of local chemical bonding. On the other hand, if the cluster morphology of g-GeO_2 is quartzlike while that of g-SiO_2 is that of the much more open cubic cristobalite network, then the difference is not only in the right direction, but it also seems quantitatively reasonable.

IX. Ring Statistics and Dynamical Force Field Constraints

As indicated in Fig. 8, all the low-pressure forms of c-SiO_2 [and GeO_2 (quartz)] are dominated by six- (cation O) membered rings. In the paracrystalline or vitreous clusters one may expect to find pairs of five- and seven-membered rings, especially near the steps on cluster surfaces, but basically one still expects the six-membered rings to overwhelmingly dominate the ring population. To emphasize this point I list in Table I the ring statistics expected for g-SiO_2 in the present model and compare them with similar distributions in the CRN model.[16]

The most direct evidence against the presence of large ring-statistical disorder is the absence of substantial infrared background absorption in the 400–600 cm^{-1} bond-bending region. As has been shown, the spectra in this region can be understood entirely as arising from the bending motion $L^1(O_s^*)$, with less than 5% background due to interior cluster bending modes F_2^1. The mechanism that quenches this latter absorption involves homogeneous disorder of the interior O ions around their linear (idealized) β-cristobalite positions and is discussed next. For the moment I mention merely that this mechanism, or any similar screening mechanism, must fail in the presence of substantial inhomogeneous ring disorder. I have used this reasoning to estimate the upper limits to ring disorder in g-SiO_2 paracrystallites in Table I.

The Raman spectrum[25] of g-SiO_2 is very strongly (a factor of 10) polarized below 700 cm^{-1} and almost depolarized above 700 cm^{-1}. Generally speaking, scattering by large symmetric molecular vibrations (such as the A_1 mode of XY_4 tetrahedral molecule) is strongly polarized and scattering by vibra-

TABLE I. SHORTEST RING STATISTICS (NUMBER OF CATION O MEMBERS PER LOOP) EXPRESSED AS PERCENTAGES IN THE CRN MODEL OF BELL AND DEAN[16] AND IN THE PRESENT PHENOMENOLOGICAL PARACRYSTALLINE MODEL[a]

Model	N_4	N_5	$(N_6 + N_7 + \cdots)$
CRN	19	28	53
Paracrystalline	$\lesssim 1$	$\lesssim 5$	$\gtrsim 95$

[a] Other CRN models discussed in Ref. 16 also produce $N_4 + N_5 \gtrsim 50$.

tions of uniaxial symmetry is depolarized. In terms of O_s^* modes, one might expect that vibrations that are confined to the internal cluster surfaces but delocalized with respect to that surface would be strongly polarized, whereas vibrations which are localized on the surface, for example, by surface step strain fields, would produce strongly depolarized Raman scattering. The interatomic force that mixes and delocalizes surface vibrations is the $O_s^*-O_s^*$ interaction, which is the same strength as the forces that determine the bending mode frequencies near and below 500 cm^{-1}. This strong polarization of the g-SiO_2 spectrum below 500 cm^{-1} is about what one would expect. [It is not very surprising that the 606-cm^{-1} $L_1(O_p^*)$ peak is strongly polarized; it is possible that this peak involves a symmetric breathing motion, e.g., around a Si or O vacancy.]

Next I will explicitly discuss the structure of β-cristobalite[36] and its relation to the vibrational spectra of g-SiO_2. Nominally the Si atoms occupy the sites of diamond lattice, while each O atom is distributed among six sites hexagonally bordering the "ideal" site midway between two Si atoms, with an O bond angle close to 146°, that is, very close to the value found in g-SiO_2. In addition to this statistical disorder of the O sites, there is considerable static disorder of the Si sites that would be present even at $T = 0$. This static disorder arises from what appears to be sixfold microtwinning of the local structure on a scale of a few unit cells,[36] which contributes a T-independent term to both the Si and O Debye–Waller factors. For the reader's convenience the overall description of this Debye–Waller disorder by Wright and Leadbetter is reproduced in Fig. 12; note that the mean amplitudes of the O and Si static disorder appear to be nearly the same.

The entire situation described by Wright and Leadbetter is a special case of the general theory of constraint-restricted configurations in solids[4] that has recently been developed and applied to c- and g-GeS_2 and SiO_2, as well as the prototypical solid electrolyte α-AgI. First, the distribution of O ions among six various sites is quite analogous to the distribution of Ag ions in

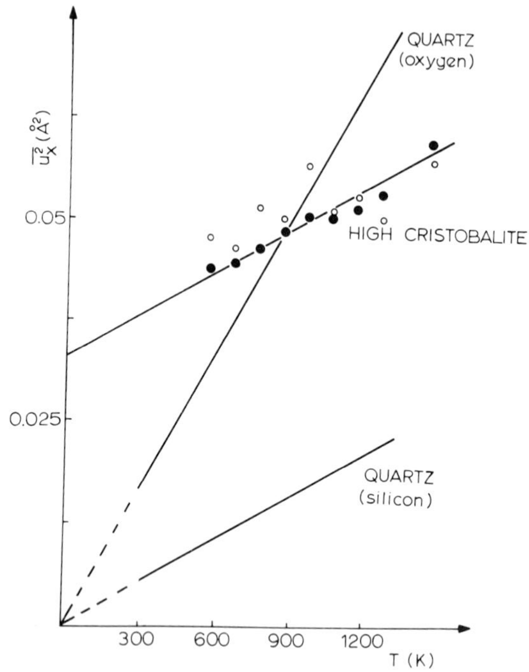

FIG. 12. Mean-square displacements of Si and O atoms (solid and open circles, respectively) in β-cristobalite as obtained from experimental Debye–Waller factors (from Ref. 36).

α-AgI, where a mean-field theory[43] has been successful in explaining the Ag–Ag radial distribution function and in calculating the activation energy for Ag diffusion. I will use these mean-field ideas shortly; meanwhile, some further analogies should be mentioned. Wright and Leadbetter developed their concepts of microdomains and microtwinning in powder cristobalite by comparing the structure of β-cristobalite to β-$AlPO_4$ and to BPO_4 and $BAsO_4$. Quite a similar comparison has been made for α-AgI, Ag_2S, and Ag_3SI, for which single-crystal data are available[44] and for which the evidence for microdomain formation is quite conclusive.[45]

As recognized by Wright and Leadbetter and in studies of Ag-based solid electrolytes, there must be substantial ordering of the statistically distributed atoms to minimize the local strain energies, which otherwise would be quite large. Obviously some kind of domain structure will have this effect, and the type of domain structure will vary according to the framework structure

[43] W. Andreoni and J. C. Phillips, *Phys. Rev. B: Condens. Matter* [3] **23,** 3098, 6456 (1981).
[44] R. J. Cava, F. Reidinger, and B. J. Wuensch, *Solid State Commun.* **24,** 411 (1977).
[45] R. J. Cava and D. B. McWhan, *Phys. Rev. Lett.* **45,** 2046 (1980).

and the nature of the interatomic forces involved. The purpose of the constraint theory is to discuss the *general* features of this energy minimization process that are common to *all* soft solids showing substantial atomic disorder. The theory is sufficiently general that many of the ideas apply equally well to crystals (such as β-cristobalite, c-GeS_2 and α-AgI) as to glasses such as g-SiO_2 and g-GeS_2.

Suppose one is able to count the number of constraints (in a mean-field sense, before domains or clusters have formed) per formula unit; in fact, this is quite straightforward when only valence force fields are involved and Coulomb forces can be neglected. Then one finds that the total number of constraints N_t is the sum of $(CN)^2/2$ over all the atoms in the formula unit, where CN is the coordination number of each atom, which itself is supposed to be a constraint. One must then compare this sum with NN_D, where N is the number of atoms in the formula unit and $N_D = 3$ is the dimensionality of the space in which the atoms are embedded.

In the case of a perfect glass, $N_t = 3N$ and this is the ideal glass-forming condition, which is satisfied by As_2Se_3. For an A(CN = 4)[B(CN = 2)]$_2$ glass such as $GeSe_2$ on SiO_2, $N_t = 18$, while $3N = 9$. Thus these glasses are *overconstrained*, and this is the reason why $GeSe_2$ is such a marginal glass former. The same reasoning shows that $N_t = N$ for $y = 1/6$ in Ge_ySe_{1-y} alloys and thus that the glass-forming tendency is maximized at this composition, in agreement with experiment.[46]

When a network glass is overconstrained, as in A(4)[B(2)]$_2$ glasses such as SiO_2, the strain frustration may be resolved in three ways. The strain energy may be equipartitioned; unexpectedly, this intuitively appealing physical condition is the least likely eventuality, because it does not reflect the connectivity and local atomic CN of the network. A second possibility is that the second-neighbor spacings may have equal widths, $\overline{\delta r^2_{AA}} = \overline{\delta r^2_{BB}}$. Unless the bond-bending force constants are equal, this geometrical condition is different from the physical one of energy equipartition. This geometrical condition is the one that is actually realized in the CRN model of Bell and Dean.[16] Finally, there is the solution that seems to be realized in g-SiO_2 and g-GeS_2; namely, the width of the smaller of the two second-neighbor spacings (O—O in g-SiO_2, Ge—Ge in g-GeS_2) is almost zero, while the width of the larger second-neighbor spacing is three to five times greater. In effect, the constraints associated with the smaller spacing are almost completely intact, while the constraints associated with the larger spacing are badly broken.[4] Thus $\overline{\delta r^2}$ has nearly a step function increase starting from near $r = 0$ and including the first neighbors and nearest second neighbors, where $N_t \ll 3N$, to above the second-neighbor spacing, where $N_t \gg 3N$. The transition occurs between the two second-neighbor peaks, where

[46] R. Azoulay, H. Thibierge, and A. Brenac, *J. Non-Cryst. Solids* **18**, 33 (1975).

$N_t \simeq 3N$. It is probably triggered by the difference in second-neighbor force constants, but (especially in g-GeS_2) there can be little doubt that this force-constant difference is not sufficiently large (it would have to be a factor of order 25) to account for the difference between $\overline{\delta r^2_{AA}}$ and $\overline{\delta r^2_{BB}}$. Instead, the network appears to relax by using all the available degrees of freedom to satisfy the stronger constraints and break all the weaker constraints.

What are the implications of these ideas for infrared oscillator strengths in a strongly ionic glass such as g-SiO_2? They suggest that it may be possible that the network topology (which is quenched in at T_g) may be such that the electric dipole oscillator strength associated with the $F_2{}^1$ internal cluster modes is almost completely screened by the statistical distribution of the oxygen atoms and by the associated displacements of the Si and O atoms from their idealized sites, as indicated by the Debye–Waller factors for cristobalite shown in Fig. 12. Approximately half the valence force field constraints are broken in this way, and thus one might expect that the infrared-active half of the paracrystalline interior modes could be screened while the Raman-active half is actually activated by disorder. Thus the oscillator strength of the $F_2^{1,2}$ bulk infrared-active modes may be screened, while that of the surface infrared modes remains largely intact. In the absence of a specific mean-field calculation including relaxation of Coulomb interactions, this suggestion is nothing more than that. I note, however, that a very strong and broad Raman-active peak associated with the $F_2{}^1$ modes is evident in the Raman spectrum shown in Fig. 5, but that no similar infrared absorption is present in Fig. 4. Thus if relaxation-facilitated screening is the explanation of this anomaly (the $F_2{}^1$ band is both Raman and infrared active according to group theory), then the screening must be very nearly complete.

X. Vibrational Spectroscopy of g-GeO_2

The infrared and Raman spectra[25] of g-GeO_2, which are reproduced for the reader's convenience in Figs. 13–15, show large quantitative and qualitative differences from the corresponding spectra for g-SiO_2. The quantitative differences represent the shifts in peak frequencies and are due to changes in the interatomic forces, which also change the corner-sharing O bond angle, and so on. These forces can be adjusted to fit experiment and are of little concern here.

The qualitative differences between the g-GeO_2 and g-SiO_2 spectra are much more interesting, and once again these occur primarily in the bond-bending region below 600 cm^{-1}. First there is a very large difference in the FWHM of the 278-cm^{-1} infrared absorption peak in g-GeO_2 compared to that of the 460-cm^{-1} peak in g-SiO_2. In the g-SiO_2 case after correcting for

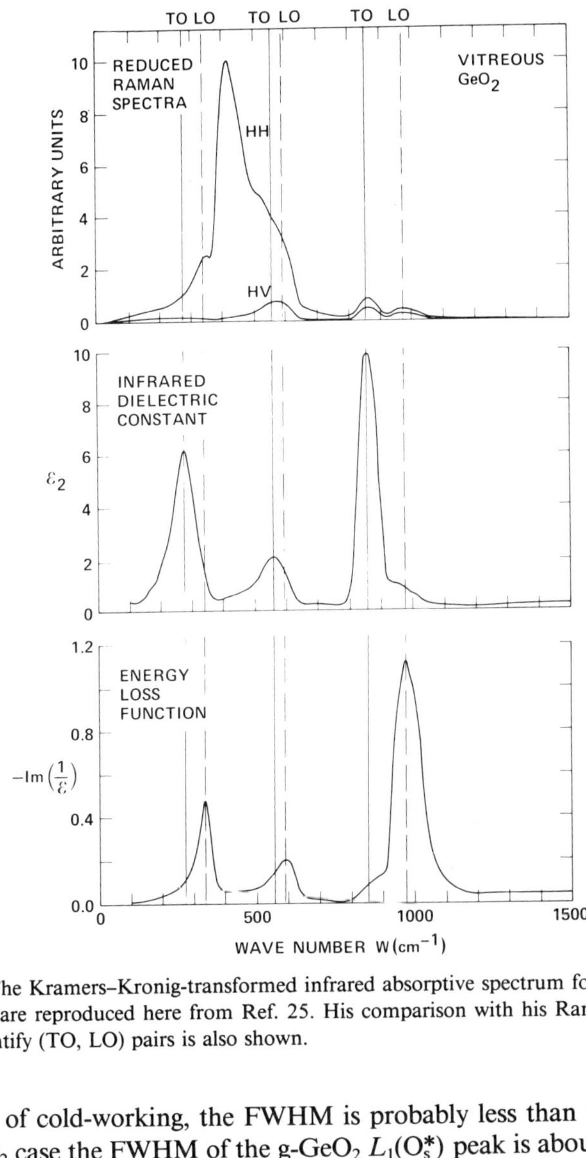

FIG. 13. The Kramers–Kronig-transformed infrared absorptive spectrum for g-GeO_2. Galeener's data are reproduced here from Ref. 25. His comparison with his Raman data in an effort to identify (TO, LO) pairs is also shown.

the effects of cold-working, the FWHM is probably less than 20 cm^{-1}. In the g-GeO_2 case the FWHM of the g-GeO_2 $L_1(O_s^*)$ peak is about 100 cm^{-1}, or about five times larger than the corresponding width in g-SiO_2. Even more striking, however, is the fact that in g-GeO_2 the bulk $F_2{}^1$ Raman peak frequency $\omega_R{}^b = 415\ cm^{-1}$ is *larger* than the surface $L_1(O_s^*)$ infrared frequency $\omega_I{}^s = 278\ cm^{-1}$; the quantitative relation is

$$\omega_R{}^b = 2^{1/2}\omega_I{}^s\,, \tag{10.1}$$

which will be explained shortly.

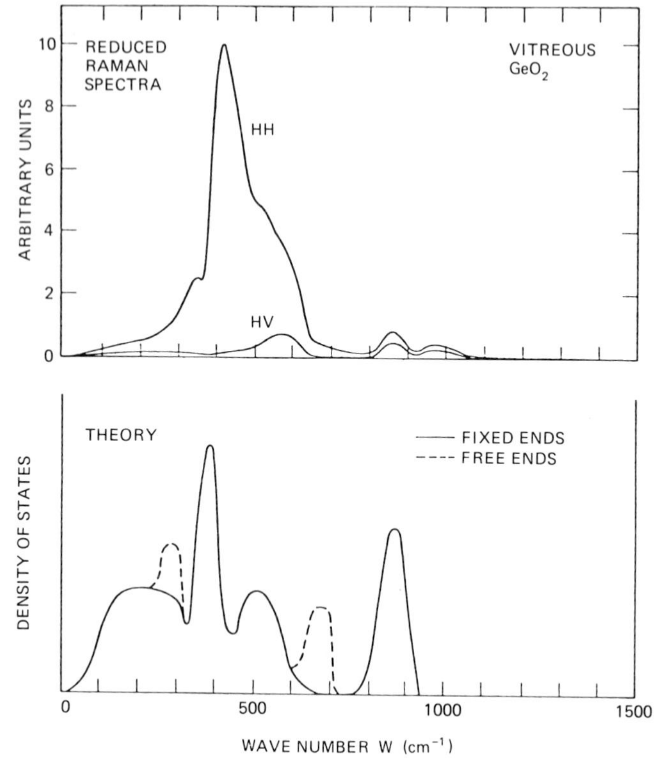

FIG. 14. Polarized and depolarized Raman spectrum of g-GeO_2, with Galeener's data from Ref. 25. The experimental data are compared with theoretical predictions from Ref. 23.

There is a large qualitative difference in the Raman spectra that is not immediately obvious but which I believe to be quite significant. At first sight there appear to be two large broad peaks in the Raman spectrum of g-GeO_2 near 415 and 595 cm^{-1}, corresponding respectively to similar peaks in g-SiO_2 near 430 and 820 cm^{-1}, respectively. In g-SiO_2 the former is very strongly polarized and the latter is almost completely depolarized. Comparison of the HH and HV polarized data for g-GeO_2 shown in Figs. 14 and 15 suggests, however, that sandwiched between the two large broad peaks, which are also respectively polarized and depolarized, is a third narrower, strongly polarized peak at Ω_R = 500–520 cm^{-1}. This third peak is quite reminiscent of the strongly polarized and very narrow 490-cm^{-1} peak in g-SiO_2. Thus it appears that there may be *two* candidates for the $L_1(O_s^*)$ peak in g-GeO_2 at 347 and 500 cm^{-1}, compared to only one in g-SiO_2.

The resolution of these substantial qualitative differences may lie in the morphological differences between the cristobalite g-SiO_2 clusters and the quartz g-GeO_2 clusters. The surface texture of the former is predominantly (100) and the surface molecules have the local siliconyl configuration $(O_{1/2})_2{-}Si{=}O$. For the hexagonal quartz structure the surface texture is more complicated. Looking at Figs. 16 and 17, as given by Bragg,[1] one sees that (001) quartz planes are analogous to cristobalite (001) planes, in the sense that two of the bonds for each surface Si atom in a given layer have a positive z component and two a negative component. Thus a (001) GeO_2 quartz plane can have surface molecules with the germanyl configuration

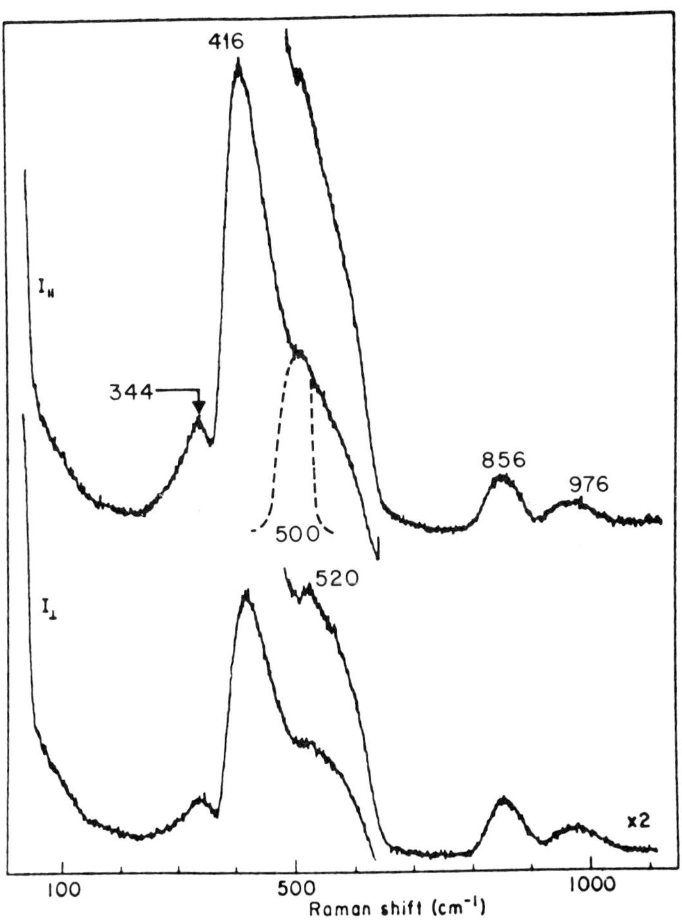

FIG. 15. Fine structure in the polarized and depolarized Raman spectrum of g-GeO_2, reproduced from Sharma's data.[25] The polarized narrow peak near 500 cm^{-1} is indicated by dashed lines. The degree of polarization in these data is less than is shown in Fig. 14, but the qualitative behavior is similar.

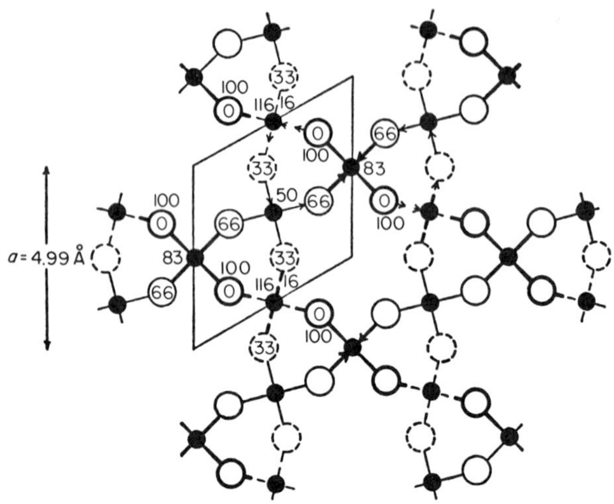

FIG. 16. A perspective sketch of the structure of α-quartz, taken from Ref. 1. The numbers indicate vertical displacements in units of $0.01c$.

$(O_{1/2})_2{-}Ge{=}O_s^*$, but the surfaces normal to the (001) plane must have a different configuration. These planes (which contain the c axis) probably form hexagonal arrays, which makes close packing in a honeycomb net possible in the x–y planes; that is, the clusters form bundles of truncated hexagonal prisms. From Fig. 16 we see that Ge atoms in these lateral planes have the same local environment as the Si atoms on the (110) faces of cristobalite; that is, there is one back bond, two surface bonds, and one dangling bond. Surfaces of this type could undergo a 2×1 reconstruction, with half the dangling bonds occupied by O_s^* atoms and half of them unoccupied. The building block of the reconstructed 2×1 surface is therefore represented by the diradical $[(O_{1/2})_3Ge]_2{-}O_s^*$. Of course, this diradical nominally contains two paramagnetic electrons, but in a partially covalent glass these electrons will form weak bonds and be nonmagnetic. The ultraviolet absorption associated with these weak bonds may have been observed, as discussed in Section XI.

One sees, then, that the locally uniaxial morphology of g-GeO_2 clusters gives rise to *both* singly and doubly bonded O_s^* atoms. Since the internal O atoms have two single bonds, the bending frequency $\omega_s[L_1(-O_s^*)] = \omega_s{}^1$ should satisfy the relation

$$(\omega_s{}^1)^2 = \omega_b{}^2/2, \tag{10.2}$$

which is essentially Eq. (10.1). Thus the 278-cm^{-1} line can be assigned to the lateral $[(O_{1/2})_3{-}Ge]_2{-}O_s^*$ surface molecules. On the other hand, for the doubly bonded (001) germanyl $(O_{1/2})_2{-}Ge{=}O_s^*$ units the double-bonding force constant is much larger. Thus $\omega_s\ [L_1({=}O_s^*)]$ is assigned to the 500-cm^{-1} peak, which has not been identified previously.

The large FWHM of the 278-cm^{-1} infrared absorption peak is now more easily understood, since it is associated with the highly deformable diradical $[(O_{1/2})_3Ge]_2{-}O_s^*$. It is also possible that the infrared line corresponding to the $=O_s^*$ atoms lies on the high-frequency side of this peak (near 340 cm^{-1}) and is not resolved as a separate peak, while the low-frequency side is broadened by large cold-working surface effects of the kind discussed in Section VI for g-SiO_2. In any case, it appears that the separation of the $(\omega_{TO}, \omega_{LO})$ pair of $=O_s^*$ modes is much larger than that of the $-O_s^*$ modes. This is what one would expect, since the vibrationally induced charge redistribution, as measured by the effective charge Z, should be larger for the higher-frequency bending mode. Moreover, the good agreement[25] between theory and experiment for the $-O_s^*$ (278, 347) cm^{-1} pair splitting may be partially accidental, the result of compensation between the effects of cold-working and the composite character of the 278-cm^{-1} peak.

As indicated by Galeener and Lucovsky,[25] the remaining features of the g-GeO_2 pair spectrum are rather similar to these of g-SiO_2 and therefore do not deserve special comment, except for the reminder that once again the condition $Z_s^* \gg Z_b^*$ [Eq. (6.4)] must hold; that is, the infrared oscillator strengths of the cluster interior modes $F_2^{1,2}$ must be strongly screened in both glasses.

XI. Electronic Structure

The cluster radii proposed in the present model are sufficiently small, and the fraction $x_s \sim 0.2$ of surface molecules sufficiently large, that ap-

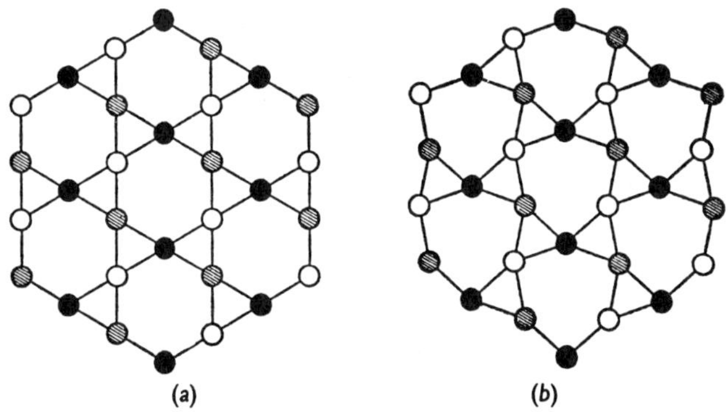

FIG. 17. The Si framework atoms only in α- and β-quartz, taken from Ref. 1.

proximately 10% of the bonds in the glass should be associated with O_s^* atoms. Moreover, in my model there is a very great qualitative difference between the surface textures of the cluster in g-GeO_2 compared to g-SiO_2; namely, in the former one has both $-O_s^*$ and $=O_s^*$ bonds, whereas in the latter only $=O_s^*$ bonds are present. In g-GeO_2 the LO Raman-active bending modes associated with the $-O_s^*$ and $=O_s^*$ bonds fall at 347 and 500 cm^{-1}, respectively. The latter frequency is higher than the bulk frequency ω_b = 415 cm^{-1}, suggesting a stronger bond whose electronic excitation energy may be larger than the bulk or cluster interior excitation energies, so that its electronic optical absorption would be superimposed on a large bulk background from which it could not easily be separated. On the other hand, the relatively low-frequency $-O_s^*$ LO vibration at 347 cm^{-1} suggests a weaker bond whose ultraviolet absorption should fall in the transparent region of the bulk glass and hence should be easily identified.

The fundamental optical spectra of many crystals have been measured and interpreted in detail in terms of transitions between conventional one-electron delocalized energy states.[47] The qualitative relations between bond strengths and electronic energy gaps are also well understood quantitatively for simpler crystals.[48] The situation is much more complex for glasses, where the theories of the optical spectra even of g-SiO_2 are still in a primitive state.[49,50] I will return to the electronic structure and ultraviolet optical spectrum of g-SiO_2 later in this section. The fundamental optical spectra of c- and g-SiO_2 and c- and g-GeO_2 were measured and compared by Pajasová.[51] Her results, which are shown in Figs. 18 and 19 for the reader's convenience, are quite surprising and they constitute one of the hitherto major unsolved problems[52] in the fundamental optical spectra of solids. Above 10 eV the spectra are rather similar, except that the excitonic widths are narrower in g- and c-SiO_2 than in g- and c-GeO_2, suggesting a more defect-free and homogeneous structure for the former. Below 9 eV, however, both c-SiO_2 and g-SiO_2 are nearly transparent, while both c-GeO_2 and g-GeO_2 exhibit an additional absorption band between 5 and 9 eV that is entirely absent from g- and c-SiO_2.

It is my contention that this additional band, whose structure changes substantially between c-GeO_2 and g-GeO_2, as shown in Fig. 19, is associated with the lateral cluster surfaces and specifically with bonding $\rightarrow$ antibonding transitions of electrons belonging to $-O_s^*$ bonds. To support this interpre-

[47] J. C. Phillips, *Solid State Phys.* **18,** 55 (1966).

[48] J. C. Phillips, *Rev. Mod. Phys.* **42,** 317 (1970); "Bonds and Bands in Semiconductors." Academic Press, New York, 1973.

[49] J. R. Chelikowsky and M. Schlüter, *Phys. Rev. B: Solid State* [3] **15,** 4020 (1977).

[50] R. B. Laughlin, *Phys. Rev. B: Condens. Matter* [3] **22,** 3021 (1980).

[51] L. Pajasovà, *Czech. J. Phys.* **B19,** 1265 (1969).

[52] J. C. Phillips, *Phys. Rev. B: Solid State* **9,** 2775 (1974).

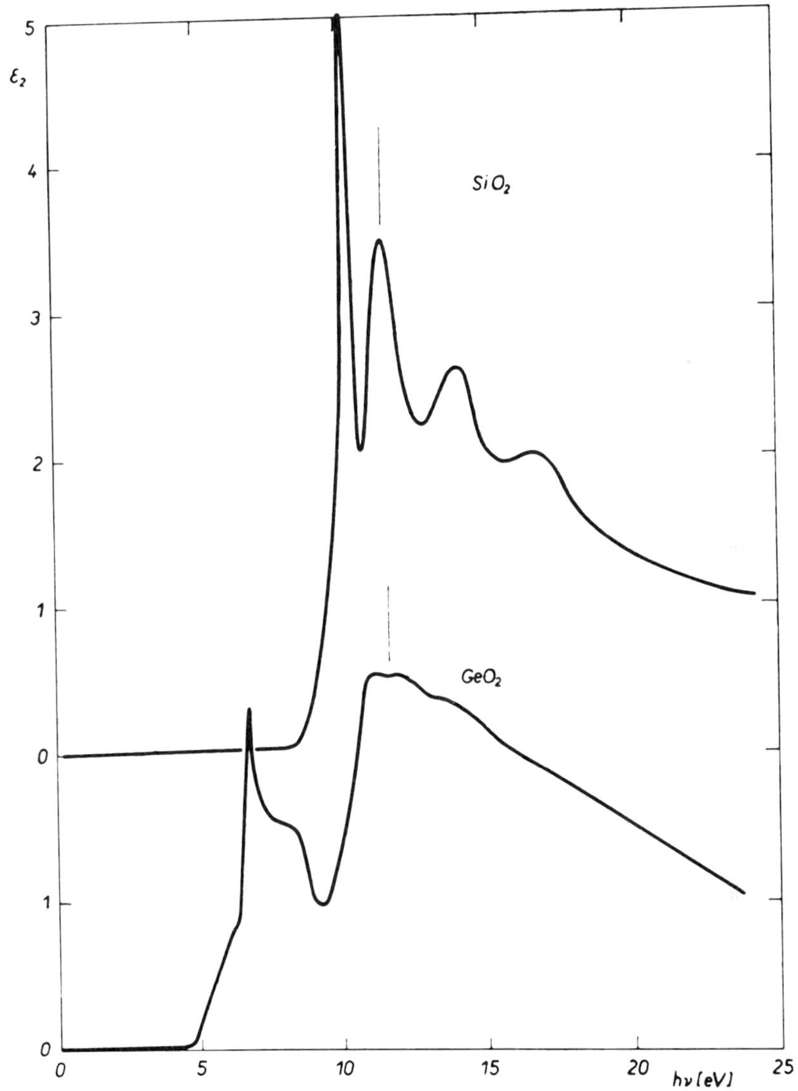

FIG. 18. Comparison of the fundamental optical spectra of g-SiO_2 and g-GeO_2 (from Ref. 51).

tation some details of the sample preparation[51] should be mentioned. The GeO_2 glass was prepared from a fused powder. The polycrystalline samples were prepared in several ways, either by evaporation, which could easily produce a microcrystalline morphology, or by hydrothermal growth. The latter samples contained[51] "a few percent of water."

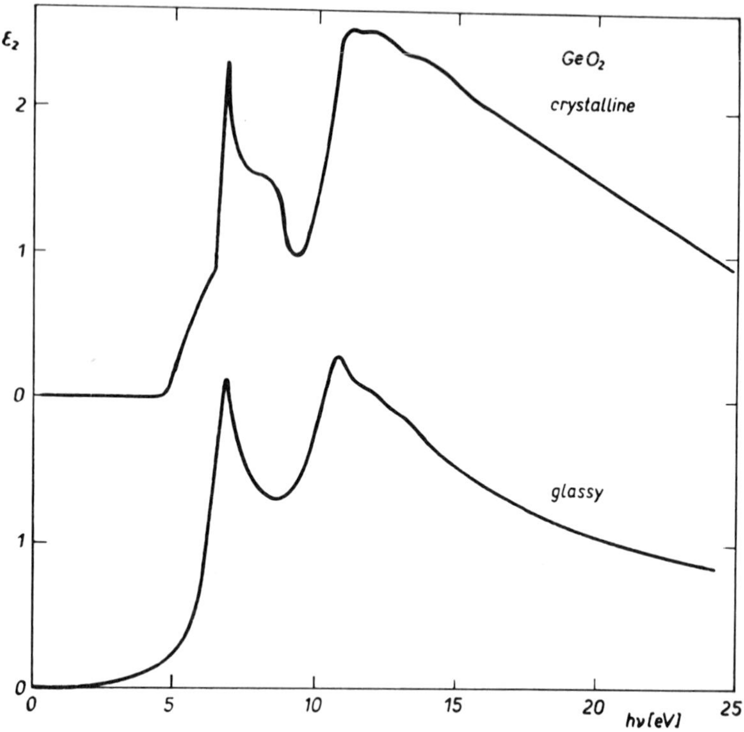

FIG. 19. Comparison of the fundamental optical spectra of c- and g-GeO_2. Note that c-GeO_2 is not only polycrystalline but also probably microcrystalline and contains "a few percent H_2O"; that is, the microcrystalline diameters are probably $\lesssim 100$ Å (i.e., the same scale as observed by Zarzycki[40]) (from Ref. 51).

Perhaps the most striking qualitative feature of the spectra shown in Fig. 19 is that although the 5–9 eV spectral structure changes shape between the crystallites and the glass, its overall integrated strength is almost constant. In my model this implies that the surface-to-volume ratio, which is determined by the cluster size, is nearly constant. This is entirely possible if the cluster growth is self-limited by the accumulation of strain energy,[53] as I have previously suggested to explain the morphology of a-Si(:H). Note that this mechanism is expected to be dominant for quartz–GeO_2 (which must be under substantial internal tension because it also exists as rutile–GeO_2, which is 50% denser) but not for g-SiO_2; that is, the presence of water relieves the domain wall or twin or grain boundary mismatch energies in quartz–GeO_2 at a microcrystallite radius of order 30 Å, whereas it is easy to grow much larger domains in quartz–SiO_2. (The shimmering of Dauphiné

[53] J. C. Phillips, *Phys. Rev. Lett.* **42,** 1151 (1979).

twin SiO_2 α–β quartz domain walls has been observed directly[54] by TEM on a scale of 10^2–10^3 Å.)

The next qualitative question concerns the difference between the $-O_s^*$ surface absorption spectra in the 5–9 eV range for microcrystallites and glass clusters. The surfaces of the former are essentially free standing (at worst, they are separated by molecular layers of H_2O), whereas the paracrystallites of the latter are fused when the melt is quenched. The resulting interface width is of the order of one molecular diameter, and since the paracrystallites are in general misoriented, this must mean a very high density of steps on the paracrystallite surfaces; that is, their surface textures are very rough. This surface roughening erases the fine structure seen in the crystalline 5–9 eV spectrum and leaves a single asymmetric peak whose functional form resembles that of the Penn model of an amorphous semiconductor.[47,48] It would be very interesting to see whether modern band theory[49,50] could calculate the surface electronic spectra of the $-O_s^*$ bands for g-GeO_2.

The correctness of this surface model can be checked quantitatively by noting that the number of valence electrons contributing to the f sum rule,

$$N_{\text{eff}}(\Omega) = \int_0^{\Omega} \omega\epsilon_2(\omega)\, d\omega, \tag{11.1}$$

in the 5–9 eV region is about 10% of the total, up to $\Omega = 30$ eV. (This can be confirmed by integrating the curves shown in Fig. 19.) This result indicates again that about 10% of the bonds in g-GeO_2 are of the $-O_s^*$ type.

Because hyper-Raman scattering[34] is sensitive to vibrations localized on internal surfaces, it would be of great interest to obtain the hyper-Raman spectrum of g-GeO_2.

XII. Structure and Vibrational Spectra of Alkali Silicate Glasses

The infrared and Raman vibrational spectra of silicate glasses of formula $(XO)_x(SiO_2)_{1-x}$ have been studied extensively by many workers. Here XO denotes one or a combination of many ionic oxides or network modifiers such as Y_2O, where Y is an alkaline element, or ZO, where Z is Sn or Pb,[55] or an alkaline earth or TO_n, where T is a transition metal and n represents full oxidation. The range of x of technological interest is generally $\frac{1}{3} \leq x \leq \frac{2}{3}$, but for the purposes of this article, the range of greatest interest in

[54] G. Van Tendeloo, J. Van Landuyt, and S. Amelinckx, *Phys. Status Solidi* A **33,** 723 (1976).

[55] J. Götz, D. Hoebel, and W. Wicker, *J. Non-Cryst. Solids* **22,** 391 (1976). A very complex phase diagram for $(PbO)_x\,(SiO_2)_{1-x}$ is discussed by R. M. Smart and F. P. Glasser, *J. Am. Ceram. Soc.* **57,** 378 (1974).

analyzing the spectra is $0 \leq x \leq \frac{1}{3}$. Here data sufficiently closely spaced to determine trends quantitatively are much less numerous, and much still remains to be done. One of the aims of this section is to suggest possible models for structural transformations with x increasing from zero in the hope that this will encourage further study of vibrational spectra in this range.

The structural progression[19] in alkali silicate crystals is one of dimensional regression of the silica framework, that is, from $N_D = 3$ at $x = 0$ (cristobalite) to $N_D = 2$ sheet corner-sharing tetrahedral structures at $x = \frac{1}{3}$ (disilicates) to $N_D = 1$ metasilicate ($x = \frac{1}{2}$) corner-sharing tetrahedral chains, to pyrosilicate anions Si_2O_7 and orthosilicate anions SiO_4, both $N_D = 0$. With increasing x a larger and larger fraction of the O atoms become (nominally) double bonded to the framework, while at the same time they are weakly associated with the metallic cations such as the alkalies[56] (see Fig. 20). This association is nondirectional so that the positions occupied by the additive cations (such as Li and Na) may vary with size according to packing considerations. The vibrational modes centered on the additive cations have little Raman scattering strength, at least in the cases of the lighter alkalies, and one must scrutinize the spectra carefully to locate features not primarily associated with the silica framework.

Gaskell's early analysis[19] of structural trends in the silicate crystalline spectra was carried out before Raman spectra were available, and so in order to compare theory with experiment he could utilize only his own infrared reflection spectra. The long-range order in the crystals makes the selection rules and the number of infrared- and Raman-active lattice modes much larger than predicted by Gaskell's molecular idealizations of the silicate crystal unit cells, and in general the correlation between the lattice bands, for example, of quartz, coesite,[57] and g-SiO_2 is at best semiquantitative with regard to frequency and at best qualitative with regard to selection rules. According to my model, the strongest infrared- and Raman-active lines in the g-SiO_2 are associated with internal surface O_s^* atoms. As is the case for all small cluster models, it is difficult to distinguish bulk and surface modes in Gaskell's models, because the molecules he considers are very small. In spite of these limitations, Gaskell did identify an increase in bond-bending frequencies with increasing x, which he attributed to increasing bond-bending force constants with increasing x and increasing concentrations of nonbridging O atoms.

Haas'[27] early Raman data on $(Na_2O)_x(SiO_2)_{1-x}$ for $x = 0, 0.14, 0.25$, and 0.50 showed that the network additive had several dramatic effects on the

[56] J. Zarzycki and F. Naudin, *Verres Refract.* **14**, 113 (1960).

[57] S. K. Sharma, J. F. Mammone, and M. F. Nicol, *Nature (London)* (1900).

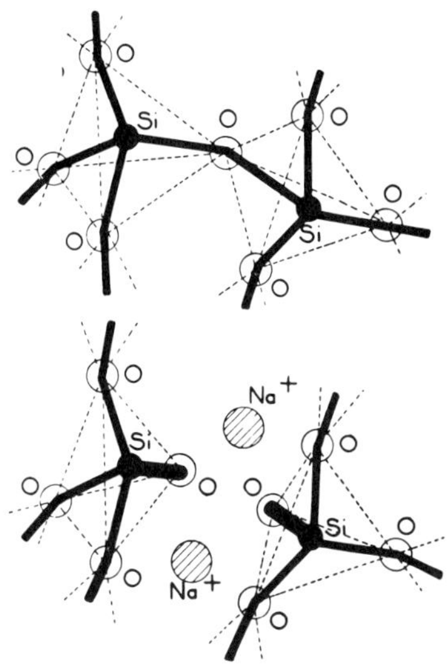

FIG. 20. Sketches of the local structure of a $(Na_2O)_x(SiO_2)_{1-x}$ glass, from Ref. 56. Upper figure: $x = 0$, corner-sharing tetrahedra, with lone-pair electrons indicated by small open circles. Lower figure: $x = 1/2$, metasilicate chains, with alkali cations intercalated between chains and associated with nonbridging O anions.

Raman spectrum. At $x = 0.14$ the largely depolarized bond-stretching (ω_{TO}, ω_{LO}) pair at (1060, 1200) cm^{-1} found at $x = 0$ became a single strongly polarized line at 1100 cm^{-1}. Small changes were observed near 500 cm^{-1}, but these were much more dramatic at $x = 0.25$. At this composition the $L_1(O_s^*)$ line at 490 cm^{-1} shifted to 540 cm^{-1}, and the broad band $F_2{}^1$ at 430 cm^{-1} at $x = 0$ nearly completely disappeared.

More closely spaced compositions for $(K_2O)_x(SiO_2)_{1-x}$ glasses were studied by polarized Raman scattering[58] at $x = 0$, 0.10, 0.20, 0.30, and 0.45. The results were similar to those reported by Haas for g-$(Na_2O)_x(SiO_2)_{1-x}$, except that the broad $F_2{}^1$ band disappeared between $x = 0.10$ and $x = 0.20$.

Polarized Raman scattering from g-$(Na_2O)_x(SiO_2)_{1-x}$ samples has recently been reported[59] for $x = 0$, 0.05, 0.10, 0.15, 0.20, 0.25, and 0.33. The broad $F_2{}^1$ band disappears abruptly at $x = x_F$ between $x = 0.20$ and $x = 0.25$. Another feature that can be studied quantitatively is $\omega[L_{1,2}(O_s^*)]$ as a function of x in $[(K_2O)_x$ or $(Na_2O)_x)](SiO_2)_{1-x}$ glasses. This is sketched in

[58] W. L. Konijnendijk and J. M. Stevels, *J. Non-Cryst. Solids* **21,** 447 (1976).

[59] B. O. Mysen, D. Virgo, and C. M. Scarfe, *Am. Mineral.* **65,** 690 (1980), and private communication (1981); T. Furukawa, K. E. Fox, and W. B. White, *J. Chem. Phys.* **75,** 3226 (1981); S. Veprek, Z. Iqbal, H. R. Oswald, F. A. Sarott, J. J. Wagner, and A. P. Webb, *Solid State Commun.* **39,** 509 (1981).

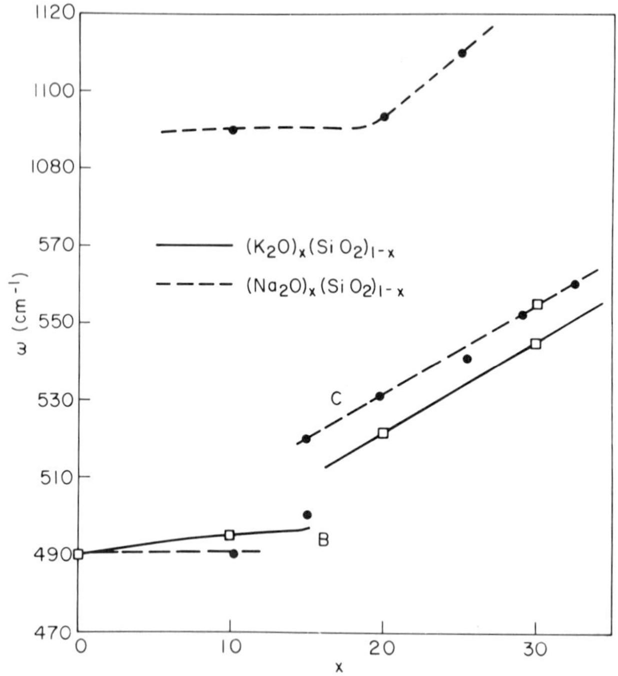

FIG. 21. Composition dependence of the narrow O* peak frequency in Na and K silicate glass Raman spectra. The □ data are taken from Ref. 58 and the ○ data from Ref. 59. Note the two branches B and C of the O* bending frequency near 500 cm^{-1}.

Fig. 21. Although the data are sparse, in both cases the frequencies are nearly constant for small x. Near $x = x_c = 0.15$ the frequencies $\omega(x)$ increase discontinuously and also show a jump in slope. In Fig. 22 I have taken advantage of the unique data of Furukawa *et al.*[59] to plot the height of this narrow peak as a function of x. Clearly there are two separate peaks (labeled B and C in Figs. 21 and 22), with peak B disappearing and peak C emerging near $x_c = 0.15$. Peak B is assigned to O_s^* bending modes, while peak C probably is merely an O* (nonbonding O) mode of a single disilicate layer.

A useful quantitative analysis of the relative intensities of the 950- and 1100-cm^{-1} bands in $(Na_2O)_x(SiO_2)_{1-x}$ glasses has been given by Furakawa *et al.*, [59] as reproduced in Fig. 23. Their theoretical discussion in terms of $L_2(O^*)$ (nonbridging oxygen stretching) modes is based on the normal-mode analysis of small molecules and suffers from the limitations of Gaskell's discussion (see before), in the sense that it is difficult to distinguish bulk and surface modes or to understand how the small molecules can be embedded in a larger network while salvaging the identities of the O* atoms. Within the context of the present model, one would thus say that the 950-

cm^{-1} line should be assigned to O* atoms attached to the metasilicate chains. The 1100-cm^{-1} line, on the other hand, is associated primarily with the $L_2(O_s^*)$ modes of disilicate clusters. Apparently the clusters of metasilicate chains are small up to $x = 0.36$. Between $x = 0.36$ and $x = 0.45$ the glass metasilicate frequency[59] increases linearly from 950 to 966 cm^{-1}, extrapolating to 974 cm^{-1} at $x = 0.50$, in excellent agreement with the crystalline powder peak[58] at 975 cm^{-1} in c-$Na_2O \cdot SiO_2$ (see Fig. 24). This shift probably represents a contraction of metasilicate microcrystallite bond lengths with increasing size; a corresponding shift with size has been observed by Veprek *et al.*[59] for Si microcrystallite optic mode frequencies.

Unfortunately, data on disilicate powders are not available to confirm the assignment of the 1100-cm^{-1} line to disilicate micro- or paracrystallites;

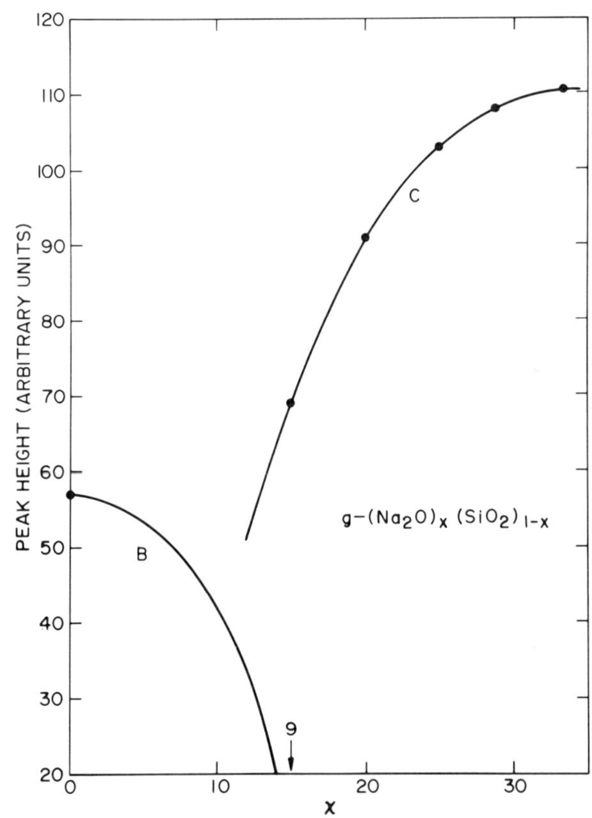

FIG. 22. The composition dependence of the B and C peak heights (Fig. 21) in Na silicate glasses. The data are taken from Ref. 59. Furukawa *et al.* are the only authors to have measured relative Raman peak heights on a series of silicate glasses of varying composition.

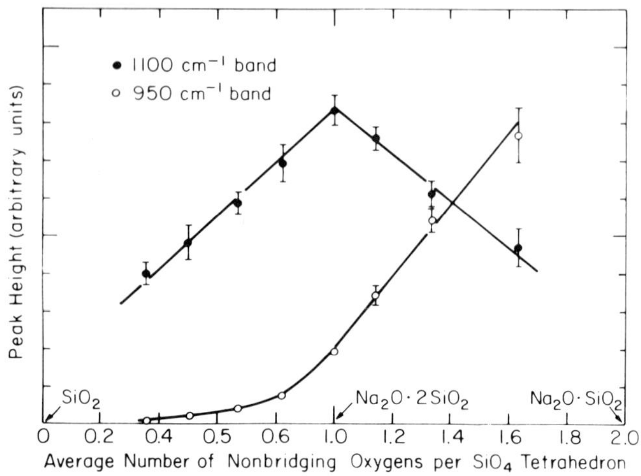

FIG. 23. Peak heights of the O_s^* bond-stretching bands in Na disilicate and metasilicate microcrystallites (see text) (from Ref. 59).

however, the composition dependence of the 1100-cm^{-1} peak strength shown in Fig. 23 shows an interesting anomaly. Although it is linear and peaks at $x = 0.33$, it does not extrapolate to zero at $x = 0$. The nonlinearity begins below $x = 0.10$ and this composition corresponds fairly closely to the maximum in the spinodal for phase separation (Section XIX). I believe that the composition dependence of peak heights shown in Fig. 23 is most easily explained by phase separation on a scale <100 Å. An alternate explanation based on cancellation (or screening) of scattering strengths by internal and surface bonds has been proposed by Furakawa *et al.*, but it seems unlikely that the small molecules they considered could produce such strictly linear behavior and still be properly polymerized.

Near the metasilicate composition ($x = 0.5$) there is evidence in the $(K_2O)_x(SiO_2)_{1-x}$ Raman spectra for the coexistence of metasilicate microcrystallites in the glassy matrix. Only one strong high-frequency Raman line at 965 cm^{-1} is seen in the $x = 0.50$ polycrystalline $K_2O \cdot SiO_2$ samples obtained by devitrifying the glass. The Raman spectrum of the glass with $x = 0.45$ is shown in Fig. 25. The $(\omega_{TO}, \omega_{LO})$ glass pair at (1055, 1075) cm^{-1} is present as a single unresolved doublet at $x = 0.30$, with a very weak satellite at 940 cm^{-1}. The latter becomes a strong narrow sideband at 950 cm^{-1} at $x = 0.45$, and it almost certainly corresponds to the 965-cm^{-1} line in the metasilicate polycrystalline samples. This supports the assignment of the 950-cm^{-1} sodium silicate line to metasilicate chains (Fig. 23). Similar comments apply to the 300-cm^{-1} line in Fig. 25 (340 cm^{-1} in the polycrystalline spectrum shown in Fig. 24).

Many authors have noted close similarities between the infrared and Raman spectra of silicates and germanates and the same spectra of polycrystalline powders. Certainly the differences between these glassy and crystalline spectra are much smaller than the differences between the glassy spectra and the predictions of CRN models of Bell *et al.*[16,21,23] Nevertheless it is quite difficult in general to say that a given line in a glass spectrum should be associated with another line in the crystalline powder spectrum. The 950-cm^{-1} line for the Na and K silicates is a very favorable case for two reasons: It is very narrow and its strength varies rapidly with $(x - 0.5)^{-1}$.

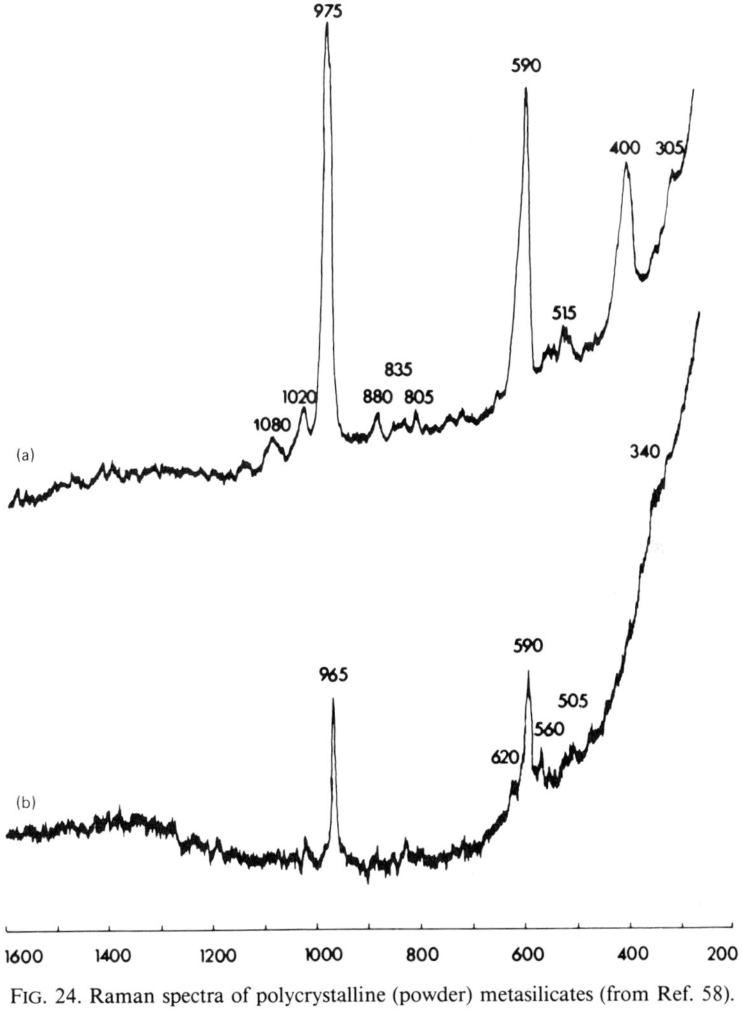

FIG. 24. Raman spectra of polycrystalline (powder) metasilicates (from Ref. 58).

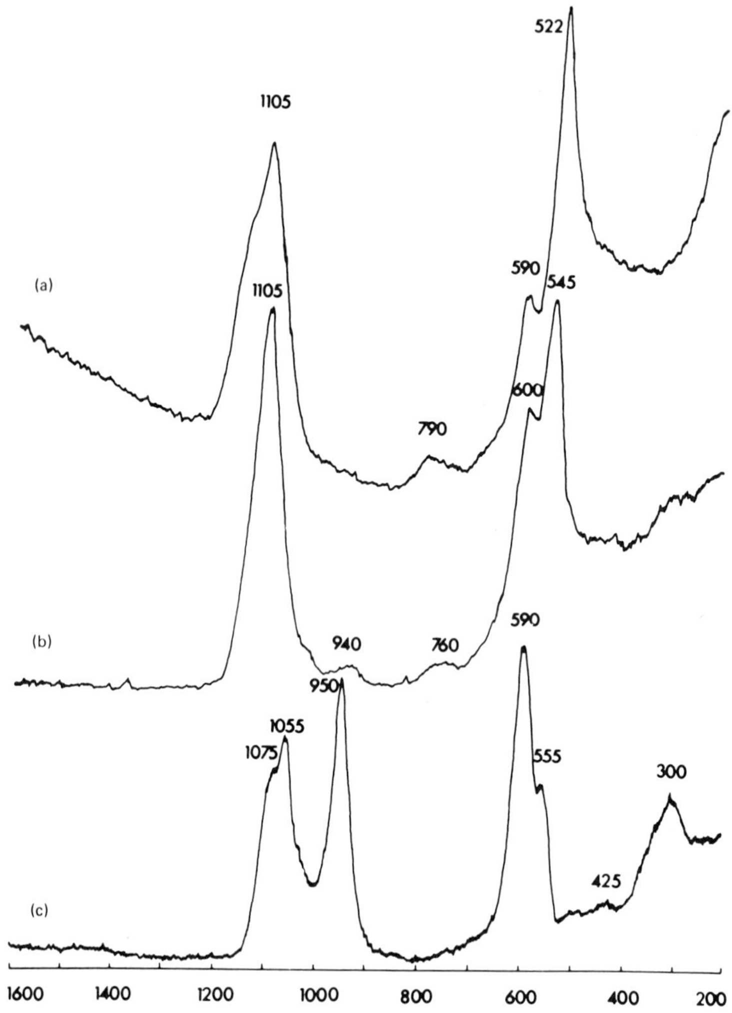

FIG. 25. Raman spectra of $(K_2O)_x(SiO_2)_{1-x}$ glasses (from Ref. 58).

Such lines are of special interest because they can serve as signatures of particular morphologies (here paracrystalline bundles of chains of corner-sharing tetrahedra). For compositions intermediate between those at which crystalline compounds are formed, one expects a mixture of morphologies. Incidentally, it is interesting that the glass line shows a 2% reduction in its bond-stretching frequency here relative to the crystal, presumably because of the lower density of the glass.

In spite of the sparsity of experimental data in the threshold regions sketched in Figs. 21 and 22, I will suggest a heuristic mechanism that may explain the quenching of $A_F(x)$ near $x = x_0 \sim 0.23$ in g-$(Na_2O)_x(SiO_2)_{1-x}$ and the abrupt increases in $\omega[L_{1,2}(O_s^*)]$ near $x = x_c \sim 0.15$. The quenching of $A_F(x)$ is related to the dimensional regression of the silica framework from $N_D = 3$ to $N_D = 2$. This alters the depolarizing effects near $\omega = 500$ cm^{-1} of the alkali ions that occupy isolated voids inside the clusters for $N_D = 3$ but which are intercalated between the silica sheets for $N_D = 2$. In the latter case near $\omega = 500$ cm^{-1} the alkali ions behave almost as free charges and therefore quench the polarized cluster interior $F_2{}^1$ bond-bending Raman scattering. In effect, the quenching of $A_F(x)$ could be regarded as a kind of insulator–metal transition for the high-frequency alkali ion motion.

An attractive feature of this mechanism is that not only does it explain the gradual quenching of the broad F_2^1 line, but it also explains how this can occur while the $L_1(O_s^*)$ line is disappearing near $x = x_c = 0.15$. The smaller alkali ions might be concentrated for low x in the cluster centers; with increasing x they spread out until they reach the Si surface layer just inside the O_s^* layer, at $x \sim x_c < x_0$. When $x \gtrsim x_c$, the silica clusters have the secondary structural role of filling the interstices between the dominant disilicate layer clusters. In this context the silica cluster surfaces may be poorly defined and the line so broad as to merge with the background. Near $x_0 \sim 0.23$ the silica clusters completely lose their identity and SiO_2 molecular complexes may even be merely "intercalated" between disilicate layers.

An interesting aspect of the frequency dependence of the layer O* bending modes for $x \gtrsim 0.15$ (which are labeled C in Fig. 21) is that they extrapolate to $\omega = 490$ cm^{-1} (the value in g-SiO_2) at about $x \approx 0.05$. This suggests a possible microscopic model for these modes. In general, one might have expected to see two peaks of this type in the disilicate composition range, one associated with O_i^* modes localized on layers interior to a cluster and another associated with O_s^* modes associated with layers composing the cluster exterior surface. Instead, one sees a single hybrid resonance, with $\omega(C)$ tending linearly to $\omega(B)$ and extrapolating to intersect it at $x \approx 0.05$.

First I would like to explain the absence of a separate surface mode. This could arise from a relaxed, self-consistent distribution of alkali ions that reduces the vibrational free energy by equalizing the silicate O_i^* and O_s^* frequencies. For smaller and smaller alkali concentrations x the difference between $\omega(O_i^* + O_s^*)$ and $\omega(O_s^*)$ may be expected to decrease linearly with x and tend to zero as $x \rightarrow 0$. In fact, the difference is proportional to $x - x_0$, with $x_0 \approx 0.05$. I believe that this, $x_0 \neq 0$, is the result of a quasi-phase transition in cluster morphology, which will be discussed in Section XIX [see also Fig. 34, where $x(A) \approx x_0$].

XIII. Structure and Vibrational Spectra of Alkali Germanate Glasses

Verweij and Buster have recently presented Raman spectra of Li, Na, and K germanate glasses and have included an extensive review of the literature on the structure of these materials.[60] In contrast to the silicates, with increasing concentration of network additive the alkali germanates exhibit maxima in density and molar refractivity.[61] The explanation[62] for these nonlinearities is that near $x = 0.2$ in $(XO)_x(GeO_2)_{1-x}$ a maximum is reached in the fraction f_6 of Ge atoms with CN = 6. Although this explanation was first advanced in the glass context, subsequent studies, for example, of X = $(Li, Na, or K)_2$ devitrified germanate powders, have confirmed the presence of octahedra in a number of alkali sheet crystalline compounds.[63]

The formation of octahedra in the germanate glasses or crystals can be regarded as arising in several ways. The network additive can be imagined to exert an internal pressure that induces an increase in CN from 4 to 6. A more satisfactory description, specific to the layer morphology, is that intercalation of the network additive induces sheet instability or buckling, and the octahedra are then able to cross-link the sheets because of their cubic symmetry. This mechanism is not available in the silicates, because the enthalpy of silica octahedra is too high, whereas in the stannates the octahedral energy is lower than the tetrahedral energy. Thus only in g-GeO_2 is the octahedral energy close to but slightly above the tetrahedral energy. These relations are reflected, of course, by the transition pressures P_t of c-SiO_2 and c-GeO_2 from CN = 4 to CN = 6, that is, ~90 and 0 kbar, respectively.[64]

Verweij and Buster, in analyzing their Raman spectra, divided the modes into five types: ν_s(Ge-O-Ge) and ν_{as}(Ge-O-Ge) of tetrahedra, an A_1-like ν_s(Ge-O-Ge) of octahedra, X—O modes, and (in the polycrystalline samples) lattice modes. Their ν_s(Ge-O-Ge) and ν_{as}(Ge-O-Ge) tetrahedral modes correspond roughly to what I have called $F_2^{1,2}$ modes, using a tetrahedral rather than a trilinear nomenclature. [A truly unambiguous approach is that of

[60] H. Verweij and J. H. J. M. Buster, *J. Non-Cryst. Solids* **34,** 81 (1979).

[61] M. Krishna Murthy and J. Ip, *Nature (London)* **201,** 285 (1964); H. Verweij, J. H. J. M. Buster, and G. F. Remmers, *J. Mater. Sci.* **14,** 931 (1979).

[62] K. S. Estropév and A. O. Ivanov, *Adv. Glass Technol.* **2,** 79–85 (1963).

[63] H. Völlenkle, A. Wittmann, and H. Nowotny, *Monatsh. Chem.* **101,** 46 (1970); **102,** 361 (1971); A. Wittmann and E. Modern, *ibid.* **96,** 581 (1965); and other references cited in Ref. 60. An attempt has recently been made to determine $f_6(x)$ in Li germanate glass by EXAFS [A. D. Cox and P. W. McMillan, *J. Non. Cryst. Solids* **44,** 257 (1981)]. The results give $f_6 \simeq 0.2$ for $0.10 \lesssim x \lesssim 0.25$, which roughly agrees with the results shown in Fig. 26.

[64] C. W. F. T. Pistorius, *Prog. Solid State Chem.* **11,** 1 (1976).

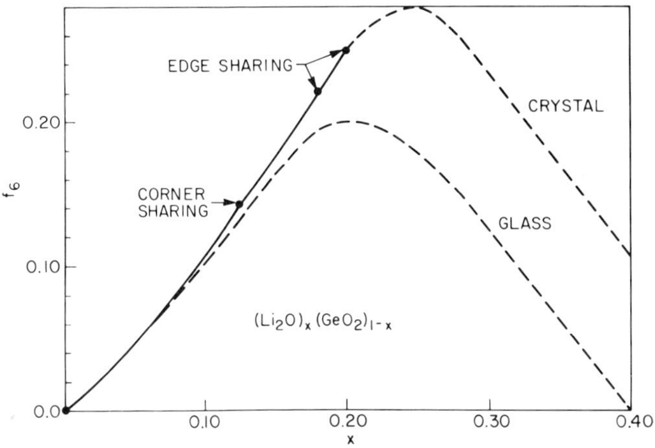

FIG. 26. A sketch of f_6, the fraction of octahedra in Li germanate glasses as a function of composition for crystals and glasses. The crystalline structures contain octahedra–tetrahedra corners or edges, as indicated.

Ref. 20, where the low- (high) frequency modes are shown to be bond-bending (stretching) modes.] In any case, I agree in general with their description of the spectra, but after some study I believe that I have identified additional features that contain considerable information concerning the glass morphology.

To begin the discussion, I summarize the structure data on octahedral populations by sketching in Fig. 26 the fraction f_6 of CN = 6 Ge atoms determined in crystals[63] and estimated for the glasses as a function of composition in g-$(LiO_2)_x(GeO_2)_{1-x}$. The crystalline curve is extrapolated to $f_6 = 0$ in the metagermanate ($x = \frac{1}{2}$, chains of corner-sharing tetrahedra) limit. Attempts to prepare $3LiO_2 \cdot 8GeO_2$ have produced a mixture of $Li_2O \cdot GeO_2$ chains and $LiO_2 \cdot 4GeO_2$ sheets.[60]

Because the glass is less dense than the crystal, f_6(glass) < f_6(crystal). Also, f_6(glass) $\rightarrow 0$ as $x \rightarrow 0.4$, according to extrapolations of the molar refractivities,[61] which also indicate a maximum in f_6 near $x = x_6 = 0.20$. This x_6 maximum shifts to smaller x with increasing alkaline ion size.[61] Finally, the cross-linking octahedra share edges with tetrahedra for the crystalline compositions near $x = 0.20$ and corners near $x = 0.10$.

The strongest lines in the powder spectra at $x = 0.125$, 0.18, and 0.27 are at 475, (566, 595), and 656 cm^{-1}. The latter two groups were assigned by Verweij and Buster to Ge(6) A_1 modes. I would prefer to assign all three lines to Ge(6) A_1-like modes, with octahedral–tetrahedral corner-sharing, octahedral–tetrahedral edge-sharing, and octahedral c-axis face sharing, respectively. I would argue that the 100-cm^{-1} shift at each step reflects the clamping effects of edge and face sharing. The last assignment, *c*-axis face-sharing rutilelike chains, does not have a c-germanate analog, but as Verweij and Buster noted, the strongest line in rutile–GeO_2 is the A_1 line at 702 cm^{-1}. It seems that microcrystallites containing *c*-axis face-sharing rutilelike chains may be present in powder $3LiO_2 \cdot 8GeO_2$. A precedent for a glass morphology that does not have a full crystalline analog can be found in the ethanelike chains[9] in g-$Ge_{0.4}Se_{0.6}$.

Concerning now the g-GeO_2 spectra of Verweij and Buster, the tetrahedral–octahedral edge-sharing doublet seen in the crystalline powder at $x = 0.18$ and 0.20 is very prominent in the polarized spectrum of the glass at $x = 0.11$ at (540, 560) cm^{-1}, corresponding to a relaxation of 25 cm^{-1} relative to the crystal. The second strongest line, at 435 cm^{-1}, may be associated with corner-sharing tetrahedra, since it seems too low to be a tetrahedral–octahedral corner-sharing line (475 cm^{-1} in the powder). This is a rather surprising result, because it suggests that CN = 6 is stable in the glass only when it shares edges with tetrahedra. The reason for this may be that the layers in the glass are more strongly warped (puckered) than in the crystal, and that therefore only the stronger cross-linking units, the edge-sharing octahedra, are formed in the glass.

For layer values of $x \gtrsim 0.15$ in the $(Li_2O)_x(GeO_2)_{1-x}$ glasses only one broad peak is seen in the polarized spectrum between 400 and 700 cm^{-1}. It seems that this peak is composed of the two groups of peaks seen at $x = 0.11$ and also includes a third broad peak near 500 cm^{-1}, which may be associated with the A_1 mode of nonbridging oxygens associated with six-membered corner-sharing tetrahedral rings in c-$Li_2O \cdot 2GeO_2$. This 500-cm^{-1} line is a germanate π-bonded analog of the collective $L_1(=O_s^*)$ mode in g-GeO_2, which was discussed in Section X.

In the polarized spectra of g-$(Na_2O)_y(GeO_2)_{1-y}$ and g-$(K_2O)_z(GeO_2)_{1-z}$ a single broad line near 530 cm^{-1} is observed for $0.11 \lesssim y \lesssim 0.25$ and for $0.11 \lesssim z \lesssim 0.15$. For $y = 0.33$ and for $z \gtrsim 0.18$ the broad line is better described as composite, with lines near (600, 650) cm^{-1} in addition to the largest peak near 540 cm^{-1}. It is tempting to describe these high-energy lines as A_1 octahedral–tetrahedral shared-edge modes, but the fact that the values of y and z at which these features appear are larger than the corresponding $x = 0.11$ value in the $(Li_2O)_x(GeO_2)_{1-x}$ glasses seems to be inconsistent with the observation that $z_6 \sim 0.15 < y_6 < x_6 \sim 0.20$ as the compositions at which the molar refractivity deviates most from linearity.[61] It is possible that this additional structure is instead associated with the disappearance of the octahedra, that is, the formation of corner-sharing tetrahedral rings with six (540-cm^{-1}) and five (600-cm^{-1}) members. The layer morphology here is far from random, but the severe warping of the layers by the larger alkali ions may cause a greater spread in the intralayer ring statistics in the glass than is found in the crystal.

The appearance of octahedra for values of x as small as 0.1 in the germanate glasses implies the presence of layer morphologies at these low concentrations. Thus the most interesting composition range for testing the quartz paracrystalline model for g-GeO_2 is the nearly pure range $0 \leq x \lesssim 0.05$, where it may be possible to observe the disappearance of the $L_1(O_s^*)$ modes *before* the layer morphology (which introduces nonbridging oxygen atoms in a different way) develops. On the other hand, at $x \sim 0.05$ some morphological separation of $N_D = 3$ quartzlike regions and $N_D = 2$ layerlike morphologies may already have taken place; that is, the alkalies may not segregate to the O_s^* regions. Further experiments in the nearly pure range with various additives would be of great interest.

XIV. Ternary Alkali Germanate–Silicate Glasses

Polarized and depolarized Raman spectra of alkali metagermanosilicate glasses $XO_2 \cdot (SiO_2)_{1-y}(GeO_2)_y$ and alkali digermanosilicate glasses $XO_2 \cdot 2(SiO_2)_{1-y}(GeO_2)_y$ have been reported by Verweij.[65,66] The main difference between the alkali silicates and germanates in the framework bending 500-cm^{-1} region is the presence of polarized octahedral lines in the germanates with molar additive concentrations $x \lesssim 0.2$, so that the spectra at these compositions $x > 0.33$ are of relatively little interest in the strong bending mode region. However, at frequencies below 300 cm^{-1} (400 cm^{-1} for X = Li) in the depolarized spectrum characteristic structure is observed that is clearly associated with the additive X ions. To my knowledge these are the first data to exhibit this structure, which is of considerable interest in the present theoretical model because of its bearing on the screening of polarized framework bending modes by depolarizing quasimetallic (i.e., low-frequency) alkali motion.

In Fig. 27 Verweij's depolarized spectra for sodium digermanosilicate glasses[66] are reproduced for the reader's convenience. Consider the spectrum for $y = 0.4$, which contains a narrow peak near 280 cm^{-1} above a broad band peaking near 150 cm^{-1}. (These are unreduced spectra, and the relative strengths of the broad and narrow bands change not only with y, but would also be modified if the spectrum were reduced by phonon thermal occupation numbers. These quantitative corrections are of secondary importance to the following discussion.) Similar depolarized structure is seen for X = Li and X = K, but the narrow band is enhanced in the K glasses and suppressed in the Li glasses except at $y = 0$. Thus in the Na digermanosilicate glasses the low-frequency spectrum nearly replicates the nar-

[65] H. Verweij, *J. Non-Cryst. Solids* **33**, 41 (1979).

[66] H. Verweij, *J. Non-Cryst. Solids* **33**, 55 (1979).

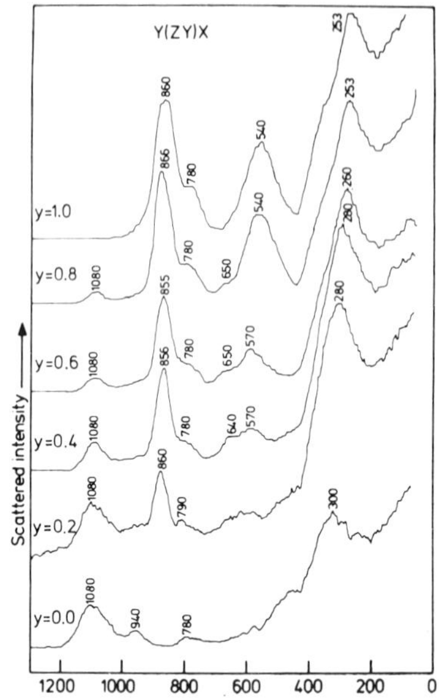

FIG. 27. Composition dependence of the depolarized Raman spectra of Na_2O-$[(GeO_2)_y(SiO_2)_{1-y}]_2$ glasses (data from Ref. 66).

row broad) framework bending structure near (480, 430) cm^{-1} in g-SiO_2 and near (500, 416) cm^{-1} in g-GeO_2 discussed in Sections VI and X.

The narrow-band frequencies ω_s for $y = 0$ and X = Li, Na, and K are 420, 330, and 300 cm^{-1}, and for $y = 1.0$, not resolved, 280, and 253 cm^{-1}, respectively. For intermediate values of y the dependence is S-shaped. The $y = 0$ and $y = 1.0$ limits of ω_s can be fitted by

$$\log \omega_s = a_X \log m_X + a_T \log m_T + \text{const}, \tag{14.1}$$

where X = Li, Na, or K and T = Si or Ge. The fit is excellent with $a_X = -0.20 = 2a_T$. This shows that ω_s is the frequency of a X—O—T bending mode; the ratio $a_X/a_T = 2$ reflects the nonbridging nature of the X–O bond and suggests that the X atom may occupy an internal surface site X_s^*.

It was suggested in Section XII that the quenching of the cluster interior broad 430-cm^{-1} F_2^1 polarized Raman band with increasing x in silicate glasses, shown in Fig. 21, might be associated with additive screening of the polarization associated with these modes. The low-frequency X modes that were observed below 300 cm^{-1} in Fig. 27 (depolarized spectra) are just these screening modes. It would therefore be of considerable interest to monitor the strengths of the screening modes in the depolarized spectrum for closely spaced compositions $x \sim x_F$ or $x \sim x_c$—the composition at which $d\omega[L_1(O_s^*)]/dx$ changes abruptly (see Fig. 22). Note that it seems quite reasonable for both dynamical and symmetry reasons that low-frequency bond-bending depolarizing X modes could be strongly coupled to and quench the Raman scattering from higher-frequency bond-bending polarized T modes, especially in the presence of X–O–T bonds.

XV. Pb Silicate and Germanate Glasses

In $(PbO)_x(GeO_2)_{1-x}$ glasses nonlinear variations in the molar density $\rho(x)$ and refractivity are observed that are similar to those observed in the alkali germanates.[67] Quantitatively there are several important differences, however. These differences are seen by comparing $\rho(x)$ in $(PbO)_x(GeO_2)_{1-x}$ with $(X_2O)_x(GeO_2)_{1-x}$ in Figs. 28 and 29. For the X alkalis $\rho(x)$ is linear only for $0 \leq x \lesssim 0.1$, whereas for Pb germanates $\rho(x)$ is extremely linear up to $x = 0.2$ or 0.3.

At small x the structure of one crystalline compound, $PbGe_4O_9$, has been determined.[68] The compound contains GeO_6 units with $f_6 = 0.25$ and its Ge_4O_9 framework structure is very similar to that reported[60,69] for $X_2Ge_4O_9$ (X = Li, Na, K), except that the unit cells of the latter are doubly reconstructed along the c axis. The pronounced break in $\rho(x)$ for the Pb germanates near $x = 0.3$ is suggestive of phase separation[70] between g-$(PbO)_y(GeO_2)_{1-y}$, with $y \lesssim 0.1$ and g-$PbGe_4O_9$. I shall return to this point in Section XIX.

Once again the compositions x of (PbO or $PbO_2)_x(SiO_2$ or $GeO_2)_{1-x}$ glasses that have been studied by Raman scattering are for the most part too large to be connected to the Raman spectra of g-SiO_2 or g-GeO_2. Thus for the lead silicates I have found recent data[71,72] only for $x \geq 0.33$, and for the lead germanates only some old unpolarized Raman data[73] obtained from a Hg arc (not laser) light source, detected photographically. However,

[67] J. A. Topping, I. T. Harrower, and M. K. Murthy, *J. Am. Ceram. Soc.* **57**, 209 (1974); M. Krishna Murthy and J. Ip, *Nature (London)* **201**, 285 (1964).

[68] C. R. Robbins and E. M. Levin, *J. Res. Natl. Bur. Stand., Sect. A* **65**, 127 (1961).

[69] J. H. Jolly and R. L. Myklebust, *Acta Crystallogr., Sect. B* **24**, 460 (1968).

[70] R. R. Shaw and D. R. Uhlmann, *J. Non-Cryst. Solids* **1**, 474 (1969); B. M. Cohen, D. R. Uhlmann, and R. R. Shaw, *ibid.* **12**, 177 (1973). In $(PbO)_x(SiO_2)_{1-x}$ melts linear variations in $\rho(x)$ with sharp kinks at $x = 0.33$ and 0.50 have been reported by M. A. Sherstobitov, *Russ. J. Phys. Chem. (Engl. Transl.)* **43**, 1784 (1969).

[71] C. A. Worrell and T. Henshall, *J. Non-Cryst. Solids* **29**, 283 (1978).

[72] T. Furukawa, S. A. Brawer, and W. B. White, *J. Mater. Sci.* **13**, 268 (1978).

[73] V. N. Morozov, *Zh. Prik. Spektrosk.* **8**, 830 (1968); *J. Appl. Spectrosc. (Engl. Transl.)* **8**, 501 (1968).

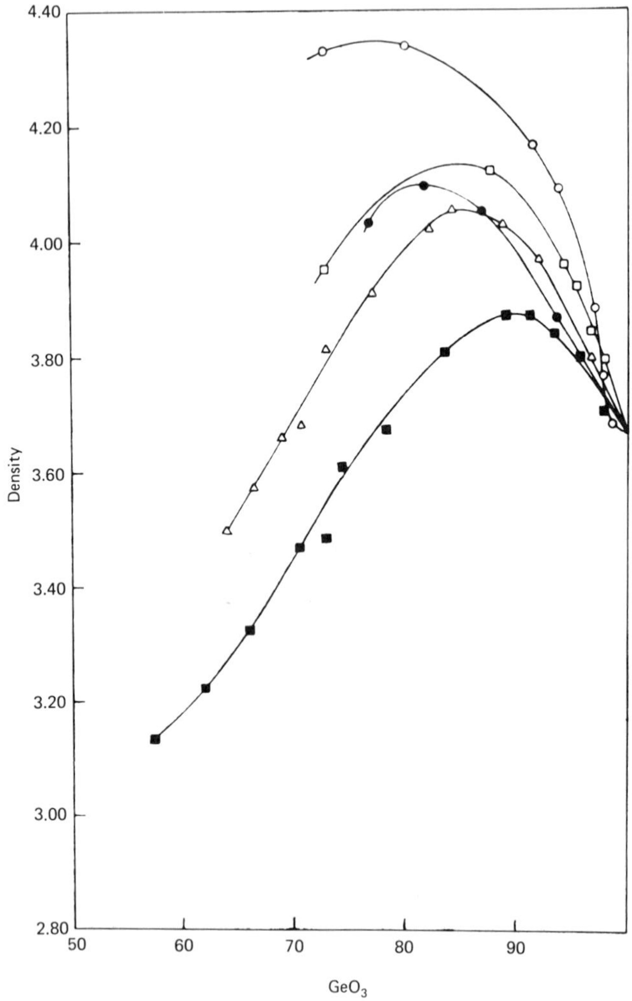

FIG. 28. Molar densities of alkali X germanates as a function of composition. Starting from left top: X = Cs, Rb, Li, Na, and K (data from Ref. 67).

the old germanate data include $x = 0.00, 0.12, 0.15, 0.25, 0.33$, and 0.45, so that in spite of the quaint experimental techniques employed, the data are of considerable interest and are shown for the reader's convenience in Fig. 30.

Once again the interest here will be focused on the F_2^1 broad band peaking near 412 cm^{-1} at $x = 0$. The quenching of this band between $x = 0.15$

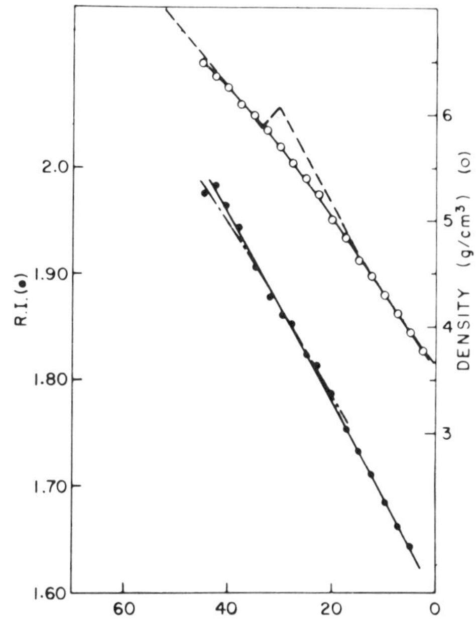

FIG. 29. Molar densities and refractive index of Pb germanates (data from Ref. 67).

(curve 4) and $x = 0.25$ (curve 3) is immediately evident; I would guess that $x_F \sim 0.20$. No data are available in this composition range for the Pb silicates, but at $x = 0.33$ there are substantial differences between the Pb digermanate Raman spectrum[73] (which, though unpolarized, is presumably dominated by its polarized spectrum).[71,72] In the latter as frequency decreases below 550 cm^{-1} the scattering strengths rises rapidly, as in Pb digermanate, but near 400 cm^{-1} there is no dip. Instead a broad, smooth rise is observed, increasing to very large scattering between 40 (unpolarized) and 140 (polarized) cm^{-1}. The latter peak is most probably[71,72] associated with the A_1 mode of OPb_n units of the type found in c-PbO. Thus there is some evidence for paracrystalline PbO clusters in the Pb silicates.

XVI. BeF_2: A Nonoxide Tetrahedral Glass

Chemically speaking, BeF_2 and $ZnCl_2$ may be described as exotic fused salts in the sense that although they are fully ionic because of the large electronegativity differences between cation and anion, the crystalline forms consist of corner-sharing tetrahedra. Indeed, g-BeF_2 is almost isostructural

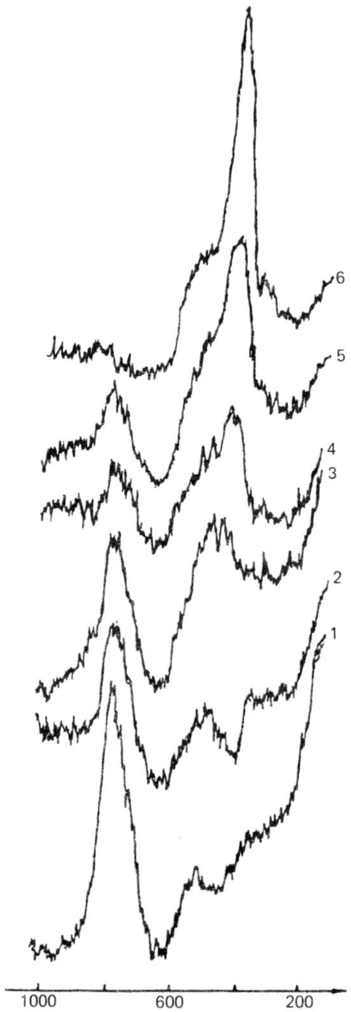

FIG. 30. Composition dependence of the Raman spectra of $(PbO)_x(GeO_2)_{1-x}$ glasses (from Ref. 73). Starting with the uppermost curve, $x = 0.00, 0.12, 0.15, 0.25, 0.33$, and 0.45.

to g-SiO_2, but a proper deconvolution[74] of the diffraction data[75] to obtain the half-widths of the two second-neighbor peaks to test the constraint theory[4] may not be possible because of the weakness of the Be–Be pair scattering contribution to the weighted radial distribution function.

[74] J. R. G. Da Silva, D. G. Pinatti, C. E. Anderson, and M. L. Rudee, *Philos. Mag.* [8] **31,** 713 (1975).

[75] A. H. Narten, *J. Chem. Phys.* **56,** 1905 (1972).

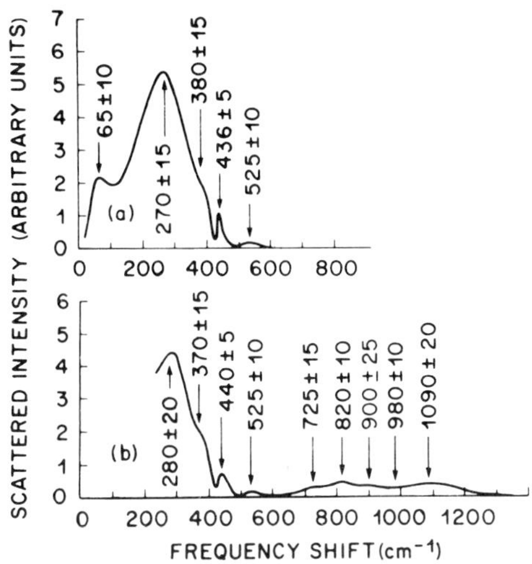

FIG. 31. Two Raman spectra of BeF_2 (from Ref. 76). Because of difficulties with the sample surface, these spectra contain a larger background than is the case for Raman spectra of oxide glasses. The main features associated with the bending mode region between 150 and 400 cm^{-1} are, however, quite similar and well defined in the two spectra.

Within the present context I wish to use g-BeF_2 as an exception that proves (i.e., tests) my model of tetrahedral oxide glass clusters. The notion that such clusters can be terminated by double-bonded oxygen surface ions and that these surface ions give rise to strong narrow bands in the Raman spectra of g-SiO_2 and g-GeO_2 can be tested by examining the Raman spectrum[76] of g-BeF_2, reproduced for the reader's convenience in Fig. 31. This shows striking similarities to the Raman spectrum of g-SiO_2, but not to that of g-GeO_2. Specifically, the bands at 490 and 606 cm^{-1} in g-SiO_2, which the present model ascribes to O_s^* and O_p^* bending modes, have analogs at 380 and 440 cm^{-1} in g-BeF_2. There is no analog in g-BeF_2 to the 347-cm^{-1} peak in g-GeO_2.

These observations are explained quite well by the microcluster model. The known crystalline phases for BeF_2 have cristobalite and quartz structures[76] that are analogous to the crystal structures of SiO_2. The quartz form of BeF_2 that would contribute the two surface bands which I ascribe to the quartz morphology of g-GeO_2 clusters is probably not thermodynamically accessible to g-BeF_2. This is attractive circumstantial evidence for the correctness of the present model and is summarized in Table II.

[76] G. E. Walrafen and R. H. Stolen, *Solid State Commun.* **21,** 417 (1977); A. J. Leadbetter and A. C. Wright, *J. Non-Cryst. Solids* **3,** 239 (1970).

TABLE II. THE NUMBER N_s^* OF NARROW SURFACE BENDING BANDS IN TETRAHEDRAL GLASSES[a]

Material	Morphology	N_s^*
SiO_2	Cristobalite	1
GeO_2	Quartz	2
BeF_2	Cristobalite	1

[a] A phenomenological correlation is observed between N_s^* and the glass cluster morphology.

XVII. Spectroscopic Analogs of Low-Temperature Thermodynamic Anomalies

Low-temperature anomalies in specific heat, thermal conductivity, and acoustic attenuation have been extensively studied in insulating glasses, including tetrahedral oxide glasses. These anomalies in general can be separated into two terms, for example, in the specific heat C_p a term C_l proportional to T^n, with n near 1.0 and an excess Debye term ΔC_p proportional to T^3. The excess Debye term is often comparable in magnitude to the crystalline Debye term.

Low-temperature measurements in general contain too little information to justify the construction of microscopic structural models that would explain the origin of these anomalies on an atomic scale. However, I have recently argued[77] that there is now enough evidence regarding the universality of internal surfaces in insulating glasses so that one can reasonably ascribe ΔC_p to collective internal surface sliding modes and C_l in oxide glasses to *extrinsic* OH groups or *intrinsic* intercluster bridging O atoms. (Notice that the cluster model, as distinguished from point defect or microvoid models, provides a natural mechanism for *intrinsic* two-level states.) The strongest evidence for intercluster interfaces in oxide glasses is associated with the low-frequency cutoff or shoulder seen in Raman and infrared spectra. This cutoff was previously discussed in detail[17,77] for layered chalcogenide crystals and glasses and I will now demonstrate similar behavior for the oxide glasses where the crystal structures may not be layered, but interfaces between clusters with three-dimensional morphologies (SiO_2, BeF_2: cristobalite; GeO_2, quartz) are still present, as sketched in Figs. 9 and 10.

Early workers found[78] a pronounced maximum in α/ω^2, where α is the

[77] J. C. Phillips, *Phys. Rev. B: Condens. Matter* [3] **24,** 1744 (1981).

[78] P. Flubacher, A. J. Leadbetter, J. A. Morrison, and B. P. Stoicheff, *J. Phys. Chem. Solids* **12,** 53 (1959).

infrared absorption constant in g-SiO_2 near $\omega = 40\ cm^{-1}$. Stolen showed[79] that a similar maximum occurred in the Raman scattering strength, with infrared (Raman) peaks at 42 (54), 38 (36), and 26 (24) cm^{-1} in glassy SiO_2, GeO_2, and B_2O_3. In the early work[78] the shoulder was attributed to low-lying optical modes, and this assignment was confirmed by studies[33] of neutron radiation-compacted samples that reduced both the scattering intensity of the low-lying modes and the excess heat capacity. Stolen concluded that the reduced strength of the shoulder with compaction indicated that the shoulder must be associated with vacancies, but he did not indicate whether the vacancies were of the point or internal surface type.

In layer crystals Zallen and Slade[79] showed that interlayer shear modes are found near 30 cm^{-1} in chalcogenide compounds and at 45 cm^{-1} in graphite, which is closer in atomic size to the O^- ions in oxide glasses. The identification of the interlayer shear modes[9] is particularly striking in c-$GeSe_2$, which, like SiO_2, is constructed of tetrahedral building blocks. The corresponding glasses exhibit maxima in α/ω^2 and Raman scattering intensity that closely resembles those seen in the oxide glasses. Thus both qualitatively and quantitatively I am well justified in assigning the low-temperature thermal anomaly ΔC_p to collective internal surface optic shear modes.

The evidence that has been discussed which indicates that the ~60-cm^{-1} "Bose" band in g-SiO_2 is an interfacial optic shear band is very substantial, although somewhat circumstantial. Thus I believe that from a phenomenological point of view this interpretation is well established. A direct test of the interpretation is possible through studying the composition dependence of the relative scattering strength of the "Bose" band from silica to disilicate glass. The first steps in this direction have already been taken by Furakawa *et al.*[59] Close study of these authors' polarized spectra of Na silicate glasses shows that the scattering strength of the Bose band is about three times larger at $x = 0.15$ than at $x = 0$. According to Fig. 22, $x = 0.15$ is the crossover composition for the scattering strengths of the 490-cm^{-1} O_s^* band of g-SiO_2 and the 560-cm^{-1} nonbridging oxygen disilicate layer band, both of which have frequencies near 490 cm^{-1} near $x = 0.15$ (see Fig. 21). While one cannot expect to be able to separate the Bose bands of SiO_2 paracrystalline surfaces and those of individual disilicate layers in frequency, one may expect an abrupt change in the slope of the composition dependence of the integrated scattering strength of the band to appear in the neighborhood $0.12 \lesssim x \lesssim 0.15$. (The apparent crossover should appear at slightly smaller x than for the O_s^* band, because the relative strengths of the latter in SiO_2 and the disilicate are nearly equal, whereas for the Bose band the disilicate strength is much larger. This presumably reflects the

[79] R. H. Stolen, *Phys. Chem. Glasses* **11,** 83 (1970); R. Zallen and M. Slade, *Phys. Rev. B: Solid State* [3] **9,** 1627 (1974).

large number of nonbridging oxygens per layer; that is, the disilicate layers create, in effect, many more internal surface modes.) This is a difficult but valuable experiment, as there is little doubt about the existence of internal surface layer sliding or shear optic modes.[9,79]

While low-temperature and low-frequency measurements cannot be used to deduce the microstructure of glass, they can be used to demonstrate the inappropriateness of structural models that treat the glass as a homogeneous continuum analogous to a normal liquid which contains point defects that are analogous to impurity molecules in a normal liquid. The very-low-frequency ($\sim$5-cm^{-1} at $T \sim 40$°K) Raman scattering,[80] for example, in such models contains a direct Raman contribution, according to Theodorakopoulos and Jäckle,[80] of the phonon-assisted tunneling type.

Winterling's experiments have recently been extended to ultralow frequencies ($\sim$1.5 cm^{-1}) and temperatures ($T \sim 1.5$°K) by Stolen and Bösch.[80] These authors showed that the predicted phonon-assisted scattering mechanism yields results that are not observed and which are more than an order of magnitude larger than the background level. The coupling constant for the theoretical mechanism is evaluated from ultrasonic attenuation measurements that involve wavelengths which are very long compared to cluster sizes (of the order of 100 Å). Probably the minimum in the reduced Raman scattering near $\omega_{min} = 7$ cm^{-1} reflects the transition from Raman scattering by optic mode motion across one intercluster interface at higher frequencies to intercluster acoustic vibrations involving many interfaces at lower frequencies. An important formal feature of the spectrum in this region omitted in the Jäckle model[80] is the requirement that the internal surface optic modes be orthogonal to the acoustic continuum modes. This requirement can lead to Fano-type "holes" in the spectrum that might explain the very low Raman scattering intensity near $\omega = \omega_{min}$.

XVIII. High-Resolution Electron Micrographs

In Section VII, in order to obtain several independent estimates of domain size I compared density deficits (assuming that g-SiO_2 paracrystallites have the crystobalite structure) with absolute internal surface $L_1(O_s^*)$ neutron scattering band strengths as discussed in Section III. Both estimates gave domain diameters of the order of 60 Å. It was then noted that the same domain size had been obtained by Zarzycki in electron microscopy studies[40] of the microstructure structure of glass filaments fractured *in situ* in an inert atmosphere. Three independent estimates thus converged on a common

[80] G. Winterling, *Phys. Rev. B: Solid State* [3] **12**, 2432 (1975); N. Theodorakopoulos and J. Jäckle, *ibid.* **14**, 2637 (1976); R. H. Stolen and M. A. Bösch, to be published.

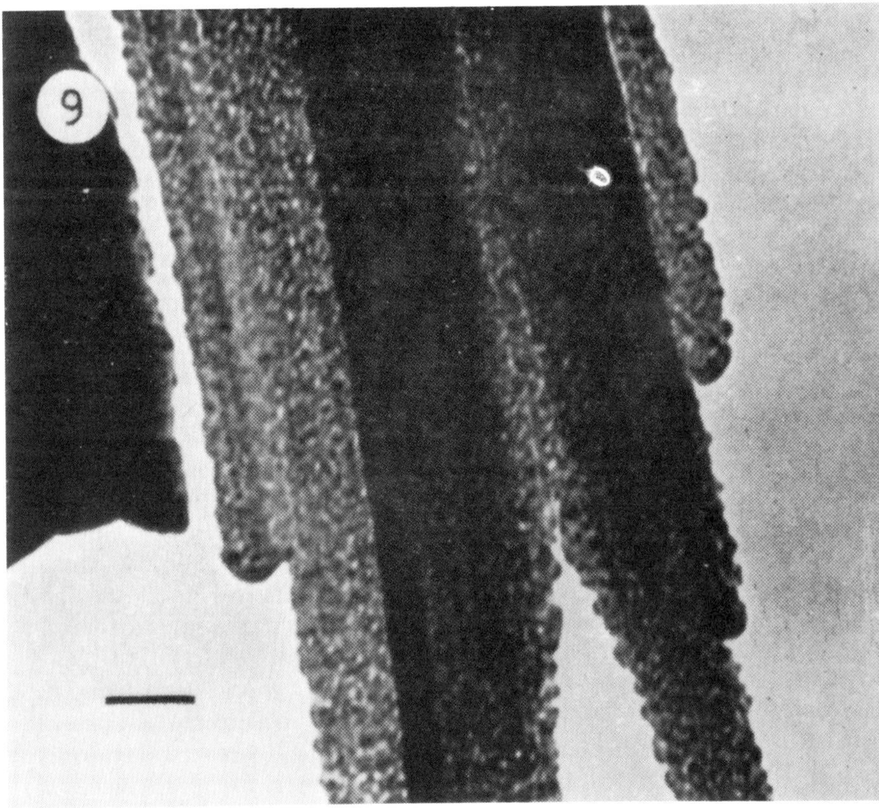

FIG. 32. Electron micrographs of filaments of commercial glass drawn and micrographed in purified Ar (from Ref. 40).

domain size. In this section I will discuss further the nature of the glass microstructure revealed by electron microscopy.

An essential feature of electron microscopy studies of insulating glass microstructure is that the experiments must be carried out on very thin samples in order to avoid charging of the sample. In Zarzycki's experiments samples were obtained by fracturing thin filaments by heating with an electron beam in an inert atmosphere. The fractured filamentary tips were found to contain very thin lamellae of thickness 100–200 Å. An electron micrograph of these lamallae is shown for the reader's convenience in Fig. 32. The domain structure gives rise to the dappled texture observed.

One of the criticisms that has been lodged[81] against these experiments

[81] T. P. Seward and D. R. Uhlman, *in* "Amorphous Materials" (R. W. Douglas and B. Ellis, eds.) p. 327. Wiley, New York, 1972.

is that Zarzycki showed that the domain structure disappeared after the samples had been exposed for a few minutes to humid air; it was therefore concluded that the observed structure was a surface feature not intrinsic to the bulk glass. In view of the evidence assembled in this article, the following interpretation seems to be much more appropriate. The domains or paracrystallites are indeed intrinsic features of the bulk glass, but during the fracture process the very snug interfacial fit discussed in Section VII and illustrated in Figs. 9 and 10 is "broken" on a scale at least of the order of 5 Å or larger. This greatly enhances the density contrast of the interfacial regions and makes possible the observation by microscopy of the domain boundaries. At the same time the "broken" interfaces become accessible, for example, to H_2O molecules instead of just He atoms, which was the situation in the original snugly fitting geometry (Section VIII and Ref. 42). Thus they become highly reactive, polymerization (especially with H_2O) erases the density deficit, and the interfaces disappear within a few minutes after exposure to moisture. This scenario is not merely possible or plausible; in fact, it is exactly what one would normally expect from the conditions of the experiment. An alternate scenario would envision the domain interfaces as being created from a CRN by the rupture process itself, but this seems quite unlikely, considering the large number of nearest-neighbor bonds that would have to be broken on a macroscopic scale.

Particulate SiO_2 particles with mean diameter ~150 Å precipitated from an aqueous gel have been examined by TEM.[82] Because the surfaces of these particles are stabilized by OH ions, they cannot be compared directly to my melt-quenched SiO_2 paracrystalline clusters with mean diameter ~60 Å stabilized by O_s^* surface atoms. Gaskell and Mistry pointed out quite properly[82] that strain energy accumulates less rapidly in g-SiO_2 than in a-Si(:H?), so that one would expect to find larger clusters in g-SiO_2 than in a-Si. This is only one factor determining cluster size, however. The second factor is the surface energy associated with terminating the cluster. It appears that because oxygen so readily forms double bonds, while Si greatly prefers to form single bonds, the surface energy associated with g-SiO_2 clusters is much smaller than that associated with a-Si. This factor at least compensates the strain factor, leaving the paracrystalline clusters in melt-quenched g-SiO_2 at least as small, if not smaller than, the clusters found in a-Si.

The lamellar microstructure expected for a disilicate glass has apparently been resolved directly in recent high-resolution electron microscopy (HREM) experiments.[83] In this case the very thin samples were prepared by plastic deformation near crack tips. The layer spacing is close to 10 Å, as expected, and the layer branching appears to occur between 50- and 200-Å spacings.

[82] P. H. Gaskell and A. B. Mistry, *Philos. Mag.* [8] **39A,** 245 (1979).
[83] V. Schmidt, J. Hopfe, and R. Scholz, *Ultramicroscopy* **5,** 223 (1980).

The average diameter estimated for a silicate paracrystalline cluster by Zarzycki[40] was about 100 Å. The chains have been oriented in the stress direction by the plastic stress, which appears to have broadened the apparent distribution of cluster diameters as measured by the branching spacing, but the average value (100 Å) does not seem to have changed greatly.

XIX. Phase Separation in Silicates

Strictly speaking, because glasses are not in thermodynamic equilibrium, one should not speak of glass "phases." However, because glass clusters contain thousands of atoms, and because these clusters change size and composition rather slowly with annealing at temperatures below the glass transition temperature T_g, this phrase has been employed extensively to describe the emergence of chemical inhomogeneities in thermally treated glasses, especially silicates.

In earlier sections of this article I have considered a wide variety of evidence that indicates that clusters in g-SiO_2, GeO_2, and even BeF_2 have paracrystalline topologies. I have also found evidence in Raman spectra of the existence of microcrystallites of metasilicates and metagermanates. What is the structure of clusters at intermediate compositions?

I will anticipate the answer to this question by saying that it appears that there are many clusters with predominantly paracrystalline structure, with relatively small amounts of nonstoichiometric material dissolved in them. Thus at composition x such that x_1 and x_2 are crystalline compositions, and $x_1 < x < x_2$, there exist clusters containing fractions f_1 and f_2 of the glass molecules, with f_1 and f_2 approximately determined by the lever rule

$$f_1(x - x_1) = f_2(x_2 - x). \tag{19.1}$$

In fact, Eq. (19.1) would be exact if the glass behaved as a regular solution[84] of clusters of composition x_1 and x_2. However, some material of the alternate composition is dissolved in the cluster of primary composition x_i; that is, the clusters actually have compositions $x'_1 > x_1$ and $x'_2 < x_2$. The interesting question is whether anything can be said about the relative magnitudes of $|x'_1 - x_1|$ and $|x'_2 - x_2|$ in specific cases.

As an example consider phase separation in sodium silicate glasses, which has been extensively studied by Porai-Koshits and Russian co-workers[85,86] in the composition range $0.05 \lesssim x \lesssim 0.20$. These authors studied the inho-

[84] R. J. Charles, *Phys. Chem. Glasses* **10**, 169 (1969).

[85] V. I. Averganov, N. S. Andreev, and E. A. Porai-Koshits, *in* "Physics of Non-Crystalline Solids" (J. A. Prins, ed.), p. 580. North-Holland Publ., Amsterdam, 1965.

[86] E. A. Porai-Koshits and V. I. Averjanov, *J. Non-Cryst. Solids* **1**, 29 (1968).

mogeneities induced by annealing by two methods, optical opalescence and electron microscopy of replicas. Their results showed that the phase separation takes place in two stages, which they describe as primary and secondary and which I explain as follows.

When the glass is originally quenched, some phase separation takes place during the quenching process itself. The minority clusters then form nuclei for heterogeneous growth during the early stages of annealing, up to a temperature T_c, at which point the opalescence reaches a maximum that is in some ways analogous to critical opalescence at a second-order phase transition (see Fig. 33). At this temperature the clusters are about 3000 Å in diameter. The opalescence declines at higher temperatures and the glass is nearly transparent at $T = T_c$, the liquifaction temperature where the glass has become a supercooled liquid.

Between T_c and T'_c the large primary clusters ($d \sim 300$ Å) disappear and are replaced by background or secondary clusters with diameters $\lesssim$500 Å. According to my estimates of paracrystalline dimensions, the large primary clusters must contain many paracrystalline subunits. The background or secondary clusters could, however, represent individual paracrystallites at high temperatures.

The composition dependence of T_l is shown[86] together with the equilibrium liquidus in Fig. 34. It is strongly asymmetric, with the disilicate-dominated compositions ($x \gtrsim 0.08$) "clearing" or liquefying at lower temperatures than the silica-dominated compositions ($x \lesssim 0.08$). Alternatively, the maximum in T_l is not at the symmetric value $[x(\text{disilicate}) - x(\text{silica})]/2 = 0.17$, but at only half this value. This means that silica dissolves to a greater extent in the disilicate than in the reverse situation. This, however, is what one would expect from the layer morphology of the disilicates; that is, silica molecules can be intercalated (together with Na_2O molecules) between the disilicate layers. In the crystobalite structure it is much more difficult to find space for the Na_2O molecules.

It seems likely that further understanding of phase separation in annealed glasses will eventually be obtained through Raman scattering studies. The behavior of these spectra in the region near T_c would be particularly interesting.

XX. Frustration of the Crystallization Path

In spite of the existence of abundant circumstantial evidence concerning the presence of paracrystallites in glass, most Western materials scientists, unlike their Russian counterparts,[2] have remained skeptical of this possibility and have instead pursued the concept of continuous, homogeneous, and largely "random" structures.[16] In this article, I have gathered together,

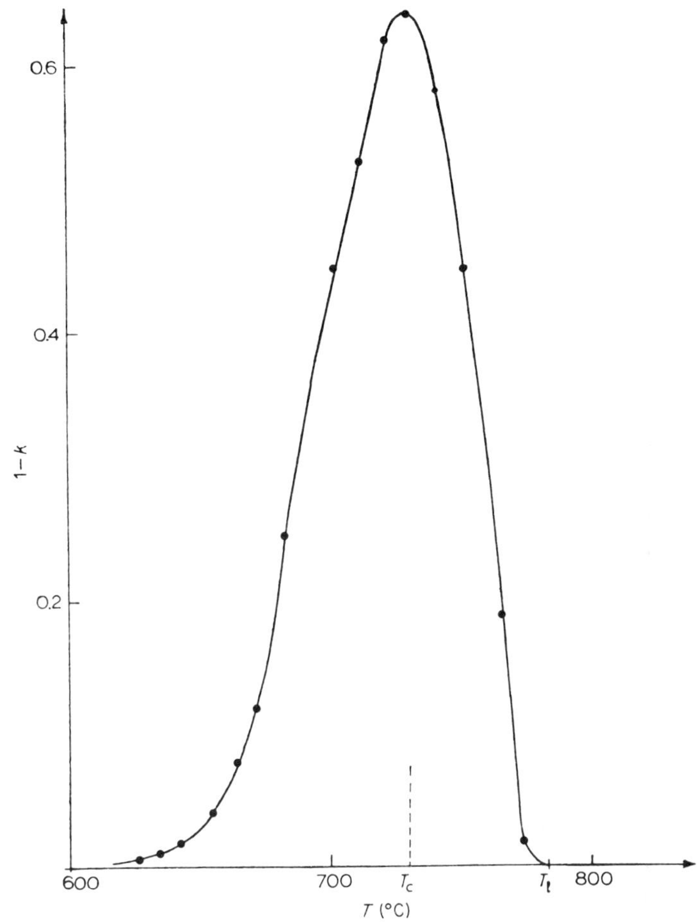

FIG. 33. Fraction of scattered 2.4-eV light (5400 Å) at $x = 0.14$ in a sodium silicate glass as a function of annealing temperature (annealing time about half an hour; T_l and T_c are little changed by longer annealing periods) (from Ref. 85).

I believe for the first time, extensive microscopic evidence, especially in the vibrational spectra, supporting the paracrystallite model even in uniphase systems such as g-SiO_2, GeO_2, and BeF_2. The basic question that many materials scientists unfamiliar with the complexities of spectral analysis will now raise is the following: "Your model *may* all be very well, and of course your analysis of narrow-band anomalies in the vibrational spectra in terms of internal surface modes is very interesting and even *may* [and of course they also mean *may not*] be correct, but if paracrystallites with diameters of the order of 60 Å are present in g-SiO_2, why do these crystallites not grow

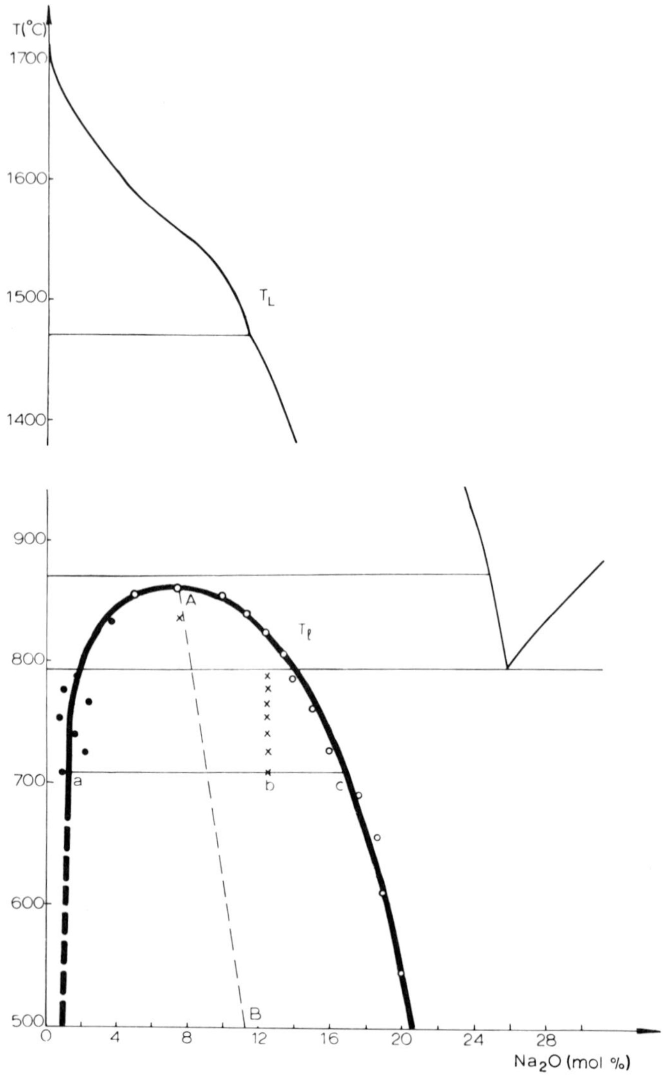

FIG. 34. Composition dependence of T_l defined in Fig. 33 (from Ref. 86).

to macroscopic dimensions? What is so special about SiO_2 that would prevent this from happening?"

The basic mechanism that frustrates crystallization of network glasses has been discussed previously[4] in terms of configurations in phase space and the constraints which are imposed upon these configurations by the interatomic bonding forces. The effect of these constraints on the widths of the second-neighbor peaks in g-SiO_2 is dramatic and is shown once again in

Fig. 35 for the reader's convenience. In a CRN model[16] these two peak widths are equal, reflecting essentially the random equipartition of strain energy. The actual peak widths differ by at least a factor of 5 (FWHM), which could be explained in a CRN model only by assuming a ratio of second-neighbor force constants greater than 25.

In place of an abstract-configurational discussion, which is instructive on a scale of 10 Å, one may consider the path to crystallization on a scale of paracrystalline diameters of the order of 50–100 Å. On this scale the dominant factors are still not the free energies of the crystal and the glass; these energies are relevant for discussing single-crystal growth. The scale of 50–100 Å is the one appropriate to nucleation theories of crystal growth.

Figure 36 shows the interface between two paracrystallites. General thermodynamic models usually consider this interface in a uniphasal material to be symmetric, which it is on the average. However, the interfacial growth kinetics are strongly influenced by a very specific type of asymmetry that arises because of the anisotropy of the surface energy. In Fig. 36 the surface of paracrystallite A at the interface is a vicinal low-index surface with a low step density and low surface energy, while the surface of paracrystallite B has a high step density and a high surface energy. The growth of paracrystallite A and shrinkage of paracrystallite B will increase the area of the A surface at the expense of the B surface and reduce the total interfacial energy.

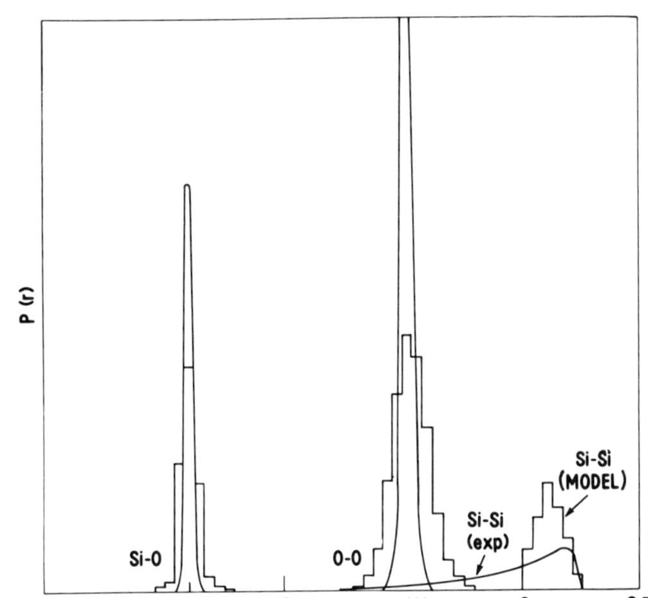

FIG. 35. Comparison between a CRN ball-and-stick model of the radial distribution function of g-SiO_2 (from Ref. 16) and the deconvoluted X-ray distribution function (from Ref. 74).

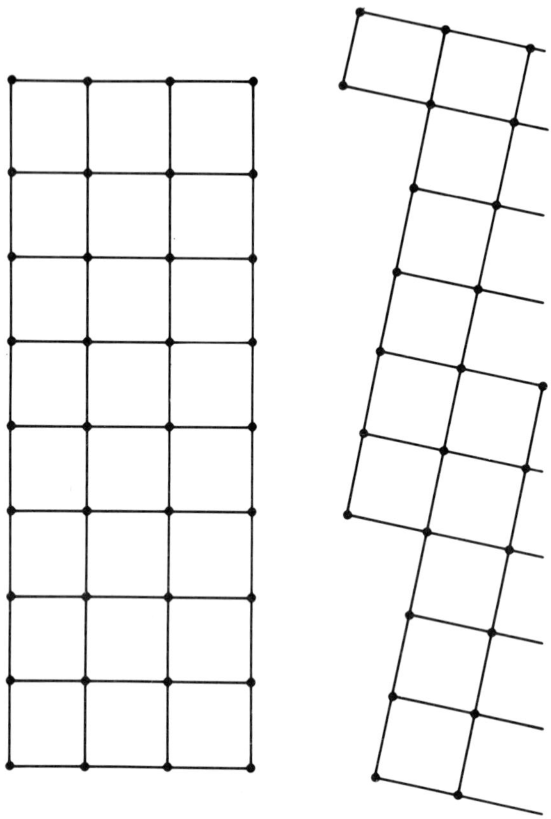

FIG. 36. At the interface between two paracrystallites the two surface energies are generally different. This is accompanied by different step densities, which are represented schematically by atomic planes that terminate at the interface. This difference in surface energy supplies the kinetic driving force for interfacial displacement. In the extreme example shown here for cross sections of simple cubic monatomic paracrystallites the step density of the left surface is zero, so that all the steps are located on the surface of the right paracrystallite.

Summing over all the surfaces of paracrystallite A, one sees that if its surface energy is below average, it will tend to grow into a macroscopic crystallite; similarly, if the surface energy of paracrystallite B is above average, it will tend to shrink and perhaps even disappear as its material is transferred to more stable crystallites such as A. This is the normal growth mechanism for polycrystalline materials.

One must now ask, if paracrystallites are present in g-SiO_2, why doesn't it develop into a polycrystalline material for $T < T_g$? In other words, why don't the low-step-density paracrystalline surfaces grow at the expense of the high-step-density surfaces? The answer to this seems to be that for T_m/

$2 \lesssim T \lesssim T_g \simeq 2T_m/3$ the anisotropy of the surface energy of a good glass-forming material is small; that is, the free energy associated with step formation is nearly zero. These are several ways in which this could happen, at least in principle.

One way to describe zero surface energy anisotropy would be to imagine that the surfaces have been thermally roughened, in the sense conjectured by Frank and Van der Merwe.[87] In the present context this description, however, does not seem to be satisfactory, for we have seen that the narrow bands in the Raman spectra are indicative of well-defined surface textures.

Another way to explain why the anisotropy of the surface energy should be very small is to return to the constraint argument[4] and to compare the number of constraints N_r per formula unit, with the number of degrees of freedom, N_d. In the bulk glass one supposes that $N_r^b = N_d$. At the surface the number of constraints is smaller, $N_r^s < N_r^b$, if only because some of the bonds are weakened (or even broken, in the case of double-bonded oxygen O_s^*, whose coordination number CN has been reduced from CN = 2 in the bulk to CN = 1 at the surface). Now with $N_r^s < N_d$ one has additional degrees of freedom $N_d - N_r^s$ available for reconstructing the surface, that is, forming steps, at almost no cost in free energy. Thus the glass-forming tendency itself provides the configurational basis for nearly isotropic surface energy.

In this discussion of the anisotropy of the surface energy I have considered the surfaces of paracrystallites on a scale of 50–100 Å. Of course, on a macroscopic scale the anisotropy of the surface energy, for example, of quartz, is very well defined. This is a totally different scale, however, and the kinetics of paracrystalline nucleation and growth are a separate issue from those of the morphological habits of macroscopic single crystals.

In conclusion, I feel that with good glass-forming materials there are strong configurational reasons for supposing that the path to crystallization will be sufficiently impeded by the presence of high-energy barriers at non-coalescing interfaces so that paracrystallites on a scale of 60 (g-SiO_2) or 100 Å (disilicates at low T) or perhaps even 300 Å (silicates at high temperatures $T_c \lesssim T < T_l$) may be metastable for laboratory times of interest. In this respect the good uniphasal glass formers that satisfy the condition $N_r \cong N_d$ differ fundamentally from poor glass formers (such as metallic glasses, $N_r > N_d$), where it is necessary to frustrate crystallization both chemically (through working at compositions close to a eutectic) and kinetically (by splat- or even laser-pulse quenching). The reason for this is that stable noncoalescing interfaces with low-surface-energy anisotropy cannot be constructed in poor glass-forming materials ($N_r > N_d$).

[87] W. K. Burton and N. Cabrera, *Discuss. Faraday Soc.* **5**, 33 (1949); W. K. Burton, N. Cabrera, and F. C. Frank, *Philos. Trans. R. Soc. London, Ser. A* **243**, 299 (1951).

XXI. Conclusions

When I began this article my intention was primarily to examine the data on vibrational spectra in order to distinguish features of a n-dimensional nature, specifically $n = 3$ (bulk), $n = 2$ (internal surface), and $n = 0$ (point-defect) vibrational modes (see, e.g., Fig. 1). I soon discovered, however, that the subject itself, as well as the available data, was much more extensive than I had previously realized; moreover, it had been almost 10 years since anyone had attempted a broad theoretical review of the subject. During this period there have been enormous advances in both the quantity and quality of the experimental data. As a result, the nature of this article changed as it developed, and finally what emerged was a cross between an original research paper and a comprehensive review; this hybrid may be called a *thematic review.*

Like its biological analog the mule, a thematic review has both merits and weaknesses. Its chief merit is that it is not restricted in scope to the development and presentation of the results of only one technique, either experimental or theoretical. In vast and still largely undefined subjects such as the molecular structure and micromorphology of inorganic glasses, this feature is very valuable; its chief drawback, on the other hand, is its incompleteness. In the course of preparing this article I have read approximately three to four times as many papers as I have cited, and I have scanned almost all the papers published in some journals (such as the *Journal of Non-Crystalline Solids*). However, I have deliberately attempted to confine my citations to papers whose results, so far as I have been able to perceive, bore directly and successfully on the themes central to this paper. These themes are by no means original and many papers that I have not cited have been written with them in mind. Some of these uncited papers were unsuccessful in spite of their authors' intentions, and in other cases I simply failed to grasp the implications of their results. At this point I therefore plead, in extenuation of these omissions, that this is a very complex subject. I also hope that this article will serve to stimulate discussion that in turn will correct in due course the shortcomings of my work.

Appendix A: Combined AX_4 and A_2X Symmetries

The problem of combining symmetries of clusters embedded in a connected network has no rigorous or even qualitative solution in general, as is obvious in the present case, where one may center the cluster either on the A atom (AX_4 cluster) or on the X atom (A_2X cluster). For this reason Bell *et al.* restricted their normal-mode analysis[6] to statistical studies of normal modes of two-force-constant models containing several hundreds of atoms. These authors then classified the normal modes by projection onto the normal modes of A_2X bent molecules,[8] ignoring entirely the alternate projection on AX_4 normal modes. A similar approach was followed in a one-force-constant model (in which the E and F_2^1 frequencies are zero) by Sen and Thorpe.[7] This procedure is rendered plausible by the relative simplicity of the A_2X normal modes and by the simple conclusion reached[7] when the A_2X bond angle is close to $\pi/2$; namely, that in this case the A-centered motions decouple, which explains why the tetrahedral modes are narrow and readily identified in the GeS_2 and $GeSe_2$ glasses.[9]

In this appendix I adopt the opposite approach; namely, I classify the normal modes by projection onto the AX_4 cluster because it has the highest symmetry. I then ask what the effect of coupling the clusters through A_2X linkages may be. Because in g-SiO_2 the A_2X bond angle is close to 150°, these coupling effects are expected to be large.[7] However, they are also systematic for reasons that I will now discuss.

The AX_4 molecule has 15 degrees of freedom, but 6 of these are associated with external translational and rotational motion, leaving 9 internal degrees of freedom divided among the $A_2(1)$, $E(2)$, and $F_2^{1,2}(6)$ modes. In the network one may write the corner-sharing tetrahedron as $A(X_4)_{1/2}$, which again gives $3 + {}^{12}\!/_2 = 9$ degrees of freedom, a consistent result that emphasizes the utility of the corner-sharing nomenclature as a means of avoiding double-counting degrees of freedom. If one now adds the four internal degrees of freedom associated with the three normal modes of bent A_2X molecules, one has an overcomplete set of modes, with 13 modes instead of the allowed 9.

As the next step one can orthogonalize the nine tetrahedral modes to the four bent A_2X modes, recovering the proper dimensionality in vibrational space. This orthogonalization process couples the A_1, E, and $F_1^{1,2}$ modes via an intertetrahedral nonorthogonality interaction g. When the A_2X bond angle is close to π, this interaction has nearly even parity and its chief effect is to mix the even-parity A_1 and E modes with each other but not with the odd-parity $F_2^{1,2}$ modes. Moreover, there is already a strong intratetrahedral interaction that has split the F_2 modes. The g interaction must also produce an acoustic band near zero frequency.

The continuum vibrational density of states for g-SiO_2 measured by neutron scattering (Section III) can now be compared with the normal molecular mode frequencies of SiF_4. It is already known[10] from comparing g-GeS_2 with $GeCl_4$ and g-$GeSe_2$ with $GeBr_4$ that there is good agreement of molecular and localized glass mode vibrational frequencies in the case where the intertetrahedral coupling g is weak, because the corner-sharing bond angle is close to $\pi/2$. (In these chalcogenide cases the glass frequencies are about 10% lower than the molecular frequencies. Since the electronic dielectric constant that screens intertetrahedral interactions is about half as large in the oxides as in the chalcogenides, most of this difference should be removed in the comparison between g-SiO_2 and SiF_4. In any case, these 10% uncertainties are much smaller then those that may be introduced by a multiparameter interatomic force field.) My reasoning suggests that $\nu(F_2^{1,2})$ should be almost the same in g-SiO_2 [(350, 1090) cm^{-1}] as in SiF_4 [(390, 1030) cm^{-1}][11] and this is indeed seen to be the case with no adjustment of parameters. On the other hand, the strong mixing of the A_1 and E bands generates a broad band with small peaks near its end at $\nu_c = 100\ cm^{-1}$ and $\nu_c' = 800\ cm^{-1}$. The upper cutoff ν_c' agrees with the A_1 frequency of SiF_4.

The reasoning given here is quite general and it can be used to describe the continuum vibrational spectrum of a network of GeO_2 tetrahedra, given the normal modes of GeF_4 ($A_1 = 740\ cm^{-1}$, $E = 205\ cm^{-1}$, $F_2^1 = 260\ cm^{-1}$, $F_2^2 = 800\ cm^{-1}$). One must also allow for the reduction in g due to the reduction of the corner-sharing bond angle θ from 150° in g-SiO_2 to 140° in g-GeO_2, as well as a concomitant increase in f, the odd-parity part of the intertetrahedral interaction.

As in the case of SiO_2, the F_2 frequencies of GeF_4 [(260, 800) cm^{-1}] are in good agreement with those obtained for g-GeO_2 by neutron scattering [(260, 860) cm^{-1}]. The lower cutoff for the A_1–E band is not identifiable, but the upper cutoff ν_c' falls at about 600 cm^{-1}. This is about 100 cm^{-1} lower then expected from the A_1 frequency of GeF_4; together with the increased value of $\nu(F_2^2)$, it results from increasing f interactions between A_1 and F_2 modes with decreasing θ.

Appendix B: Clusters and Small-Angle X-Ray Scattering

In Section VII it was mentioned that in the older literature the absence of measurable small-angle X-ray scattering (SAXS) was often cited as conclusive evidence against the presence of clusters in glasses such as g-SiO_2 with diameters $d \gtrsim 20$ Å. More recent experience has suggested that SAXS is not a reliable guide to fine-grained structure in some cases, but that it is quite useful in others. The literature on this subject has largely been produced by experimentalists using oversimplified mathematical models that make no allowance for the actual physical structure of intercluster interfaces. While a complete discussion of this question is beyond the scope of this paper, a few comments to clarify the problem seem to be in order.

All small-angle scattering calculations have been based on a superposition of scattered intensities from independent scatterers with spherical symmetry. In some cases the scatterers are regarded as spherical microvoids or spherical shell voids.[88,89] Similar scattering problems arise in solutions of the Schrödinger equation, for example, in liquid metals. In these cases care is taken,

[88] A. Bienenstock and B. G. Bagley, *J. Appl. Phys.* **37**, 4840 (1966).
[89] J. Zarzycki, *Int. Congr. Glass* [*Pap.*], *10th, 1974* pp. 12–28 (1974).

for example, in pseudopotential theory, to obtain self-consistent (multiple-scattering) results in the long-wavelength limit[90] $q \rightarrow 0$. The importance of making allowances for such corrections in the SAXS problem has also been discussed, but not actually implemented.[88]

The first and most important point to notice is that spherical models of the void distribution place only *upper limits* on the SAXS magnitude.[88] Any geometry that is less symmetrical then a sphere will produce less scattering. Consider, for example, a cubic array of cubic crystobalite paracrystallites with perfectly planar surfaces. (As has been shown in Sections VII and XVI, this is quite a good model not only for g-SiO_2, but also for g-BeF_2.) The SAXS for this model arises primarily from scattering from edges, which can be regarded as cylindrical microvoids with diameters $\lesssim 3$ Å. The fraction of the total void density concentrated in these edges is of the order of 3%. A comparison with Figs. 4 and 5 of Ref. 88 shows that this scattering is at least 50 times smaller than that obtained by Weinberg[91] near $k = 0.2$ Å^{-1}. Even if allowances are made for the microvoid volume associated with the presence of steps on the paracrystallite surfaces, the additional scattering will still be that associated with microvoids with diameters $\lesssim 2$ Å, and even many such steps will make a contribution below that which is experimentally detectable.[88]

A key feature of the foregoing discussion is the adaptability of the SiO_2 surfaces, which, according to the topological constraint theory, is something that is intrinsically related to the glass-forming tendency itself.[4] If one looks instead at GeSe, which has a deformed rock salt structure (coordination number = CN = 6, which is very large compared to the CNs found in network glasses), then the interfaces are expected to be much less adaptable and the interstitial voids between paracrystallites more nearly spheroidal. In this case SAXS may be much more informative, and recent SAXS experimental data[92] seem to indicate the existence of crystallites or voids with diameters near 30 Å. Note, however, that in Ge_xSe_{1-x} alloys for $x \geq 0.42$ the materials no longer form glasses, but are always polycrystalline, even on this small scale.[93]

Acknowledgments

I am grateful to F. Galeener, P. H. Gaskell, A. J. Leadbetter, G. Lucovsky, K. Nassau, R. H. Stolen, D. Virgo, W. L. White, and J. Zarzycki for discussions and correspondence concerning much of the material described here.

[90] V. Heine, *Solid State Phys.* **24,** 1 (1970).

[91] D. L. Weinberg, *J. Appl. Phys.* **33,** 1012 (1962).

[92] H. Terauchi, H. Kawamura, H. Macda, and K. Osamura, *J. Non-Cryst. Solids* **45,** 293 (1981).

[93] H. Kawamura and M. Matsumura, *Solid State Commun.* **32,** 83 (1979).

SOLID STATE PHYSICS, VOLUME 37

Extended X-Ray Absorption Fine Structure Spectroscopy

T. M. HAYES AND J. B. BOYCE

Xerox Palo Alto Research Center, Palo Alto, California

I. Introduction

The availability of intense sources of continuous X radiation has stimulated the development of a technique for structure determination which is particularly suited to studies of short-range interatomic correlations in complex systems. The absorption cross section for the photoexcitation of an electron from a deep core state to a continuum state exhibits oscillations as a function of photon energy known as the extended X-ray absorption fine structure, or EXAFS. These oscillations are a final-state electron effect.

ISBN 0-12-607737-1

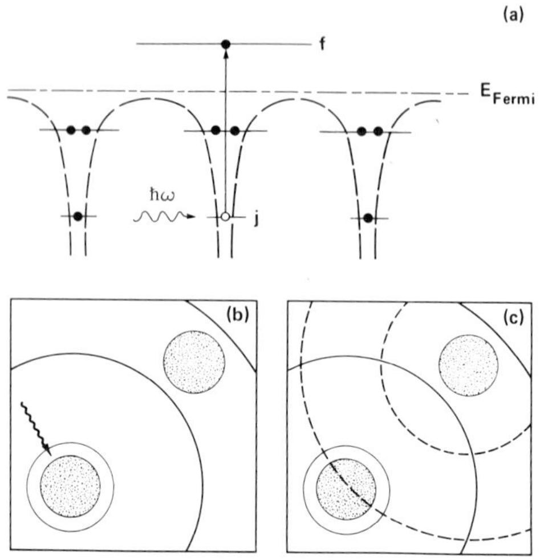

FIG. 1. The origin of the extended X-ray absorption fine structure, or EXAFS. (a) An X-ray photon is absorbed by an atom through the excitation of an electron from a core state j to an unoccupied continuum state f (i.e., $E_f = E_j + \hbar\omega \geq E_{\mathrm{FERMI}}$). (b) The photoexcited electron waves propagate outward from the excited atom (nodes denoted by solid lines). (c) A scattered wave from a neighboring atom interferes with the outgoing wave at the excited atom (nodes denoted by dashed lines), modulating the absorption cross section.

The photoelectron wave function in the core region, and hence the transition rate, is modulated by interference between the outgoing portion of that wave function and that small fraction of the wave which is scattered back from near-neighbor atoms. This phenomenon is illustrated schematically in Fig. 1. The primary objective of studies of EXAFS is to determine the local environment of the excited atom species by analyzing the measured oscillations. The interference pattern reflects directly the net phase shift of the backscattered electron wave, which is predominantly proportional to the product of the momentum of the electron k and the distance traveled. The atomic identity of both the excited and backscattering atoms has a more subtle but nonetheless significant effect on the interference. As a consequence, analysis of the EXAFS can yield not only the distance but also the type and number of the nearest neighbors of the excited atom.

Our first purpose in writing this article is to describe EXAFS spectroscopy with a breadth sufficient to communicate to the nonexpert the potential of this technique. Our second purpose is to discuss the theoretical underpinnings and experimental state of the art with depth sufficient to delineate clearly to a practitioner the capabilities and limitations of this spectroscopy. To this end, we shall first describe the phenomenon of EXAFS in qualitative terms. Then we will derive an especially useful expression for the EXAFS from basic considerations, delineating explicitly the approximations required at each step. The more significant of these approximations are examined in the context of both theory and experiment. The calculation of EXAFS spectra is discussed. The structural information content of the EXAFS is set forth explicitly and compared with the information available from a diffraction measurement. Turning to experiment, we present in some detail the considerations which are important to this spectroscopy. Various techniques for acquiring EXAFS data are examined and their relative advantages and disadvantages delineated. The analysis of EXAFS data is discussed extensively. Finally, a survey is presented of the wide and growing variety of applications of this technique, with particular emphasis on those in condensed matter physics. For supplemental information and alternative viewpoints, the reader is referred to other reviews and summary treatments of this spectroscopy.[1-9]

The development of EXAFS spectroscopy as a tool to examine short-range interatomic correlations has occurred principally since 1970. The phenomenon of EXAFS had, however, been known and understood qualitatively for some time before that.[10] The first published observations[11] of fine structure above an X-ray absorption edge occurred in 1920, 7 years after de Broglie first published[12] a measured absorption spectrum. In 1931, Kronig formulated two alternative explanations for this fine structure based on the photoexcitation of a single electron. For the case of photoabsorption in solids,[13] he neglected the energy dependence of the transition probability and attributed the fine structure to variations in the final-state electron density of states [or, more specifically, to the structure of allowed and forbidden zones which arises from long-range order (LRO)]. Since this theory

[1] E. A. Stern, *Sci. Am.* **234,** 96 (1976).

[2] T. M. Hayes, *J. Non-Cryst. Solids* **31,** 57 (1978).

[3] P. Eisenberger and B. M. Kincaid, *Science* **200,** 1441 (1978).

[4] E. A. Stern, *Contemp. Phys.* **19,** 289 (1978).

[5] D. R. Sandstrom and F. W. Lytle, *Annu. Rev. Phys. Chem.* **30,** 215 (1979).

[6] G. S. Knapp and F. Y. Fradin, *in* "Electron and Positron Spectroscopies in Materials Science and Engineering" (O. Buck, J. Tien, and H. Marcus, eds.), p. 243. Academic Press, New York, 1979.

[7] H. Winick and S. Doniach, eds., "Synchrotron Radiation Research." Plenum, New York, 1980.

[8] B. K. Teo and D. C. Joy, eds., "EXAFS Spectroscopy: Techniques and Applications." Plenum, New York, 1981.

[9] P. A. Lee, P. H. Citrin, P. Eisenberger, and B. M. Kincaid, *Rev. Mod. Phys.* **53,** 769 (1981).

[10] The reader is referred to the detailed accounts of the historical development of this field by L. V. Azároff, *Rev. Mod. Phys.* **35,** 1012 (1963) and by Sandstrom and Lytle.[5]

[11] For example, see H. Fricke, *Phys. Rev.* **16,** 202 (1920).

[12] M. de Broglie, *C. R. Hebd. Seances Acad. Sci.* **157,** 924 (1913).

[13] R. de L. Kronig, *Z. Phys.* **70,** 317 (1931); **75,** 191 (1932).

predicts far too much fine structure, Kronig grouped the peaks suitably to explain experiment. The resulting positions of the maxima and minima are in qualitative agreement with the fine structure data. Hayashi later formulated an alternative model, also based on LRO, in which the maxima in the fine structure correspond to electron excitations to "quasi-stationary" final states.[14] The transition probability to these states is enhanced by their increased wave function amplitude at the excited atom due to coherent reflections from planes of atoms in the solid. The quality of the early experimental data made it very difficult, however, to decide whether or not either of these approaches could explain the observed EXAFS quantitatively.

At the same time, in recognition that a LRO theory could not explain fine structure observations in molecules, Kronig began the formulation of a short-range order (SRO) theory of the EXAFS based on scattered waves.[15] In this approach, the fine structure is attributed to variations in the transition probability arising from modification of the final-state electron wave function by waves scattered back from the other atoms in the molecule.[16] Petersen extended this theory,[17] showing that the contributions of neighboring atoms are additive. It was applied with reasonable success by Hartree *et al.*[18] to explain the fine structure in $GeCl_4$. These authors drew attention to one of the principal virtues of EXAFS spectroscopy: A measurement of the fine structure probes the environment of each atom species individually. By eliminating contributions from some elements of structure, this selectivity can simplify substantially the interpretation of data on a sample containing more than one atom species. Though it was reasonably successful for molecules and remarkably close to the current model for EXAFS in solids, this theory was not believed at the time to be applicable to condensed matter.

The absence of a convincing verification of the LRO model did, however, stimulate later work on a SRO theory for solids. Kostarev extended the Kronig–Petersen model by including in the photoelectron wave function the phase shift due to the excited atom potential.[19] The last of the major elements needed for a correct SRO theory of EXAFS was introduced into the literature when Shiraiwa and co-workers[20,21] included the effects of

[14] T. Hayashi, *Sci. Rep. Tohoku Univ.* **33,** 123, 183 (1949); **34,** 185 (1950).
[15] R. de L. Kronig, *Z. Phys.* **75,** 468 (1932).
[16] Note the similarity with the LRO model of Hayashi.[14]
[17] H. Petersen, *Z. Phys.* **80,** 258 (1933).
[18] D. R. Hartree, R. de L. Kronig, and H. Petersen, *Physica (Amsterdam)* **1,** 895 (1934).
[19] A. I. Kostarev, *Zh. Eksp. Teor. Fiz.* **11,** 60 (1941); **19,** 413 (1949); **21,** 917 (1951); **22,** 628 (1952).
[20] T. Shiraiwa, T. Ishimura, and M. Sawada, *J. Phys. Soc. Jpn.* **13,** 847 (1958).
[21] T. Shiraiwa, *J. Phys. Soc. Jpn.* **15,** 240 (1961).

multiple-atom scattering and accounted for the attenuation of the final-state electron wave due to elastic and inelastic electron–electron scattering. It is unfortunate that they neglected the effect of the excited atom phase shift in their formulation.

Experimental efforts to decide between the alternative LRO and SRO approaches for solids continued to be disappointing in spite of these theoretical advances. As observed by Koslenkov,[22] one difficulty is that the two approaches lead to similar predictions for the position and temperature dependence of the maxima and minima in the fine structure. This is not entirely surprising, given that the LRO density of electronic states arises from the scattering of electrons by individual atoms. Another difficulty is that none of the explicit formulations then in existence was very successful at predicting the *shape* of the EXAFS oscillations.

In 1962, however, the key measurements of Nelson, Siegel, and Wagner[23] provided a convincing qualitative argument in favor of the SRO model. These authors measured the fine structure on the Ge K-shell absorption out to 350 eV in amorphous GeO_2 and in two crystalline polymorphs, hexagonal and tetragonal. These spectra are shown in Fig. 2. Each Ge atom has four nearest-neighbor O atoms in the hexagonal polymorph, and six in the tetragonal one. The experiment revealed significant differences between the EXAFS from the hexagonal and tetragonal GeO_2 samples, which would be expected on the basis of either LRO or SRO theories. On the other hand, the EXAFS spectrum from the amorphous sample is very similar to that from the hexagonal crystal. Since only one of these samples possesses any LRO, one must conclude that only the *near* neighbors affect the EXAFS. These measurements suggested strongly that the appropriate model for EXAFS in solids would be based on SRO, although the range of the neighbors which affect the spectrum was not established. This conclusion was reinforced by the work of Sayers *et al.*,[24] who showed that EXAFS spectra from Fe, Cu, and Ge compare quite favorably with the predictions of a point-scattering SRO expression which incorporates most of the previously introduced effects. Finally, Schaich demonstrated in 1973 that the LRO and SRO approaches lead to identical results if electron mean-free-path effects are properly taken into account.[25]

The widespread realization that EXAFS spectroscopy could become a useful structural probe required two further developments. Firstly, Sayers, Stern, and Lytle demonstrated in 1971 that a Fourier analysis of the measured EXAFS spectrum greatly facilitates interpretation of the data in terms

[22] A. I. Kozlenkov, *Izv. Akad. Nauk SSSR* **25,** 957 (1961).

[23] W. F. Nelson, I. Siegel, and R. W. Wagner, *Phys. Rev.* [2] **127,** 2025 (1962).

[24] D. E. Sayers, F. W. Lytle, and E. A. Stern, *Adv. X-Ray Anal.* **13,** 248 (1970).

[25] W. L. Schaich, *Phys. Rev. B: Solid State* [3] **8,** 4028 (1973).

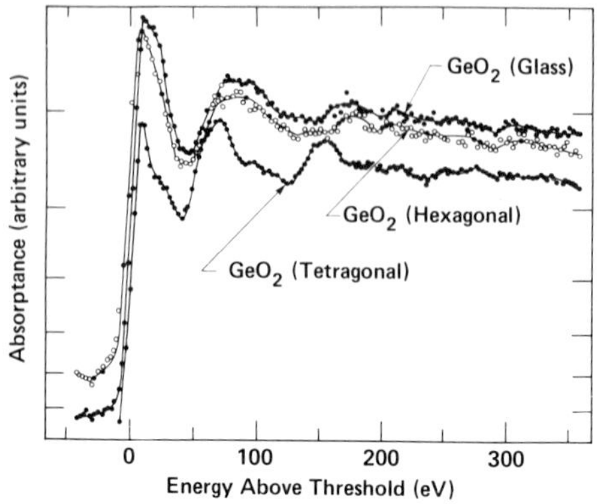

FIG. 2. The absorptance of one glassy and two crystalline samples of GeO_2 as a function of X-ray photon energy relative to the threshold for Ge K-shell absorption. Note that the EXAFS from the glassy form of GeO_2 is very similar to that from the hexagonal polymorph, but is significantly different from the EXAFS from the tetragonal polymorph. As discussed in the text, this observation provides the basis for a convincing argument in favor of SRO models for the EXAFS. [Taken from Fig. 1 in W. F. Nelson, I. Siegel, and R. W. Wagner, *Phys. Rev.* [2] **127,** 2025 (1962).]

of a SRO model.[26] Secondly, the development of a source of synchrotron radiation at Stanford University in the early 1970s provided the high-intensity continuous X-radiation necessary to measure the EXAFS to sufficient accuracy. During the next 10 years, our understanding of EXAFS evolved considerably. We present it below in its simplest terms, and in greater detail in succeeding sections.

1. Qualitative Description of EXAFS

The microscopic origin of the EXAFS is very well understood by now, having been discussed at length in the literature.[27–29] Of the many nearly equivalent single-scattering expressions for the EXAFS, we will use one in which the atomic pair correlation functions p appear explicitly. The absorption cross section for the photoexcitation of an electron from the K shell of atom species α can be written as

$$\sigma_\alpha(\omega) = \sigma_\alpha^0(\omega)[1 + \chi_\alpha(\omega)], \tag{1.1}$$

[26] D. E. Sayers, E. A. Stern, and F. W. Lytle, *Phys. Rev. Lett.* **27,** 1204 (1971).
[27] E. A. Stern, *Phys. Rev. B: Solid State* [3] **10,** 3027 (1974).
[28] C. A. Ashley and S. Doniach, *Phys. Rev. B: Solid State* [3] **11,** 1279 (1975).
[29] P. A. Lee and J. B. Pendry, *Phys. Rev. B: Solid State* [3] **11,** 2795 (1975).

where $\hbar\omega$ is the X-ray photon energy. The σ^0 factor in Eq. (1.1) is similar to the absorption cross section observed for free atoms, but is essentially featureless except for the threshold. χ expresses the modulation of the photoexcitation rate arising from changes in the photoelectron wave function in the core region caused by interference between the outgoing portion of that wave function and that small fraction of the wave which is scattered back from near-neighbor atoms. This modulation yields the oscillations in σ with increasing photon energy which are known as the EXAFS. The oscillations in the interference result from the energy dependence of the phase difference between the outgoing and backscattered waves. The principal contribution to this phase difference is simply the product of the photoelectron momentum k and the round-trip distance $2r$ to a near neighbor. There are also energy-dependent contributions to the phase arising from the excited atom and backscattering atom potentials. Finally, the amplitude of the backscattered electron wave depends on the backscattering strength of the near-neighbor atom potential and on the attenuation of the electron wave in traveling the distance $2r$. Combining these elements, the EXAFS for randomly oriented local environments can be expressed as[30,31]

$$k\chi_\alpha(\omega) = \sum_\beta \int_0^\infty dr\, r^{-2}\, p_{\alpha\beta}(r)\, 2\mathrm{Re}[e^{2ikr}\Lambda_{\alpha\beta}(k, r)], \tag{1.2}$$

where

$$\Lambda_{\alpha\beta}(k, r) \simeq -(2i\pi^2 m/\hbar^2)\, t_\beta^+(-\mathbf{k}, \mathbf{k}) \exp[-2r/\lambda(k) + 2i\eta_\alpha(k)]. \tag{1.3}$$

In these expressions, $E = E_{1s} + \hbar\omega = E_0 + \hbar^2k^2/2m$ is the final-state electron energy, E_0 is the final-state energy corresponding to $k = 0$,[32] and m is the mass of the electron. The term which represents the EXAFS, χ, has been divided into two distinct contributions: direct structural information in $p(r)$ and complicated energy dependences in Λ. $p_{\alpha\beta}(r)$ is the radial distribution of atom species β about the excited species α, defined so that $\int_0^\infty dr\, p_{\alpha\beta}(r)$ equals the total number of β atoms in the sample. Each atom species in the system is represented in the sum over β in Eq. (1.2). The angular distribution of neighbors does not enter except for a single crystal or an oriented polycrystalline sample. In $\Lambda_{\alpha\beta}(k, r)$ are included all those factors which express the complicated interactions of the final-state electron with the excited atom, the backscattering atoms, and the intervening solid: the t matrix of the scattering atom t_β^+, the electron mean-free-path length λ, and the $l = 1$ phase shift due to the potential of the excited atom η_α. Note that Eq. (1.2) does not involve the correlations between all atom pairs, as does the analogous expression for a diffraction study (in which there is a sum over

[30] T. M. Hayes and P. N. Sen, *Phys. Rev. Lett.* **34,** 956 (1975).

[31] T. M. Hayes, P. N. Sen, and S. H. Hunter, *J. Phys. C* **9,** 4357 (1976).

[32] The assignment of a value to E_0 is discussed in Sections 5 and 10.

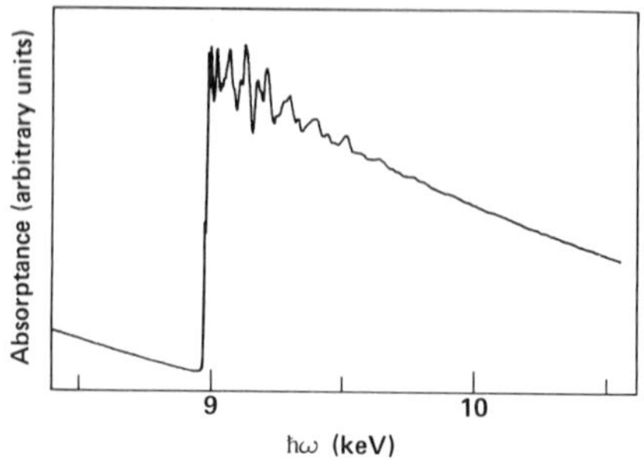

FIG. 3. The absorptance of Cu metal at 77°K as a function of X-ray photon energy, including the onset of Cu K-shell absorption at 8.98 keV.

all α as well). As is discussed in Section 6, this simplifies the analysis substantially and is a principal attribute of the EXAFS technique.

The most direct and commonly used method to acquire EXAFS data is a simple transmission experiment. The intensity of a beam of monochromatized X rays is measured before and after passing through a sample. The logarithm of the ratio of incident to transmitted intensity is the absorptance (the reflectance being negligible) and is proportional to the absorption cross section for that material. The absorptance is shown as a function of X-ray photon energy in Fig. 3 for Cu metal at 77°K. The substantial increase in absorptance as the incident photon energy is increased past 8.98 keV marks the onset of absorption due to the photoexcitation of electrons from the Cu K shell. For our purposes, this experimental absorption cross section is best regarded as the sum of two distinct contributions: the Cu K-shell absorption cross section σ, which vanishes below 8.98 keV, and a slowly varying background absorption. This background absorption results from the photoexcitation of those electrons in the sample which are less tightly bound than the Cu K-shell electrons and is of no interest to us at this point.

Examining the Cu absorptance from the perspective of Eq. (1.1), we identify with σ^0 the sharp increase in absorption at the K-shell threshold (8.98 keV) and the slow falloff above it. The sharp oscillations in the K-shell absorption cross section correspond directly to the EXAFS, represented by χ in Eq. (1.1). These have been extracted from the spectrum of Fig. 3 and are shown in Fig. 4. Note the clear presence of more than one frequency in this spectrum. As is implied in Eq. (1.2), a peak at r_i in a pair correlation function $p(r)$ introduces into $k\chi$ a sinusoid of argument similar to $2kr_i$. This is revealed explicitly in the limit where the contribution of the neighbors of the excited atom arises from only one peak in $p(r)$, containing N_i atoms confined in a narrow Gaussian distribution about r_i. In that limit,

$$k\chi(k) \to N_i r_i^{-2}|\Lambda(k, r_i)| \exp[-2(\sigma_i k)^2]\, 2 \cos\{2kr_i + \text{phase}[\Lambda(k, r_i)]\}, \quad (1.4)$$

where σ_i is the Gaussian half-width of the distribution. It is clear that each peak in $p_{\alpha\beta}$, the radial distribution function of neighbors about the excited atom, will contribute to the fine structure a sinusoid in k space of wavelength $\approx \pi/r_i$, where r_i is the mean radial position of that peak. The envelope of this sinusoid is determined by the number and position of the atoms and by their identity to the extent that their t matrix is distinct. The envelope decreases with increasing k at a rate determined at large k by the width of the distribution, σ_i. This tends to decrease the contribution of more distant peaks, especially in very disordered systems, since they are usually also more broad. The multitude of frequencies present in the EXAFS of Fig. 4 suggests, nonetheless, the contributions of many peaks in p. The principal (and lowest) frequency in Fig. 4 corresponds to the 12 nearest neighbors of each excited Cu atom. The higher frequencies correspond to more distant neighbors.

In quantifying these observations, it is often convenient to work with the EXAFS data after it has been "frequency analyzed" by a Fourier transform. The contributions from the various peaks in p are separated in favorable situations, greatly simplifying the subsequent analysis. This advantage was first pointed out by Sayers *et al.*[26] If $2k$ and r are taken to be the complementary variables, the Fourier transform FT of $k\chi(k)$ can be expressed as[2,31]

$$\varphi_\alpha(r) \simeq \sum_\beta \int_0^\infty dr'\, r'^{-2}\, p_{\alpha\beta}(r')\, \xi_{\alpha\beta}(r - r'), \tag{1.5}$$

for $r \geq 0$. As expressed by Eq. (1.5), φ is the convolution of the pair correlation function p with a peak function ξ, where $\xi_{\alpha\beta}(r, r'') = \mathrm{FT}[W(k) \times \Lambda_{\alpha\beta}(k, r'')]$. As a result, φ is essentially a linear combination of ξ's, one for each peak in p. ξ is analogous to the peak function of X-ray or neutron diffraction. W is a window function introduced to account for the finite range of the EXAFS data and is discussed in Section 10. The r'' variable

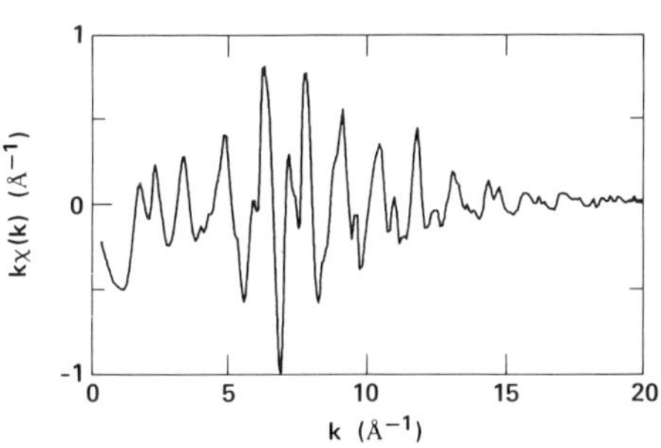

FIG. 4. The EXAFS oscillations $k\chi(k)$ on the Cu K-shell absorption cross section in Cu metal at 77°K as a function of photoelectron momentum k.

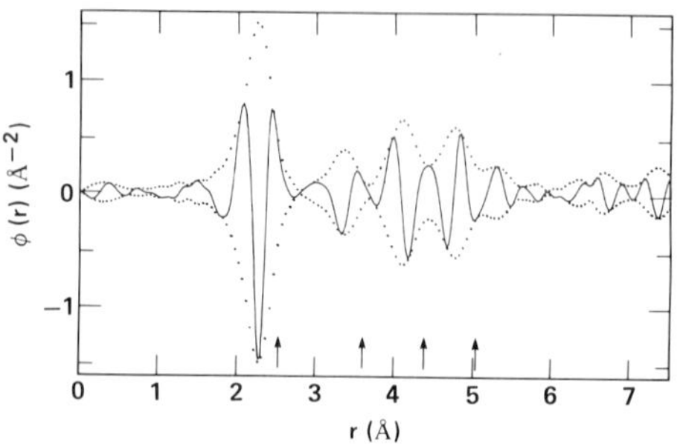

FIG. 5. The real part (solid line) and the magnitude (dotted line) of the Fourier transform $\varphi(r)$ of the EXAFS on the Cu K-shell absorption in Cu metal at 77°K. The data were transformed using a square window with k between 2 and 20 $Å^{-1}$. The near-neighbor distances in Cu metal are shown by vertical arrows.

in $\xi(r, r'')$, which arises from the electron mean-free-path factor in Λ, has been suppressed in Eq. (1.5) (and r replaced by $r - r'$). The inelastic electron scattering processes which underlie this factor must be taken into account, however, in comparisons among EXAFS signals from peaks at substantially different radii and in comparisons between data and calculated spectra.

The Fourier transform of the Cu data is shown in Fig. 5. The first four near-neighbor separations in Cu are represented by the four prominent peaks in Fig. 5. The positions of these peaks in φ are shifted from the near-neighbor separations, however, owing to the k dependence of the phase of Λ (or, equivalently, owing to the displacement of the maximum in $|\xi(r)|$ from $r = 0$). Interference with the contributions from neighboring peaks in φ can also shift the peak positions. As a consequence, analysis of data such as these requires more than merely noting the peak positions in φ. In that analysis, there are two major decisions to be made. Firstly, will the desired structural information be extracted from experimental spectra through comparisons with calculated spectra, or through comparisons with spectra derived from measurements on known systems? Secondly, will the comparisons be made in r space (as shown in Fig. 5), or after isolated peaks in r space have been transformed separately back into k space? There are significant advantages and disadvantages to be considered in making each of these decisions, as we will discuss in some detail in the section on data analysis. First, however, let us examine in detail the derivation of theoretical expressions for the EXAFS, stating explicitly the approximations required at each step.

II. Photoabsorption Theory

As discussed in the Introduction, there have been many alternative formulations of the theory of the extended X-ray absorption fine structure in the 50 years since Kronig[13,15] first attempted to describe this effect. In the following sections, we derive an especially useful form of the theory, detailing each approximation with some care. The more significant of these approximations are examined in the context of both theory and experiment. The calculation of EXAFS spectra is discussed. The structural information content of the EXAFS is set forth explicitly and compared with the information available from a diffraction measurement. Finally, we discuss briefly the structure in the immediate vicinity of the absorption edge.

2. Photoexcitation Formalism

The transmission of photons of energy $\hbar\omega$ through matter is usually described by the following expression:

$$I(z) = I(0) \exp[-(\sum_{\alpha} n_\alpha \sigma_{\alpha;\mathrm{total}}(\omega))z]. \tag{2.1}$$

In this expression, I is the intensity (or energy flux) of the photons, n_α is the density of photon scatterers of species α (e.g., a given species of atom), $\sigma_{\alpha;\mathrm{total}}$ is the total cross section (including both elastic and inelastic interactions) for that type of scatterer, z is the thickness of the material, and the sum includes all possible scattering species in the path of the photons. An expression for the total cross section is obtained most easily by use of the optical theorem, or Bohr–Peierls–Placzek relation, which relates σ_{total} to the imaginary part of the elastic scattering amplitude in the forward direction[33]:

$$\sigma_{\mathrm{total}}(\omega) = (4\pi/k_{\mathrm{X}}) \operatorname{Im} f_{\mathrm{gs}}(\mathbf{k}_{\mathrm{X}}, \mathbf{k}_{\mathrm{X}}), \tag{2.2}$$

where $\mathbf{k}_{\mathrm{X}}$ is a vector in the direction of photon propagation with magnitude ω/c (c being the velocity of light) and "gs" represents the ground state of the system. $f_{\mathrm{gs}}(\mathbf{k}, \mathbf{k}')$ is, of course, the usual scattering amplitude.[33] Alternatively, given that the Hamiltonian is Hermitian, σ_{total} may be expressed in terms of the transition matrix for the particle–photon interaction,

$$\sigma_{\mathrm{total}}(\omega) = -(2\omega/I) \operatorname{Im}\langle \mathrm{gs}|T_{\text{particle–photon}}(E_{\mathrm{gs}})|\mathrm{gs}\rangle. \tag{2.3}$$

Explicitly, $T_{\text{particle–photon}}$ is the transition matrix (or operator) of the system characterized by a Hamiltonian $H = H_{\mathrm{particle}} + H_{\mathrm{photon}}$ and an additional interaction $H_{\text{particle–photon}}$:

$$T_{\text{particle–photon}}(E) \equiv H_{\text{particle–photon}} + H_{\text{particle–photon}} G^{+}(E) T_{\text{particle–photon}}(E), \tag{2.4}$$

where the Green's function (or resolvent) for H is

$$G^{\pm}(E) \equiv (E - H \pm i\epsilon)^{-1} \tag{2.5}$$

[33] For example, see A. Messiah, "Quantum Mechanics," Vol. II, Chapter 19. North-Holland, Amsterdam, 1961.

and E_{gs} is the energy of the system in its ground state. ϵ is a positive infinitesimal constant introduced to define integrals involving the operators $G^{\pm}(E)$. In the following treatment, G will always imply G^{+}, and ϵ will be suppressed. The properties of $G^{\pm}(E)$ are discussed in many textbooks.[33]

The reduction of this expression to a more familiar form is quite straightforward. Substituting Eq. (2.4) into Eq. (2.3), we notice that the first term vanishes for a Hermitian $H_{\text{particle-photon}}$, leaving

$$\sigma_{\text{total}}(\omega) = -(2\omega/I)\,\text{Im}\langle \text{gs}|H_{\text{particle-photon}}G(E_{\text{gs}})T_{\text{particle-photon}}(E_{\text{gs}})|\text{gs}\rangle. \quad (2.6)$$

Since the particle–photon interaction is in general a weak perturbation on the system, we can evaluate Eq. (2.6) in the Born approximation, in which T is replaced with $H_{\text{particle-photon}}$, yielding

$$\sigma_{\text{total}}(\omega) = -(2\omega/I)\,\text{Im}\langle \text{gs}|H_{\text{particle-photon}}G(E_{\text{gs}})H_{\text{particle-photon}}|\text{gs}\rangle. \quad (2.7)$$

Equation (2.7) is a general expression, capable of describing the effects of photon diffraction as well as absorption and emission. Let us now confine our treatment to the absorption of a photon accompanied by an excitation of the particle field through the promotion of one or more electrons. Recall that G may be expressed in terms of the eigenfunctions of H as

$$G(E) = \sum_j |j\rangle(E - E_j)^{-1}\langle j|. \quad (2.8)$$

Consider the specific ground state consisting of a particle field in its ground state g in the presence of m photons of energy $\hbar\omega$, as represented by (g, m). We are interested in those final states j with $m - 1$ photons but the same total energy E_{gs}, represented by $(f, m - 1)$, where f denotes the excited assemblage of particles. The evaluation of Eq. (2.7) will involve matrix elements of the following form:

$$\langle f, m-1|H_{\text{particle-photon}}|g, m\rangle = (-e/c)(2\pi c I/\omega^2)^{1/2}\langle f|\exp(i\mathbf{k}_X\cdot\mathbf{r})\hat{\epsilon}_{\mathbf{k}s}\cdot\mathbf{v}|g\rangle, \quad (2.9)$$

where e is the charge of the electron, $\hat{\epsilon}_{\mathbf{k}s}$ is the unit electric field vector associated with propagation vector $\mathbf{k}_X$ and polarization s, and $\mathbf{v}$ is the velocity operator. Note that the eigenstates involved in the right-hand side of Eq. (2.9) are now only those of the particle Hamiltonian. If the electron to be promoted is in a deep core level (i.e., small radius) and the photon energy is not too high (i.e., small k_X), as is true in the cases of interest here, then $\mathbf{k}_X\cdot\mathbf{r} \ll 1$ over the range of the integral in Eq. (2.9) and $\exp(i\mathbf{k}_X\cdot\mathbf{r})$ can be replaced by unity. This is the electric dipole approximation.[34] We may now

[34] Note that $\hbar\omega = 10$ keV corresponds to $k_X \approx 5$ Å^{-1}, while a core electron bound by 10 keV has a mean radius $r_0 \approx 0.02$ Å. The first correction to the electric dipole expression for σ is proportional to $(k_X r_0)^2$ and therefore negligible.

use

$$\mathbf{v} = i/\hbar(H_{\text{particle}}\mathbf{r} - \mathbf{r}H_{\text{particle}}), \tag{2.10}$$

where $\mathbf{r}$ is the position operator, to obtain

$$\langle f, m-1|H_{\text{particle-photon}}|g, m\rangle = -ie(2\pi I/c)^{1/2}\langle f|\hat{\epsilon}_{\mathbf{k}s}\cdot\mathbf{r}|g\rangle. \tag{2.11}$$

The equation for the total cross section can now be written in the simple form

$$\sigma_{\text{total}}(\omega) = -4\pi\alpha_{\text{fs}}\hbar\omega\,\text{Im}\langle g|\hat{\epsilon}_{\mathbf{k}s}\cdot\mathbf{r}\,G_{\text{particle}}(E_g+\hbar\omega)\,\hat{\epsilon}_{\mathbf{k}s}\cdot\mathbf{r}|g\rangle, \tag{2.12}$$

where $\alpha_{\text{fs}} = e^2/\hbar c$ is the fine structure constant, G_{particle} is the Green's function of H_{particle}, and E_g is the ground-state energy of the particle field. To the extent that many-body interactions are included in G and g, Eq. (2.12) incorporates all possible excitations of the particle field: one-electron transitions, shake-up and shake-off effects (such as are discussed in Section 4,a), and truly multielectron transitions. Contact with Fermi's Golden Rule can be made by recalling that

$$\text{Im}\langle f|G_{\text{particle}}(E)|f'\rangle = -\pi\delta(E-E_f)\delta_{ff'}\,. \tag{2.13}$$

Equation (2.12) can now be written in the familiar form

$$\sigma_{\text{total}}(\omega) = 4\pi^2\alpha_{\text{fs}}\hbar\omega\sum_f|\langle g|\hat{\epsilon}_{\mathbf{k}s}\cdot\mathbf{r}|f\rangle|^2\delta(E_g+\hbar\omega-E_f), \tag{2.14}$$

where $|f\rangle$ are the true (many-body) eigenstates of the actual particle Hamiltonian H_{particle}. Equations (2.12) and (2.14) can be considered exact for the purposes of this treatment, since the only approximations needed to derive them are the Born approximation in $H_{\text{particle-photon}}$ and the electric dipole approximation. Note that the final electron eigenstate must be unoccupied prior to the transition *if* these equations are interpreted in terms of a one-electron model. This implies a factor of $1 - F_{\text{FD}}(E_g + \hbar\omega)$ in $\sigma_{\text{total}}(\omega)$, where $F_{\text{FD}}(E)$ is the Fermi–Dirac distribution function.

3. Specialization to EXAFS

Many models have been proposed for the EXAFS based on Eq. (2.12).[27–29] Building on these efforts, let us now consider a set of simplifying approximations[31] which reduces Eq. (2.12) to a form sufficiently accurate for many applications yet convenient for analysis of EXAFS data. In later sections, we will analyze further each of these approximations in the light of experiment.

If one believed that long-range structural information was contained in the EXAFS data, then an expression for that data might best be obtained by starting with Eq. (2.12) and using the techniques to calculate G which have been developed for use in the analysis of low-energy electron diffraction

(LEED), for example. It has been demonstrated often, however, that the lifetime of the final-state electron is quite brief (or, equivalently, that its mean free path is short). Thus it is more direct to use a short-range order expansion of G in terms of T such as was developed originally by Lax[35] and by Beeby and Edwards.[36,37]

To this end, let us divide H_{particle} into two contributions:

$$H_{\text{particle}} = H_0 + V. \tag{3.1}$$

Recall that

$$G_{\text{particle}}(E) = (E - H_{\text{particle}})^{-1}. \tag{3.2}$$

The first contribution, $H_0 = -\hbar^2\nabla^2/2m + V_0$ (where m is the mass of the electron), should be such that its resolvent (or Green's function) can be calculated without great difficulty. It could, for example, be the free-space Hamiltonian ($V_0 = 0$). For the purposes of this discussion, however, H_0 should represent H_{particle} quite accurately in the region of the core of the excited atom (i.e., that atom from which the photoemitted electron is expelled during the photoabsorption event). To that end, V_0 will include the excited atom potential in an effective medium designed to simulate the presence of the remainder of the solid or molecule absorbing the X rays. Inelastic processes are included through an imaginary part of the potential of the medium. Finally, H_0 is taken to be spherically symmetric about the center of the excited atom, which is located at the origin of the coordinate system. This is extremely convenient in that which follows. The remainder of the particle Hamiltonian, represented by V, should have been minimized in the region of the excited atom core by our choice of V_0. Our notation is established with the following equations:

$$G(E) = G_0(E) + G_0(E)T(E)G_0(E), \tag{3.3}$$

where

$$G_0(E) \equiv (E - H_0)^{-1} \tag{3.4}$$

and

$$\begin{aligned} T(E) &\equiv V + VG(E)V \\ &= V + VG_0(E)T(E). \end{aligned} \tag{3.5}$$

The substitution of Eq. (3.3) into Eq. (2.12) allows the separation of σ_{total} into two parts:

$$\sigma_{\text{total}}(\omega) = \sigma^0_{\text{total}}(\omega) + \sigma^1_{\text{total}}(\omega), \tag{3.6}$$

[35] M. Lax, *Rev. Mod. Phys.* **23,** 287 (1951).

[36] J. L. Beeby and S. F. Edwards, *Proc. R. Soc. London, Ser. A* **274,** 395 (1963).

[37] The choice of starting point would not be important if we could carry out the calculations exactly, since Schaich[25] has demonstrated the formal equivalence of the two approaches to solving this problem.

where

$$\sigma^0_{\text{total}}(\omega) \equiv -4\pi\alpha_{\text{fs}}\hbar\omega \ \text{Im}\langle g|\hat{\epsilon}_{\mathbf{k}s}\cdot\mathbf{r}\ G_0(E)\ \hat{\epsilon}_{\mathbf{k}s}\cdot\mathbf{r}|g\rangle, \tag{3.7}$$

$$\sigma^1_{\text{total}}(\omega) \equiv -4\pi\alpha_{\text{fs}}\hbar\omega \ \text{Im}\langle g|\hat{\epsilon}_{\mathbf{k}s}\cdot\mathbf{r}\ G_0(E)T(E)G_0(E)\ \hat{\epsilon}_{\mathbf{k}s}\cdot\mathbf{r}|g\rangle, \tag{3.8}$$

and $E = E_g + \hbar\omega$. In these expressions, σ^0 is a slowly varying atomlike contribution to the total cross section, essentially featureless except for the steps in absorption at each of the thresholds for the excitation of a core level. The oscillations in σ_{total} arise from σ^1. Note that this expression for σ^1 is exact to all orders in the true particle Hamiltonian, but is correct to only second order in the particle–photon interaction.

The derivation of a more explicit expression for the photoabsorption cross section requires an explicit formula for $G_0^{\pm}(E)$. Let us consider only those approximations of the many-body electron states in which the eigenfunctions can be expressed as products of one-electron eigenfunctions. It is then possible to expand G_0 in terms of the one-electron eigenstates of H_0. The behavior of V_0 at asymptotically large r is important in that which follows. Let E_0 be the asymptotic value of V_0 as $r \rightarrow \infty$.[32] E_0 must be spatially invariant, but may be energy dependent and/or complex. For a general spherically symmetric potential V_0 for which $r(V_0 - E_0) \rightarrow 0$ as $r \rightarrow \infty$, it is easily shown[33] that $G_0^{\pm}$ can be written as

$$\langle\mathbf{r}|G_0^{\pm}(E)|\mathbf{r}'\rangle = -\frac{2m}{\hbar^2 k}(rr')^{-1} \times \sum_{lm} \exp[\pm i\eta_l(k)]F_l(k;r_<)u_l^{\pm}(k;r_>)Y_{lm}(\Omega_{\mathbf{r}})Y^*_{lm}(\Omega_{\mathbf{r}'}), \tag{3.9}$$

where $E = E_0 + \hbar^2k^2/2m$ and $r_<$ and $r_>$ represent the smaller and larger of r and r', respectively. F_l is the solution of the radial portion of H_0 which is regular at the origin and proportional to $\sin(kr - \tfrac{1}{2}l\pi + \eta_l)$ as $r \rightarrow \infty$. η_l is the increase in the phase of the regular solution of angular momentum l and energy $E_0 + \hbar^2k^2/2m$ at large r due to the presence of V_0 (or, equivalently, of energy $\hbar^2k^2/2m$ in the presence of $V_0 - E_0$). $u_l^{\pm}$ are the "outgoing" and "incoming" wave solutions, proportional to $\exp[\pm i(kr - \tfrac{1}{2}l\pi)]$ as $r \rightarrow \infty$. Note that

$$\text{Im}[\exp(\pm i\eta_l)u_l^{\pm}(k;r)] = \pm F_l(k;r). \tag{3.10}$$

$Y_{lm}(\Omega_{\mathbf{r}})$ is the usual spherical harmonic of argument equal to the angle of $\mathbf{r}$. Note that our choice of a spherically symmetric H_0 has led to a G_0 which is diagonal in lm. The inclusion of inelastic effects through an imaginary part of V_0 would result in an imaginary part in k, corresponding to attenuation of the electron wave. The generalization of Eq. (3.9) for the case where $r(V_0 - E_0)$ approaches a constant is straightforward.[33] In the more familiar free-space limit (i.e., $V_0 = 0$), $\eta_l = 0$, $F_l(k;r)$ is replaced by $kr\,j_l(kr)$, and $u_l^{\mp}(k;r)$ becomes $kr\,h_l^{\mp}(kr)$, the usual spherical Bessel function and Hankel functions of the first and second kinds, respectively.[38]

Let us now consider that subset of all the contributions to $\sigma_{\alpha;\text{total}}$ in which the absorption of a photon corresponds to the promotion of one electron from the nl shell of an atom of species α. This is the only type of excitation channel rigorously available in a truly one-electron model. We replace $\langle \mathbf{r}|g\rangle$ with $R_{\alpha;nl}(r)Y_{lm}(\Omega_\mathbf{r})$ in Eqs. (3.7) and (3.8) and sum the resulting contributions to σ_{total} over all m and s (assuming a full nl shell). Using the rules of LS (Russell–Saunders) coupling, the sum over the spin quantum number s yields a factor of 2. With these changes, $\sigma^0_{\alpha;nl}$ is proportional to

$$\operatorname{Im} 2\sum_m \int d\mathbf{r}\int d\mathbf{r}'\, R_{\alpha;nl}(r)Y^*_{lm}(\Omega_\mathbf{r})\,\hat{\epsilon}_{\mathbf{k}s}\cdot\mathbf{r}\,(rr')^{-1}\sum_{l'm'}\exp(+i\eta_{l'})$$

$$\times F_{l'}(k;r_<)u^+_{l'}(k;r_>)Y_{l'm'}(\Omega_\mathbf{r})Y^*_{l'm'}(\Omega_{\mathbf{r}'})\,\hat{\epsilon}_{\mathbf{k}s}\cdot\mathbf{r}'\,R_{\alpha;nl}(r')Y_{lm}(\Omega_{\mathbf{r}'}). \tag{3.11}$$

The angular integrals are all of the form

$$M_\Omega(l'm'lm) \equiv \int d\Omega_\mathbf{r}\; Y^*_{l'm'}(\Omega_\mathbf{r})\,\hat{\epsilon}_{\mathbf{k}s}\cdot\hat{\mathbf{r}}\; Y_{lm}(\Omega_\mathbf{r}), \tag{3.12}$$

where $\hat{\mathbf{r}}$ is the unit vector in the direction of $\mathbf{r}$. These integrals are independent of E, and related simply to the Clebsch–Gordan coefficients and to the "$3j$" symbols of Wigner. Note that $M_\Omega(l'm'lm)$ vanishes unless $l' = l \pm 1$ and $m' = m,\ m \pm 1$, the customary dipole selection rule. Furthermore, the sum of $|M_\Omega|^2$ over m and m' in $\sigma^0_{\alpha;nl}$ can be evaluated explicitly:

$$2\sum_{mm'}|M_\Omega(l'm'lm)|^2 = (l' + l + 1)/3. \tag{3.13}$$

It is convenient to include the relevant coefficients of the cross section in defining the analogous radial matrix element:

$$M_r(l'lE) \equiv \left((4\pi\alpha_{\text{fs}}\hbar\omega)\,\frac{2mk}{\hbar^2}\,\frac{l'+l+1}{3}\right)^{1/2}$$

$$\times \int dr\, r^2\,\frac{1}{kr}\,F_{l'}(k;r)rR_{\alpha;nl}(r). \tag{3.14}$$

With these definitions, given Eq. (3.10) and the facts that R and F are real, we can write the following expression for the slowly varying portion of the absorption cross section:

$$\sigma^0_{\alpha;nl}(\omega) = \sum_{l'=l\pm1} M_r^2(l'lE). \tag{3.15}$$

[38] A. Messiah, "Quantum Mechanics," Vol. I, Appendix B. North-Holland, Amsterdam, 1961.

Turning our attention to σ^1, we see that G_0 is always operated upon by a core wave function on one side and T on the other. Specifically, σ^1 consists of terms involving factors like

$$\int d\mathbf{r} \int d\mathbf{r}' \, R_{\alpha;nl}(r) Y^*_{lm}(\Omega_\mathbf{r}) \, \hat{\epsilon}_{\mathbf{k}s} \cdot \mathbf{r} \, (rr')^{-1} \sum_{l'm'} \exp(+i\eta_{l'}) F_{l'}(k\,;\, r_<)$$
$$\times\, u^+_{l'}(k\,;\, r_>) Y_{l'm'}(\Omega_\mathbf{r}) Y^*_{l'm'}(\Omega_{\mathbf{r}'}) \langle \mathbf{r}'|T(E)|\mathbf{r}''\rangle \times \cdots. \tag{3.16}$$

Since the potential of the excited atom is included in V_0, the true particle potential can be divided between V_0 and V in such a way that the potential V [and hence T by means of Eq. (3.5)] will vanish at $\mathbf{r} = 0$. To the extent that V can be chosen to vanish wherever the core-state charge density is nonnegligible, then r is always less than r' in expression (3.16). Accordingly, F_l is always associated with the core-state portion of the integral, and u_l with the T portion. We may then write

$$\sigma^1_{\alpha;nl}(\omega) = \sum_{l',l''=l\pm 1} M_r(l'lE) K_1(l'l''lE) M_r(l''lE), \tag{3.17}$$

where

$$K_1(l'l''lE) \equiv -\frac{2mk}{\hbar^2}(4\pi)^{-1} \,\mathrm{Im} \int d\mathbf{r} \int d\mathbf{r}' \, \frac{1}{kr} \exp(+i\eta_{l'}) u^+_{l'}(k\,;\, r)$$
$$\times \langle \mathbf{r}|T(E)|\mathbf{r}'\rangle K_2(l'l''l\Omega_\mathbf{r}\Omega_{\mathbf{r}'}) \frac{1}{kr'} \exp(+i\eta_{l''}) u^+_{l''}(k\,;\, r') \tag{3.18}$$

and

$$K_2(l'l''l\Omega_\mathbf{r}\Omega_{\mathbf{r}'}) \equiv 4\pi \times 3[(l'+l+1)(l''+l+1)]^{-1/2}$$
$$\times\, 2 \sum_{mm'm''} M_\Omega(lml'm') Y^*_{l'm'}(\Omega_\mathbf{r}) Y_{l''m''}(\Omega_{\mathbf{r}'}) M_\Omega(l''m''lm). \tag{3.19}$$

Expressions (3.15) and (3.17) are quite general, restricted principally by the assumption of one-electron final states. In particular, their sum constitutes an expression for the photoabsorption cross section σ which is correct to all orders in the particle Hamiltonian. The effect on σ of the scattering of the photoelectron by atoms other than the excited atom is contained in the K_1 factor of Eq. (3.17), while the effect of the excited atom itself is contained in the M_r factors of Eqs. (3.15) and (3.17). In deriving these expressions, we have used the rules of *LS* (Russell–Saunders) coupling. Note, however, that spin–orbit coupling is important in many cases of interest. The resulting eigenstates would obey the rules of intermediate or even *jj* coupling, which would necessitate straightforward changes in this derivation.

The principal quantity of interest in EXAFS spectroscopy is the structural information contained in T. It is therefore convenient to define an EXAFS function χ as σ^1 normalized to the smooth atom-like absorption represented by σ^0. Using Eqs. (3.15) and (3.17) for σ^0 and σ^1, respectively, we can write

$$\chi_{\alpha;nl}(\omega) \equiv \sigma^1_{\alpha;nl}(\omega)/\sigma^0_{\alpha;nl}(\omega)$$
$$= \frac{\sum\limits_{l',l''=l\pm1} M_r(l'lE)K_1(l'l''lE)M_r(l''lE)}{\sum\limits_{l'=l\pm1} M_r^2(l'lE)} \tag{3.20}$$

With this definition for χ, the total photoabsorption cross section arising from the nl shell of atom α may be written as

$$\sigma_{\alpha;nl}(\omega) = \sigma^0_{\alpha;nl}(\omega)[1 + \chi_{\alpha;nl}(\omega)]. \tag{3.21}$$

These expressions for the photoabsorption cross section and for the EXAFS function χ are relatively explicit, with the exception of T. M_Ω and M_r may be calculated in a straightforward fashion, although the latter does require determining the eigenfunctions of H_0. The remaining approximations in this section are directed toward obtaining a more easily evaluated form for K_1. They are of two distinct and largely orthogonal types: treatments of the full multiple-scattering problem to obtain an easily evaluated but approximate expression for T and approximations to the "outgoing" wave u_l^+ to simplify evaluation of the particular matrix elements of T which are found in K_1. We shall treat the multiple-scattering problem first since it is the more fundamental of the two.

In the spirit of the short-range order expansion, we denote the single-site transition matrix of the atom at $\mathbf{r}_i$ by t_i, where

$$t_i(E) \equiv v_i + v_i G_0(E) t_i(E) \tag{3.22}$$

and v_i is that portion of V arising from the atom at $\mathbf{r}_i$. Lax[35] and Beeby and Edwards[36] have shown that T can be expressed as

$$T(E) = \sum_i t_i(E) + \sum_{i\neq j} t_i(E)G_0(E)t_j(E) + \cdots, \tag{3.23}$$

where all sums exclude any terms where any particular t_i occurs twice with only one intervening G_0 (i.e., all terms containing a factor like $t_iG_0t_i$). In this instance, we also exclude the atom at the origin (which is included in H_0 in our model). Each successive term in Eq. (3.23) involves an additional electron scattering event and an additional transit between two atoms, separated by at least a nearest-neighbor distance. The expansion is expected to converge quickly if the electron mean free path is short, as is the case here. For the present, we neglect all multiple-atom scattering and use only the first term in Eq. (3.23). It is convenient at this point to replace the sum of t_i over all scattering sites with an integral over the pair correlation function

$\rho_{\alpha\beta}(\mathbf{r})$ which expresses the positions of each atom of species β with respect to the excited atom species α at $\mathbf{r} = 0$:

$$\begin{aligned}\langle \mathbf{r}|T(E)|\mathbf{r}'\rangle &\rightarrow \sum_i \langle \mathbf{r}|t_i(\mathbf{r}_i; E)|\mathbf{r}'\rangle \\ &= \sum_i \langle \mathbf{r} - \mathbf{r}_i|t_i(0; E)|\mathbf{r}' - \mathbf{r}_i\rangle \\ &= \sum_\beta \int d\mathbf{r}''\, \rho_{\alpha\beta}(\mathbf{r}'')\, \langle \mathbf{r} - \mathbf{r}''|t_\beta(0; E)|\mathbf{r}' - \mathbf{r}''\rangle. \end{aligned} \tag{3.24}$$

In this expression, the sum includes all atom species β. $\rho_{\alpha\beta}$ has been averaged over all atoms of the excited atom species α, just as is the measured absorption cross section.

Let us turn to the "outgoing" wave u_l^+. The separation of the full particle potential into V_0 and V was designed to produce a V and hence a T which nearly vanish in the region of the excited atom core, where the spatially varying portion of V_0 (i.e., $V_0 - E_0$) is nonzero. If the residual overlap of T with $V_0 - E_0$ has been made sufficiently small, then u_l^+ can be approximated in Eq. (3.18) with a phase-shifted Hankel function of the first kind, h_l^+. The effect of inelastic scattering of the "outgoing" electron wave in the vicinity of an atom can be incorporated into the complex phase shifts associated with that atom's potential, as will be discussed in Section 5. The remaining inelastic processes in the interatomic regions lead to an additional factor in the final-state wave function similar to $\exp[-r/\lambda(k)]$, where λ is a mean-free-path length. Finally, we take the limit where the solid angle subtended at the excited atom by the scattering atom potential is vanishingly small, the so-called "small-atom" approximation. In this limit, the spatial dependence of the "outgoing" wave at the site of the scattering atom, $\mathbf{r}''$, can be approximated with a plane wave with equal amplitude and phase. In summary,

$$\begin{aligned}&(1/kr)\exp(+i\eta_l)u_l^+(k; r)Y_{lm}(\Omega_\mathbf{r}) \\ &\quad\rightarrow \exp(+i\eta_l)h_l^+(kr)Y_{lm}(\Omega_\mathbf{r}) \\ &\quad\rightarrow \exp[+i\eta_l - r/\lambda(k)]h_l^+(kr)Y_{lm}(\Omega_\mathbf{r}) \\ &\quad\rightarrow \exp[+i\eta_l - r''/\lambda(k)]h_l^+(kr'')Y_{lm}(\Omega_{\mathbf{r}''})\exp[i\mathbf{k}\cdot(\mathbf{r} - \mathbf{r}'')], \end{aligned} \tag{3.25}$$

where k is now taken to be real and $\mathbf{k}$ has the direction of $\mathbf{r}''$. Recall that

$$h_l^+(x) = (-i)^l[\exp(ix)/x]v_l^+(x), \tag{3.26}$$

where $v_l^+(x) \rightarrow 1$ as $x \rightarrow \infty$.[38] The validity of each of these substitutions is discussed further in Section 4.

Let us return to Eq. (3.18) for K_1 and introduce into it the single-atom scattering approximation to T given by expression (3.24) and the plane wave approximation to the outgoing wave given by expression (3.25). The $\mathbf{r}$ and $\mathbf{r}'$ integrations of Eq. (3.18) yield the matrix element of t_β between plane waves. Only the $\mathbf{r}''$ integration in expression (3.24) survives, so that K_1 can be written as

$$kK_1(l'l''lE) = (-i)^{l'+l''+2} 2\mathrm{Re} \sum_\beta \int dr \exp(+2ikr)\Lambda_{\alpha\beta}(l'l''kr) \times \int d\Omega_{\mathbf{r}}\, \rho_{\alpha\beta}(\mathbf{r}) K_2(l'l''l\Omega_{\mathbf{r}}\Omega_{\mathbf{r}}), \tag{3.27}$$

where

$$\Lambda_{\alpha\beta}(l'l''kr) \equiv (i/2)(-2\pi\hbar^2/m)^{-1}(2\pi)^3 t_\beta^+(-\mathbf{k}, \mathbf{k}) \times v_{l'}^+(kr) v_{l''}^+(kr) \exp[+i\eta_{l'} + i\eta_{l''} - 2r/\lambda(k)]. \tag{3.28}$$

In these expressions, $t_\beta^+(-\mathbf{k}, \mathbf{k})$ is the matrix element of t between wave vector normalized plane waves for an atom of species β located at the origin. It is related to the electron scattering amplitude $f_\beta(k, \theta)$ for a scattering angle of 180°:

$$(2\pi)^3 t_\beta(-\mathbf{k}, \mathbf{k}) \equiv (2\pi)^3 \langle -\mathbf{k} | t_\beta(0; E) | \mathbf{k} \rangle = -(2\pi\hbar^2/m) f_\beta(k, \pi), \tag{3.29}$$

where $f_\beta(k, \pi) = f_\beta(-\mathbf{k}, \mathbf{k})$ and $\langle \mathbf{r} | \mathbf{k} \rangle \equiv (2\pi)^{-3/2} \exp(i\mathbf{k} \cdot \mathbf{r})$.

Equations (3.20) and (3.27) together constitute a single-scattering expression for the EXAFS which can be evaluated without great difficulty and which is, at the same time, sufficiently accurate for many applications. The "small-atom" approximation has resulted in a particularly simple form for K_2 [appearing in Eq. (3.27)], provided that we choose the z axis for our coordinate system to lie parallel to the electric field vector of the incident photons $\hat{\epsilon}$, which we now do. Accordingly, we may write explicit expressions for the EXAFS oscillations $\chi = \sigma^1/\sigma^0$. The photoexcitation of an s-symmetry core electron ($l = 0$), such as from the K shell, is the most common situation in EXAFS. In that case, the above approximations lead to the following expression:

$$k\chi_{\alpha;n0}(\omega) = 2\mathrm{Re} \sum_\beta \int dr \exp(+2ikr)\Lambda_{\alpha\beta}(11kr) \int d\Omega_{\mathbf{r}}\, \rho_{\alpha\beta}(\mathbf{r}) 3\cos^2\theta, \tag{3.30}$$

where θ is the angle between $\hat{\epsilon}$ and $\mathbf{r}$. For an amorphous or randomly oriented polycrystalline sample, we can replace $\rho_{\alpha\beta}$ with its spherical average:

$$p_{\alpha\beta}(r) = 4\pi r^2 \langle \rho_{\alpha\beta}(\mathbf{r}) \rangle_{\Omega_{\mathbf{r}}}, \tag{3.31}$$

where $p_{\alpha\beta}$ is the radial correlation function of β atoms about α atoms, normalized so that

$$\int dr\, p_{\alpha\beta}(r) = N_\beta\,, \tag{3.32}$$

the total number of β atoms in the sample. The angular integration in Eq. (3.30) can be performed to yield

$$k\chi_{\alpha;n0}(\omega) = \sum_\beta \int dr\, r^{-2}\, p_{\alpha\beta}(r)\, 2\mathrm{Re}[\exp(+2ikr)\Lambda_{\alpha\beta}(11kr)]. \tag{3.33}$$

Note that this expression is equivalent to Eq. (1.2) in the limit where $v_1{}^+(kr) \to 1$ (i.e., $kr \to \infty$).

The expressions are somewhat more complicated for the photoexcitation of a p-symmetry core electron ($l = 1$). For a general $\rho_{\alpha\beta}$,

$$\begin{aligned} k\chi_{\alpha;n1}(\omega) = -2\mathrm{Re} \sum_\beta \int dr \exp(+2ikr) \int d\Omega_{\mathbf{r}}\, \rho_{\alpha\beta}(\mathbf{r}) \\ \times [M_r^2(01E)\Lambda_{\alpha\beta}(00kr) + M_r^2(21E)\Lambda_{\alpha\beta}(22kr)(3\cos^2\theta + 1)/2 \\ - 2M_r(01E)M_r(21E)\Lambda_{\alpha\beta}(02kr)(3\cos^2\theta - 1)/2^{1/2}] \\ \times [M_r^2(01E) + M_r^2(21E)]^{-1}. \end{aligned} \tag{3.34}$$

[The function M_r is defined by Eq. (3.14).] Specializing to the spherically symmetric case, we obtain

$$\begin{aligned} k\chi_{\alpha;n1}(\omega) = -\sum_\beta \int dr\, r^{-2}\, p_{\alpha\beta}(r)\, 2\mathrm{Re}\{\exp(+2ikr) \\ \times [M_r^2(01E)\Lambda_{\alpha\beta}(00kr) + M_r^2(21E)\Lambda_{\alpha\beta}(22kr)] \\ \times [M_r^2(01E) + M_r^2(21E)]^{-1}\}. \end{aligned} \tag{3.35}$$

Note that the cross term has vanished. This term represents, for example, the contribution of an "outgoing" s wave which is scattered back toward the excited atom as a d wave. The contribution of this scattering mechanism averages to zero in the randomly oriented polycrystalline case, because our expression is only to first order in $\rho_{\alpha\beta}$ (i.e., we neglected multiple-atom scattering). Expressions (3.34) and (3.35) can be simplified further by neglecting the contribution of the p $\to$ s excitation relative to that of the p $\to$ d excitation.[39–41] Note that spin–orbit coupling is often significant in determining the L-shell eigenstates in cases of interest, leading to a splitting

[39] U. Fano and J. W. Cooper, *Rev. Mod. Phys.* **40,** 441 (1968).

[40] B.-K. Teo and P. A. Lee, *J. Am. Chem. Soc.* **101,** 2815 (1979).

[41] Fano and Cooper[39] argued that $|M_r(01E)/M_r(21E)|$ is typically of order 0.3. Teo and Lee[40] have calculated it in several cases and found it to be ≈ 0.15.

of the 2p excitations between L_{II} and L_{III} absorption edges. The dominant effect in the EXAFS region is to alter the relative strengths of their respective σ^0's.[42]

Expressions similar to Eqs. (3.30) and (3.33)–(3.35) have been obtained by others.[27–29,43] The most significant difference is that we have made no unnecessary assumptions regarding the form of the interatomic pair correlation function $\rho_{\alpha\beta}(\mathbf{r})$. In this connection, it should be stressed that in the evaluation of Eqs. (3.30) and (3.34) the z axis in the coordinate system used to determine $\rho_{\alpha\beta}(\mathbf{r})$ is oriented parallel to the electric field vector of the incident photons $\hat{\epsilon}$, so that θ is the angle between $\hat{\epsilon}$ and $\mathbf{r}$.

Finally, it is interesting to note that the experimental data can be manipulated in most instances to yield a quantity equivalent to Eqs. (3.30) and (3.33)–(3.35), *except* that the real part operator Re is absent. This manipulation utilizes Fourier transforms and depends on the intrinsic structure of the EXAFS equations in a interesting way. Let us adopt the half-space Fourier transform

$$f(r) = (2\pi)^{-1/2} \int_0^\infty d(2k) \exp(-2ikr) F(k),$$

$$F(k) = (2\pi)^{-1/2} \int_{-\infty}^\infty dr \exp(+2ikr) f(r). \tag{3.36}$$

It is convenient to restrict this discussion to K-shell excitation in a polycrystalline sample, using $\chi_{\alpha;n0}$ as defined by Eq. (3.33). Since ω, E, and k are interrelated by $E = E_{\alpha;n0} + \hbar\omega = E_0 + \hbar^2k^2/2m$, $\chi_{\alpha;n0}(\omega)$ may be treated as an implicit function of k.[44] The Fourier transform of $k\chi_{\alpha;n0}(k)$ may then be expressed as

$$\varphi_{\alpha;n0}(r) = \sum_\beta \int_0^\infty dr'\, r'^{-2} \rho_{\alpha\beta}(r') [\xi_{\alpha\beta}(r - r', r') + \xi^*_{\alpha\beta}(-r - r', r')], \tag{3.37}$$

where

$$\xi_{\alpha\beta}(r, r') = (2\pi)^{-1/2} \int_0^\infty d(2k) \exp(-2ikr) W(k) \Lambda_{\alpha\beta}(11kr'). \tag{3.38}$$

As represented in Eq. (3.37), the Fourier transform of the EXAFS is the convolution of the pair correlation function $p(r)$ with a peak function $\xi(r)$. Analogous to the peak function of a diffraction experiment, ξ is the Fourier

[42] For a discussion, see Lee *et al.*[9]

[43] For p-symmetry core electron excitations, see S. M. Heald and E. A. Stern, *Phys. Rev. B: Solid State* [3] **16,** 5549 (1977).

[44] See Sections 5 and 10 for a discussion of this interrelationship.

transform of the product of Λ with a window function $W(k)$ representing the limited energy range of the experiment. Although the presence of W is irrelevant to the present discussion, we have introduced it into the definition of ξ so that these expressions will be applicable in the discussion of data analysis. Let us denote the peaks in $p_{\alpha\beta}$ by the set $\{r\}_{\alpha\beta}$. If these peaks are sufficiently sharp, Eq. (3.37) implies that $\varphi(r)$ is essentially a linear combination of $\xi_{\alpha\beta}$ peaks at $r = \{r\}_{\alpha\beta}$ and $\xi^*_{\alpha\beta}$ peaks at $r = -\{r\}_{\alpha\beta}$. In general, however, the contributions of these two sets of peaks to φ are separate from one another. This occurs because $|\xi(r, r')|$ reaches a maximum at $r \simeq -0.25$ Å and is typically only 0.5–0.7 Å wide. Since the position of the first-neighbor peak in $p(r)$ is often ≈ 2.5 Å, it is usually a good approximation to assume that the contribution of the peaks at $r = \{r\}_{\alpha\beta}$ is confined to $r > 0$, while that of the peaks at $r = -\{r\}_{\alpha\beta}$ is confined to $r < 0$. In this limit, Eq. (3.37) reduces to Eq. (1.5) for $r > 0$.

We can take advantage of this essentially complete separation of the peaks in φ into two groups to transform back into k space the contributions from only one of them. If the back-transform includes only $r > 0$ and there is no overlap of ξ^* peaks into this area, then the result is given by

$$\sum_{\beta} \int dr\, r^{-2}\, p_{\alpha\beta}(r) \exp(+2ikr)\Lambda_{\alpha\beta}(11kr). \tag{3.39}$$

A comparison with Eq. (3.33) reveals that the real part operator Re is no longer present. In other words, the phase of $e^{+2ikr}\Lambda$ has been "recovered." One important practical significance of this is increased ease in extracting the phase and amplitude of Λ, which are useful in comparisons between experimental and calculated spectra.

Inspection of Eq. (3.28) reveals that the r' variable in $\xi(r, r')$ arises from two sources: the asymptotic form of the outgoing electron wave function, approximated by v and the inelastic processes approximated by the mean-free-path length λ. While both of these effects are properly functions of the position of the scattering atom, r', neither of them is expected to affect substantially comparisons among experimental EXAFS signals arising from scattering atoms at similar distances (e.g., nearest neighbors only). Accordingly, the r' variable is often suppressed as in Eq. (1.5). Note, however, that these effects will be important in comparisons between signals from atoms at different distances, or between theory and experiment, and are discussed in more detail in Section 4,c.

Equations (3.30) and (3.33)–(3.35) are relatively general expressions for the EXAFS oscillations, applicable in *most* cases. Occasionally, however, one or more of the underlying approximations will not be appropriate. We now turn to an examination of these approximations.

4. Underlying Approximations

The derivation of Eqs. (3.30) and (3.33)–(3.35) required that several assumptions be made regarding the importance of various physical phenomena. The neglect of many-body effects is the first approximation to require further discussion. We will examine in this section those modifications of our model for the photoabsorption cross section which would be necessary to account correctly for many-body phenomena, relying heavily on the extensive literature of atomic photoabsorption. Next we examine those approximations peculiar to photoabsorption in molecules and condensed matter, including our treatments of multiple-atom scattering, of the overlap of atomic potentials, of the inelastic processes which limit the lifetime of the final electron eigenstate, and of the spherical nature of the outgoing electron wave.

a. Many-Body Effects

In a proper many-body treatment of the particle field, the Coulomb interactions between electrons lead to eigenstates in which all of the electrons are intrinsically coupled. Photoabsorption will result in transitions among these many-electron eigenstates. As might be expected, there exist many channels for excitation of the particle field which are absent from a one-electron treatment such as we presented in Section 3. The primary channel for excitation corresponds in most instances to the process which we have already considered—the promotion of a single core electron to a continuum final state. Even so, the photoabsorption cross section is often modified substantially by the other excitation channels. In discussing these modifications, we rely heavily on the treatment of these many-body effects in the context of a one-electron model for atomic photoabsorption by Fano and Cooper.[39] These authors discussed several distinct consequences of many-body interactions. For example, resonances at the threshold of excitation from a core shell can be modified strongly by Coulomb interactions between the continuum electrons, often resulting in collective excitations (e.g., plasmons). Additionally, there can be localized eigenstates which are degenerate with the continuum states in the vicinity of threshold, resulting in intermixing and autoionization. Both of these effects give rise to spectral features which are conceptually distinct from the EXAFS phenomena on which this treatment is focused. Furthermore, they do not affect the photoabsorption cross section in the "high-energy" region (beginning 20–30 eV above threshold), where a single-scattering analysis of the EXAFS is most likely to yield reliable structural information. Accordingly, while of great interest in their own rights, both of these phenomena are outside the scope of this treatment. Other many-body effects arise from the response of the electrons of the excited atom core to the departure of the primary photoelectron. They can have a substantial influence on the photoabsorption spectrum in the EXAFS region and will be considered here in some detail.

The relaxation of the core electrons of the excited atom becomes an issue even in a one-electron treatment of X-ray photoabsorption, because the effective one-electron potentials are modified substantially by the creation of a hole in the atomic core. The calculated photoelectron eigenstate (and therefore the one-electron transition rate) will vary significantly with the potential used to calculate it, which could be derived from the atom in its initial state or in a state which reflects partial or complete relaxation to the presence of the core hole.[45] Arguments in support of the choice of one of these potentials are often based on considerations of time scale, comparing that during which the photoexcited core electron "diffuses" away from the excited atom core with those required for the response of the core electrons to the corresponding change in potential. If the photoelectron diffuses away very rapidly (relative to the response times of the core electrons), then it has been argued that a potential derived from the initial state of the atom would be appropriate (e.g., see Ref. 46). If the photoelectron's departure is slow, then the proper final-state wave function might well reflect a complete or partial relaxation of the core electrons of the excited atom to the presence of the core hole. In addition to modifying the one-electron transition rates in this way, the departure of the primary photoelectron can lead to the excitation of additional electrons, forbidden transitions in a strictly one-electron treatment of photoabsorption.[47] In the terminology of atomic photoabsorption spectroscopy, these additional excitations are referred to as "shake-up" or "shake-off" transitions, depending on whether the additional electron is excited to a bound state or to a continuum state, respectively. The physical mechanism which underlies "shake" transitions is examined in this subsection together with the conditions under which the overall response of the core electrons could have significant effects on the photoabsorption cross section. The implications of these effects for the interpretation of EXAFS data are discussed.

The mechanism by which the departure of the primary photoelectron can excite additional electrons is brought out in the following application

[45] Note that experimental linewidths suggest that the filling of the core hole occurs after a much longer time, and does not affect the EXAFS except in providing an upper bound to the final-state electron lifetime.

[46] P. A. Lee and G. Beni, *Phys. Rev. B: Solid State* [3] **15,** 2862 (1977).

[47] Note that many-body interactions also give rise to a class of multiple-electron excitations that *cannot* be described in terms of core relaxation. This process is essential to explain the double-electron excitations in the He spectrum, for example, but is presumably not as important as core relaxation in cases such as we are concerned with here, where the primary photoelectron comes from a deep core level.[39] We shall not discuss it in any event.

of time-dependent perturbation theory. The Coulomb potential carried away by the escaping photoelectron contributes a time-dependent term to the effective one-electron potential acting on the remaining core electrons of the excited atom. The response of the core electrons to this stimulus will modify in turn the effective one-electron potential acting on the photoelectron. A proper self-consistent treatment of all these changes in the one-electron Hamiltonian as a function of time will clearly be quite difficult. It is possible to explore some of the interesting ramifications of the time dependence itself, however, in the much simpler problem of a system described by H_0 to which is applied (by some external source) a time-dependent perturbing potential $V(t)$. Consider a potential which vanishes for $t \leq 0$ and approaches a constant as $t \longrightarrow \infty$. The evolution of the nth eigenstate as a function of t, $\varphi_n(t)$, can be examined by expanding it in the eigenfunctions $\varphi_k^{(0)}$ of the unperturbed Hamiltonian H_0:

$$\varphi_n(t) = \sum_k a_{kn}(t)\varphi_k^{(0)}. \tag{4.1}$$

It may be shown[48] that $a_{kn}(t)$ is given by

$$a_{kn}(t) = -\frac{i}{\hbar}\int_0^t dt'\,\exp(i\omega_{kn}t')V_{kn}(t'), \tag{4.2}$$

where $V_{kn}(t) \equiv \langle k|V(t)|n\rangle$ and $\hbar\omega_{kn} \equiv E_k - E_n$, the difference in energy between the levels. Integrating by parts, one obtains

$$a_{kn}(t) = \frac{-\exp(i\omega_{kn}t)V_{kn}(t)}{\hbar\omega_{kn}} + (\hbar\omega_{kn})^{-1}\int_0^t dt'\,\exp(i\omega_{kn}t')\frac{\partial V_{kn}(t')}{\partial t'}. \tag{4.3}$$

Equations (4.1) and (4.3) describe the response of the nth eigenstate to the changing potential $V(t)$.

The first term on the right-hand side of Eq. (4.3) corresponds to the expression which one would have obtained for a_{kn} at some instant t by ignoring the time dependence of $V(t)$ and using time-independent perturbation theory to first order in the instantaneous value of V. It represents the instantaneous relaxation of the nth eigenstate to the potential in effect at any given time. The second term is an additional contribution to a_{kn} which depends directly on the rate at which the potential changes with time. Since the correct nth eigenstate of $H_0 + V(t)$ corresponds to the first term by itself, the second term on the right-hand side of Eq. (4.3) is properly regarded as describing an actual transition from n to k, driven by the time dependence of V. As an actual transition, it involves an energy transfer of $\hbar\omega_{kn}$ to the system from the external source of the change in potential $V(t)$.

[48] L. D. Landau and E. M. Lifshitz, "Quantum Mechanics: Non-Relativistic Theory," 3rd ed., Sect. 41. Pergamon, Oxford, 1977.

This is the mechanism by which the outgoing primary photoelectron can excite other electrons. The transition probability as $t \longrightarrow \infty$ due to this second term alone is

$$w_{kn} = (\hbar\omega_{kn})^{-2}\left|\int_0^\infty dt' \exp(i\omega_{kn}t')\frac{\partial V_{kn}(t')}{\partial t'}\right|^2. \tag{4.4}$$

It can be evaluated easily in two limiting situations, determined by the relative time scales of variations in $\partial V_{kn}/\partial t$ and $\exp(i\omega_{kn}t)$. If $\partial V_{kn}/\partial t$ varies little over a time of order $2\pi/\omega_{kn}$, then the integral in Eq. (4.4) will be very small. In the limit in which V_{kn} varies arbitrarily slowly, the probability of any transition with a change in energy (i.e., $\omega_{kn} \neq 0$) tends to zero. If $w_{kn} = 0$ for all k, the only effect of V on the nth eigenstate is a modification to reflect the instantaneous value of V. This constitutes an *adiabatic* response to $V(t)$ by the nth eigenstate. The opposite limit corresponds to a very rapid change in the potential. If $\partial V_{kn}/\partial t$ is nonzero only for some very short time ($\ll 2\pi/\omega_{kn}$), the exponential can be removed from the integral in Eq. (4.4) to yield a transition probability of

$$w_{kn} = |V_{kn}(t \longrightarrow \infty)/\hbar\omega_{kn}|^2. \tag{4.5}$$

This is the *sudden* limit, in which the rate for transitions from n to k is relatively easy to calculate. Note that an expression for w_{kn} in the sudden limit can be obtained for arbitrarily large V_{kn} as well[48]:

$$w_{kn} = \left|\int d\mathbf{r}\, \varphi_k^* \varphi_n^{(0)}\right|^2, \tag{4.6}$$

where φ_k are the eigenfunctions of $H = H_0 + V$. Between the sudden and adiabatic limits are a wide range of situations in which actual transitions are induced (i.e., $w_{kn} \neq 0$), but the sudden approximation is not adequate to calculate the transition rates. In all cases, the eigenstates will reflect the instantaneous value of the Hamiltonian by virtue of the mechanism represented by the first term on the right-hand side of Eq. (4.3).

In extending this discussion to encompass the photoabsorption process, it is clear that the answers to two questions should be sought. Is the response of the core electrons to the departing photoelectron simply an adiabatic relaxation of the eigenstates to ones recognizing the presence of the core hole, or will core electrons also be excited via the mechanism implicit in Eq. (4.4)? How will the departing photoelectron be affected by these changes in the core electrons? Unfortunately, a detailed analysis of photoexcitation based on Eqs. (4.3) and (4.4) is made quite difficult by the fact that the relevant time-dependent potential is not externally applied. It arises from the escape of a photoelectron, which, unlike the external source of $V(t)$ in the simple example, is not capable of more than a well-defined amount of work (i.e., it cannot give up more than a well-defined amount of energy). Furthermore, the potential must reflect self-consistently the response of the photoelectron to any changes in the core electrons caused by its escape. As a consequence, the appropriate time-dependent potential is rather difficult to specify. It *is* possible to gain some insight into the photoabsorption process, however, by examining directly the electron–electron interactions which underlie the response of the core electrons. Fano and Cooper[39] have derived approximate expressions for the matrix elements which couple the one-electron final states of interest in this context: that final state with a single excited photoelectron and those with an additional excited "shake" electron. These matrix elements are many-body corrections to the one-electron eigenstates which are calculated neglecting all changes in the effective Hamiltonian on photoexcitation. They account for changes in the mutual screening of the electrons due to the primary photoexcitation event, as well as those due to any accompanying "shake" transitions. In particular, the first of two terms in each matrix element reflects the effect on the core electrons (of the excited atom) of the difference between the wave function of the photoelectron eigenstate and that of the core eigenstate from which it came. The second term reflects the effect on the photoelectron of the difference between the wave function of a "shake" electron and that of the eigenstate from which *it* came.

The quasi-adiabatic limit corresponds to the absence of "shake" transitions.[49] In the treatment of Fano and Cooper,[39] this requires that all of the matrix elements linking the single-excited-electron final state to multiple-excited-electron final states be negligible. Given the substantial difference between the wave function of the photoelectron and that of the core eigenstate from which it came, the vanishing of the first term in each matrix element appears to be highly unlikely, *except* when no multiple-excited-electron states exist with the proper energy. In one-electron terms, this situation occurs whenever the least energy required to excite an additional electron is greater than the most energy which the photoelectron can give up, and is likely only for a final-state electron energy just at the threshold for photoexcitation from the *nl* core level.[50] Even in the absence of coupling to multiple-excited-electron final states, however, there will be nonvanishing matrix elements connecting the various one-electron final states corresponding to a single excited electron, leading to the quasi-adiabatic relaxation of the core electrons to the creation of the photoelectron. The final state would

[49] The term *quasi-adiabatic* is used in this context as a reminder that the relaxation of the core will in general require an exchange of energy between the departing photoelectron and the core electrons, even in the absence of a "shake" transition.

[50] "Shake" transitions may occur even at threshold, of course, if they are capable of giving energy to the photoelectron rather than taking energy from it.

therefore correspond to a single outgoing photoelectron with energy $E = E_{nl} + \hbar\omega$ and a wave function which reflects the relaxation of the core electrons to the transfer of charge from the nl core level to the photoelectron.

As the photon energy increases, the matrix elements corresponding to quasi-adiabatic relaxation will be joined by those corresponding to proper "shake" transitions as soon as energy conservation allows. These transitions result in energy being taken from (or given to) the photoelectron and therefore have a profound effect on the final state from the viewpoint of EXAFS analysis. It is necessary in that analysis to assign a wave vector k to the fine structure at each final-state energy E, typically through the relation $\hbar^2k^2/2m = E - E_0$, where E_0 is the zero of conduction electron energy. If a single "shake" transition occurs with a probability p_s less than unity, transferring energy E_s from the photoelectron to an electron in the $n'l'$ core state, then a final-state energy E corresponds to $1 - p_s$ electrons with $k^2 \propto E - E_0$, p_s electrons with $k^2 \propto E - E_s - E_0$, and p_s electrons with $k^2 \propto E_s + E_{n'l'} - E_0$. The simultaneous presence of several different final-state wave vectors could alter dramatically the interference effects which underlie the EXAFS, depending as they do on the phase change in the final-state wave function(s) over interatomic distances (i.e., $2kr_i$). This will be discussed below in the context of experiment. It should be emphasized that the quasi-adiabatic relaxation terms continue to operate in the presence of "shake" transitions, so that the proper outgoing electron wave functions will reflect the relaxation of the system to the transfer of charge from the nl core level to the photoelectron. The magnitude of the relaxation effect is a function of the residual screening of the core electrons by the photoelectron and will depend accordingly on photoelectron energy.

At still higher photon energies, one expects to enter the regime of the sudden approximation, in which the effects of quasi-adiabatic relaxation and of the "shake" transitions are especially amenable to calculation. Fano and Cooper[39] observed that this corresponds to neglecting completely the screening of the core electrons by the escaping photoelectron, as well as the effect of the "shake" transitions on the photoelectron. In a quasi-classical limit, they showed that the screening effect of the photoelectron on the nth core electron may be neglected completely only if

$$\tau(E, r_n) \ll \tau(E_n, r_n), \tag{4.7}$$

where r_n is the classical turning point for the nth electron,

$$\tau(E, r) = \int_0^r dr' \, [v(E, r')]^{-1}, \tag{4.8}$$

and

$$v(E, r) = \{(2/m)[E - V_0(r) - \hbar^2 l(l+1)/2mr^2]\}^{1/2}. \tag{4.9}$$

In these expressions, $\tau(E,r)$ is the time required for a quasi-classical particle of energy E and angular momentum l to traverse from the origin to r in the presence of some appropriate effective potential V_0. Accordingly, inequality (4.7) requires that the primary photoelectron be much "faster" than the nth electron for $r < r_n$. A more explicit statement is difficult to formulate, since the appropriate effective potential for a given core electron will be different in general from that for the departing photoelectron, which is presumably that appropriate to the core level from which it came. Nonetheless, it is apparent that inequality (4.7) will hold only for $E \gg |E_n|$, the binding energy of the nth electron. This condition is considerably more stringent than the requirement that the departing photoelectron be able to transfer enough energy to excite the nth core electron to an unoccupied level, which is approximately that $E \geq |E_n|$ if both energies are measured with respect to the lowest unoccupied electron states. We conclude accordingly that the strength of the shake transitions is amenable to description in the sudden approximation only if $E \gg |E_n|$. In other words, while many shallow levels may be candidates for shake transitions, few of these transitions lend themselves to treatment in the sudden approximation.

We conclude from the foregoing discussion that the final-state electron wave function must always reflect the relaxation of the excited atom core electrons to the change in mutual screening caused by the promotion of a core electron to a continuum final state. "Shake" transitions are expected to take place as soon as the departing electron is capable of providing the necessary energy, and may well have a profound effect on the EXAFS spectrum. Finally, these processes will be amenable to calculation in the sudden approximation only for photoelectrons with such high energy that their screening of the core electrons may be totally neglected. These conclusions are in qualitative agreement with the results of atomic spectroscopy. In addition, Williams and Shirley[51] have observed that the effects of shake-up and shake-off appear to saturate by 150–200 eV above threshold in the case of Ne. This suggests that the contributions to the shake spectrum of the more deeply bound states may be neglected, at least in the case of a light element.

The practical implications of the foregoing discussion for EXAFS spectroscopy are as follows. Let us examine first the effect of multiple-electron excitations, taking the case of Br_2, whose molecular photoabsorption spectrum was discussed at length by Rehr *et al.*[52] At final-state energies greater than 200 eV, an outgoing photoelectron is believed to generate shake-up

[51] R. S. Williams and D. A. Shirley, *J. Chem. Phys.* **66,** 2378 (1977).

[52] J. J. Rehr, E. A. Stern, R. L. Martin, and E. R. Davidson, *Phys. Rev. B: Solid State* [3] **17,** 560 (1978).

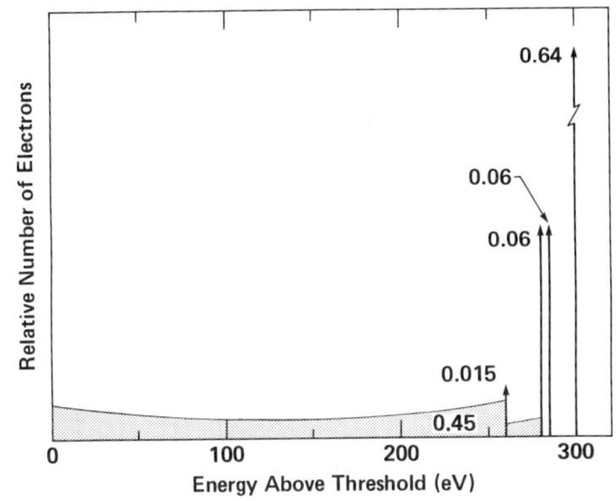

FIG. 6. Schematic spectrum of outgoing primary and "shake" electrons for Br_2 for a primary electron energy 300 eV above threshold. Primary electrons whose energies have been lowered by a "shake-up" transition are represented by δ functions at 15, 20, and 40 eV below 300 eV (the last resulting from a multiple-electron shake-up). "Shake-off" and multiple-electron shake-off processes lead to a continuum of outgoing primary and "shake" electrons, represented by shaded regions. Approximate intensities are indicated relative to that expected for the primary electron in a one-electron theory. [Based on Fig. 1 of J. J. Rehr, E. A. Stern, R. L. Martin, and E. R. Davidson, *Phys. Rev. B: Solid State* [3] **17,** 560 (1978).]

transitions involving energies of $\approx$15, 20, and 40 eV, with strengths of 6, 6 and 1.5%, respectively. Then there is a long tail of shake-off transitions extending from 20 to $\approx$200 eV, with a total strength of 22.5%. The resulting spectrum of outgoing primary *and* "shake" electrons is indicated schematically in Fig. 6 for a final-state energy E approximately 300 eV above the threshold for K-shell excitation. Thus, for a single photon energy, only approximately 64% of the outgoing "primary" photoelectrons will have the full final state E. The remainder of the outgoing electrons will have an energy (and hence momentum) reduced by the various amounts of energy transferred away during the shake-up and shake-off transitions. As discussed earlier, each of these electrons will contribute to the EXAFS spectrum a signal determined by its own value of wave vector. This leads, in principle, to a substantial complication in the analysis of EXAFS data. Rehr *et al.*[52] argued that the contributions of the continuum of shake-off transitions exhibit such a wide spread in phase (by virtue of the spread in electron momenta) that they will cancel one another to a great extent and can therefore be neglected. The contributions of the shake-up–modified primaries will each have a well-defined momentum but will be of small intensity ($\approx$13%). If both of these are neglected, the remaining effect of "shake" transitions on the analysis of the EXAFS is an apparent reduction in the EXAFS amplitude, since only $\approx$64% of the photoexcitation events contributing to σ^0 will also contribute to the EXAFS in σ^1. A detailed study by Stern and co-workers[53,54] has reinforced this model in general terms, but has led to the conclusion that atoms provide a better model system for investigating shake transitions in condensed matter than do the molecules examined by Rehr *et al.*[52] The much greater polarizability of molecules apparently leads to a significantly higher occurrence of shake transitions. Furthermore, Stern *et al.*[53,54] concluded that the effects of atomic environment on the amplitude reduction (i.e., chemical effects) are slight. This explains the common observation that approximately 80% of the absorption strength goes into the channel with one electron of unreduced energy (i.e., such that $\hbar^2k^2/2m = E_{nl} - E_0 + \hbar\omega$), independent of the material under investigation.[40,46,55] As a result, it is widely assumed that the total effect of multiple-electron excitations on the EXAFS spectrum for $E > 200$ eV can be approximated closely by decreasing the calculated EXAFS amplitude by a factor of $\sim$0.8, the precise number depending somewhat upon the excited atom species but little on the chemical environment. It may be inferred further that the amplitude reduction effect is of little consequence in comparisons among measured EXAFS spectra (owing to the weakness of chemical effects), but may be important in comparisons between theory and experiment.

It is apparent from this discussion that the effects of multiple-electron excitation on EXAFS spectra ought to be subjected to further quantitative investigation. In particular, it should be noted that the conclusions of Stern *et al.*[53,54] concerning these excitations are based on an analysis which neglects the curvature of the departing photoelectron wave front at the scattering atom (i.e., they make the usual "small-atom" approximation). As will be discussed in Section 4,c, this approximation is rather questionable in general. With reference to this particular study, Pettifer[56] has observed that the proper treatment of wave front curvature results in substantial changes in the k dependence of the calculated backscattering amplitude at low k, which is just the region where Stern *et al.*[54] found a strong k dependence which they attributed to multielectron excitations. It is possible, then, that the conclusions of Stern and co-workers will require some modification. If "shake"-modified outgoing electrons are ultimately shown to contribute significantly to EXAFS spectra, it will be necessary to devise methods for

[53] E. A. Stern, S. M. Heald, and B. A. Bunker, *Phys. Rev. Lett.* **42,** 1372 (1979).

[54] E. A. Stern, B. A. Bunker, and S. M. Heald, *Phys. Rev. B: Condens. Matter* [3] **21,** 5521 (1980).

[55] R. F. Pettifer, *Trends Phys., Pap. Gen. Conf. Eur. Phys. Soc., 4th, 1978*, Chapter 7, p. 522 (1979).

[56] R. F. Pettifer, *in* "X-Ray Processes in Solids and Innershell Ionization of Atoms" (D. J. Fabian, L. M. Watson, and H. Klienpoppen, eds.). Plenum, New York, 1981.

the analysis of EXAFS data which can cope with the contributions of more than one wave vector at each final-state energy E.

Independent of the effects of multiple-electron excitation on EXAFS spectra, it is clear from our earlier discussion based on the work of Fano and Cooper[39] that the final-state eigenfunctions ought to reflect the relaxation of the core electrons of the excited atom to the transfer of charge from the nl core level to a photoelectron state. A proper many-body treatment of this would be very difficult, involving the diagonalization of a matrix such as that discussed by Fano and Cooper.[39] In the one-electron model which has been developed for the EXAFS spectrum, the effects of core relaxation on the photoelectron are expressed through the phase shift due to the excited atom potential. The calculations of this phase shift by Lee and coworkers have been enabled by the following arguments. First, Lee and Beni[46] argued that the proper criterion to determine whether or not the nth core level will respond quasi-adiabatically to the departing photoelectron is the following. Compare the core-state relaxation time $\hbar/|E_n|$ with the time required for the photoelectron to travel twice the nearest-neighbor distance. Lee and Beni used $v = \hbar k/m$ for the photoelectron velocity, rather than the local velocity given by Eq. (4.9). This corresponds to a time scale for the change in potential substantially longer than that deduced by Fano and Cooper,[39] leading to the prediction of a predominantly quasi-adiabatic response to the departure of the photoelectron. Specifically, the criterion of Lee and Beni[46] leads to the conclusion that the response is quasi-adiabatic for all core electrons bound by more than 30 eV, given a photoelectron momentum of $k = 16$ Å^{-1} (or 15 eV, given $k = 8$ Å^{-1}). In contrast, our analysis based on the work of Fano and Cooper[39] suggests that the core electron response is typically intermediate between quasi-adiabatic and sudden (i.e., "shake" transitions occur but are not amenable to calculation in the sudden approximation). Next, Lee and Beni neglected the more shallowly bound electrons and argued that the appropriate potential for low k is that of an atom which has relaxed completely to the presence of the hole in the core corresponding to the vacated core level (called the "screened $Z+1$ atom"[57]). This implies that even the conduction electrons have relaxed to screen the resulting Coulomb potential. For large k, they argued that there has been *no* relaxation so that the Z atom with one electron missing is appropriate (called the "unscreened Z ion"). The absence of relaxation is inconsistent with the observation of "shake" transitions at high energies; that is, the existence of "shake" transitions implies the importance of matrix elements of the type discussed by Fano and Cooper.[39] Similar matrix elements will

[57] Use of the "$Z + 1$ atom" implies that the effect on the core electron eigenstates due to a missing core electron can be simulated with an additional charge on the atom nucleus. Thus the photoexcitation of a core electron transforms the Z atom to a $Z + 1$ atom.

lead to the relaxation of the core to the presence of the core hole. In addition, the appropriate potential for a high-k photoelectron ought to be tending toward the result expected in the "sudden" limit, in which the core electrons not only reflect the hole in the core but are also unscreened by the photoelectron. From any perspective, it seems apparent that the appropriate potential at high k will reflect the presence of the core hole. This suggests that the eigenfunctions of the $Z + 1$ atom would provide a better basis for the potential at high k than those of the Z atom. Lee and Beni[46] chose to compromise between their two limiting cases by using the potential of the $Z + 1$ atom with an outer electron missing (called the "unscreened $Z + 1$ ion") throughout the energy region of interest (E = 80–2000 eV). This compromise fortunately incorporates some aspects of the expected relaxation of the core electrons.[58] In the end, Lee and Beni[46] have come to a model for core relaxation which is not qualitatively at variance with the work of Fano and Cooper,[39] although it certainly neglects many of the subtleties inherent in that treatment, and which is amenable to calculation. It should be noted that Teo and Lee have departed from this model in phase-shift calculations for the heavier elements,[40] using the potential of the Z atom but omitting the contribution of the photoexcited core electron. This model neglects the expected effects of relaxation as discussed above. Finally, it should be emphasized that the core relaxation effects discussed here apply only to that portion of the EXAFS spectrum arising from a single outgoing electron with the full final-state energy E. A "shake"-modified outgoing electron will be subject to a different excited atom potential. This would obviously complicate any calculation of the contribution of "shake"-modified photoelectrons to an EXAFS spectrum.

b. Multiple-Atom Scattering

In deriving Eqs. (3.30) and (3.33)–(3.35), we emphasized that substantial simplification results from the neglect of multiple-atom scattering. In fact, EXAFS analysis would be essentially as complicated as the analysis of low-energy electron diffraction (LEED) if it were necessary to treat multiple-atom scattering effects to the same degree. Numerous authors have worked out the implications of multiple scattering for EXAFS.[25,28,29,59,60] These treatments are quite complete and will not be reproduced here. It is important to recall, however, that the proper treatment of multiple-atom scattering effects was the key issue which separated the earliest long-range and

[58] The question of whether or not to omit the potential due to the outer electron is unrelated to core relaxation and will not be addressed in this context.

[59] W. L. Schaich, *Phys. Rev. B: Solid State* [3] **14,** 4420 (1976).

[60] J. J. Rehr and E. A. Stern, *Phys. Rev. B: Solid State* [3] **14,** 4413 (1976).

short-range order (LRO and SRO) models for the EXAFS. The LRO models implicitly assumed an infinite mean free path for the final-state electron, leading to the dominance of band structure effects. The SRO models assumed rather short mean free paths owing to inelastic electron–electron scattering, leading to the dominance of near neighbors in the EXAFS spectra and substantially smaller multiple-atom scattering contributions (all of which correspond to long path lengths). Schaich[25] has shown that the LRO and SRO approaches are equivalent for identical mean-free-path lengths. This conclusion was supported through comparison with exact solutions in one dimension.[59]

Ashley and Doniach[28] observed that multiple-atom scattering should be important for close-packed structures like Cu metal, even if it is not important for open structures like tetrahedrally bonded crystalline Ge. Lee and Pendry[29] carried out an explicit calculation of multiple-atom scattering effects for face-centered cubic (fcc) Cu metal and concluded that these were small except for the contribution of the fourth shell of neighbors. This may be understood as follows. While quite complicated in principle, the contribution from a multiple-atom scattering path will contain a factor which can be represented (in oversimplified form) by

$$\exp(ikR - R/\lambda) \times t_{\beta}(\mathbf{k}, \mathbf{k}') \times t_{\beta'}(\mathbf{k}', \mathbf{k}'') \times \cdots, \tag{4.10}$$

where $R = |\mathbf{r}_1 - \mathbf{r}_0| + |\mathbf{r}_2 - \mathbf{r}_1| + \cdots + |\mathbf{r}_n - \mathbf{r}_0|$ is the total path length. The shorter single-atom and multiple-atom scattering paths in the fcc lattice are listed in Table I. Four of these are represented in Fig. 7. Note that the shortest multiple-atom scattering path is longer than the single-scattering contribution of the nearest neighbors of the excited atom. Thus, except for complicated structures, multiple-scattering effects should not interfere with the analysis of the first-neighbor contribution to the EXAFS.[61] Furthermore, as pointed out by Lee and Pendry,[29] the problem is simplified substantially by being able to identify and treat only those few multiple-atom scattering paths which will contribute to a given region of r space. There are few of these at short path lengths, as can be seen from Table I. Note also from expression (4.10) that the electron wave is attenuated according to the total path length, just as for single scattering.

There is, however, an amplitude effect peculiar to multiple-atom scattering. At high energies, electron scattering is very strong in the forward direction, and weak otherwise. Most short multiple-atom scattering paths involve at least two large-angle (and therefore weak) scattering events and are, accordingly, of small amplitude. Certain multiple-atom scattering paths

[61] This implies, of course, that a procedure such as a Fourier transform will be used to separate the various "frequencies," or spatial separations, observed in the EXAFS.

TABLE I. SINGLE-ATOM AND MULTIPLE-ATOM SCATTERING PATHS[a]

Path label	Path length (a_0)	Number of paths
1–1	1.414	12
2–2	2.000	6
1–1–1	2.121	48
1–2–1	2.414	72
3–3	2.449	24
1–3–1	2.639	144
4–4	2.828	12
1–4–1	2.828	36
1–1–1–1	2.828	144

[a] Those paths relevant to EXAFS spectroscopy in fcc crystals are listed, up to a total path length of 2.9 in units of a_0, the cube edge length. The paths are labeled by the length of the segments of which they are composed. For example, a path from the excited atom at (0, 0, 0) to one at (−½, +½, 0) to one at (−½, −½, 0) and back to (0, 0, 0) involves two nearest-neighbor distances, and one second-nearest-neighbor distance and is accordingly labeled 1–2–1. The *apparent* interatomic separation that would be inferred from an EXAFS spectrum on the assumption of only single-atom scattering is one-half of the total path length shown here. [Based on Table I of P. A. Lee and J. B. Pendry, *Phys. Rev. B: Solid State* [3] **11,** 2795 (1975).]

consist, however, of two forward scatterings and one backscattering. This contribution can be comparable to the single-scattering contribution at that path length. This is the case for the fourth-nearest-neighbor atoms in a face-centered cubic system, which lie exactly behind the first-nearest-neighbor atoms. Thus the signal from those neighbors is changed substantially by contributions from the two multiple-atom scattering paths with identical lengths shown in Table I: 1–4–1 [e.g., (0, 0, 0) to (+½, +½, 0) to (−½, −½, 0) to (0, 0, 0)] and 1–1–1–1 [e.g., (0, 0, 0) to (+½, +½, 0) to (+1, +1, 0) to (+½, +½, 0) to (0, 0, 0)]. Specifically, the experimentally observed signal from the region of the fourth-nearest-neighbor atoms in Cu metal can be as much as 90° out of phase in Fourier transform space when compared with the expected single-scattering contribution (see Fig. 5). In their treatment,[29] Lee and Pendry were able to calculate just this sort of correction due to multiple-atom scattering. Other examples of this effect are given by Lee *et al.*[9] In general, however, multiple-atom scattering effects are believed to be small at high final-state electron energies ($E > 200$ eV), where

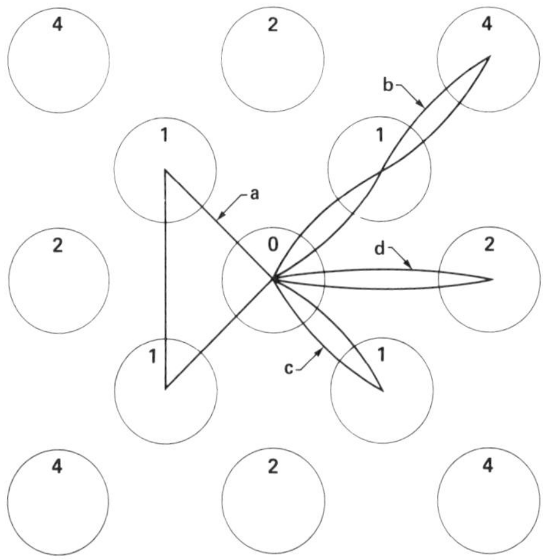

FIG. 7. Two of the shorter multiple-atom scattering paths in the (001) plane of an fcc lattice are shown together with two single-atom scattering paths. The number on each atom denotes the group of neighbors of the excited atom (denoted by "0") to which it belongs. For example, "2" denotes a second nearest neighbor to "0". Path *a* involves two nearest-neighbor atoms which are separated from each other by a cube edge. This path is denoted by 1–2–1 in Table I. Path *b* involves backscattering from a fourth neighbor and two forward scatterings from a nearest neighbor. Since each path segment connects atoms that are nearest neighbors *to each other*, this path is labeled 1–1–1–1 in Table I. Note that path *b* is precisely as long as that path involving only a fourth-neighbor atom. Two single-atom scattering paths, *c* and *d*, are shown for comparison. They are labeled 1–1 and 2–2, respectively, in Table I.

the enhanced scattering is confined to a very narrow region about the forward direction. The situation is different at low energies, where two effects combine to enhance the contribution of multiple-atom scattering. Firstly, all of the long path length contributions are more substantial in the energy region 10–50 eV above threshold, where electron mean free paths lengthen substantially (to 10–20 Å at 10 eV, rather than 5–10 Å at 50–150 eV[62]). Furthermore, electron–atom scattering becomes more isotropic at lower energies. This increases the intensity for the large-angle scattering which is necessary for those short multiple-atom scattering path lengths which interfere with near-neighbor structural information. The low-energy region, which is also complicated by the onset of shake transitions, is largely inaccessible to the simplest single-scattering EXAFS analysis.

It should be noted that Pettifer[56] has concluded from a careful comparison with theory that his EXAFS data from second nearest neighbors in the Zn

[62] For example, see I. Lindau and W. E. Spicer, *J. Electron Spectrosc. Relat. Phenom.* **3,** 409 (1974).

chalcogenides cannot be explained without resorting to the inclusion of multiple-atom scattering effects. First-neighbor scattering presumably reduces the outgoing electron amplitude anomalously at the second neighbors in this open structure, thereby reducing the scattering signal from them. It is clear that the importance of multiple-atom scattering should be reassessed as we demand more precise information from EXAFS data.

c. Outgoing-Wave Approximations

We have made several approximations regarding the outgoing electron wave in deriving Eqs. (3.30) and (3.33)–(3.35). The principal effect of these approximations is to yield an expression for the EXAFS χ such that the electron transport terms can be grouped into a Λ which is only trivially a function of the position of the scatterer. As will be appreciated later, this leads to a substantial (but not essential) simplification of the analysis. Our first approximation is that the excited atom potential does not overlap those of its neighbors, or, more precisely, that the effect of the overlap region can be approximated well by adding the phase shifts due to the individual potentials. A similar approximation is frequently made in constructing "muffin-tin" potentials for band structure calculations. The situation is complicated in this case, however, by the Coulomb potential of the core hole, which might not be completely screened at half the nearest-neighbor distance. This approximation has not been examined in detail in the literature, but is nonetheless commonly made.

The second approximation involves the inelastic effects which shorten the electron lifetime. Pendry[63] has argued that the predominant inelastic scattering mechanism in the interatomic region is plasmon excitation, which yields a constant imaginary contribution to the self-energy on the order of $V_{\text{inelastic}} \approx 3$–$6$ eV, as verified by LEED studies. The corresponding final-state electron mean-free-path length is given by $\lambda(k) = 2k/V_{\text{inelastic}}$. Approaching the same phenomenon from an atomic viewpoint, Beni *et al.*[64] and Lee and Beni[46] have studied electron–electron correlation effects in electron scattering from atoms. Their work suggests that inelastic events in the region of the atomic core reduce the magnitude of the backscattering t matrix by as much as 50% for $k < 4$–6 Å^{-1} and alter substantially the phase shifts for those values of k which contribute significantly to the EXAFS, leading to errors ≈ 0.1 Å in nearest-neighbor position if neglected. In the Lee and Beni treatment,[46] a plasmon pole approximation is used to incorporate these inelastic events into the complex phase shifts associated with the excited and scattering atoms (see Section 5 for a further discussion).

[63] J. B. Pendry, "Low Energy Electron Diffraction." Academic Press, New York, 1974.
[64] G. Beni, P. A. Lee, and P. M. Platzman, *Phys. Rev. B: Solid State* [3] **13,** 5170 (1976).

The remaining events are due to electron–electron scattering in the interatomic regions. Their effect can be approximated through an effective mean-free-path length λ applied only to that portion of the electron path outside the excited and backscattering atoms. This suggests that the factor of $\exp(-2r/\lambda)$ in Λ [see Eq. (3.28)] be replaced by $\exp[-2(r - d)/\lambda]$, where d is similar to the nearest-neighbor distance. This substitution has been verified experimentally by Stern *et al.*,[54] who also concluded that λ increases by a factor of 2 with increasing ionicity in tetrahedrally bonded semiconductors.[65] Thus the treatment of electron lifetime suggested in the derivation of Eqs. (3.30) and (3.33)–(3.35) is reasonable *if* EXAFS analysis proceeds by determining Λ experimentally (as is discussed in Section 10), since the measured Λ's will incorporate the dominant effects of inelastic scattering. On the other hand, calculations of the EXAFS phase shifts and t matrices must include a proper treatment of these effects.

Finally, we introduced the so-called "small-atom" approximation, in which the matrix element of t appearing in the expression for the EXAFS is evaluated using plane waves rather than the actual spherical wave function of the outgoing electron as indicated in Eq. (3.18). The complex amplitude and wave vector of the plane wave are chosen to reproduce the outgoing wave function at one point, the center of the scattering atom. This approximation is reasonable only if the curvature of the spherical outgoing wave function and the r dependence of $v_l^+(kr)$ can be neglected over the spatial extent of the scatterer. If the curvature or the r dependence is significant, the proper matrix element of t will depend nontrivially on the distance r from the excited atom to the scattering atom, which will necessitate a more sophisticated treatment. For the case of the nearest-neighbor contribution in Cu, Lee and Pendry[29] showed that the "small-atom" approximation leads to only small errors in the amplitude of t for $k \geq 5$ Å^{-1}, but introduces errors in the phase of t which grow more important with decreasing k, reaching ≈ -0.6 rad by 5 Å^{-1}. Use of the asymptotic value for v [see Eq. (3.26)] to obtain the amplitude of the plane wave leads to an additional error ≈ -0.6 rad. Lee and Pendry concluded that the small-atom approximation is reasonable only for $k > 10$ Å^{-1}. Pettifer and coworkers have pointed out that this situation becomes much worse as the atomic number of the backscatterer increases.[55,56,66] From studies of ZnSe and ZnTe, they argued that the proper treatment of the spatial dependence of the outgoing wave at the first neighbor is as important in calculations as the proper treatment of inelastic effects. The errors in the calculated

[65] Note that the analysis of Stern *et al.*[54] neglected the possibility of non-Gaussian peak shapes, which might complicate their determinations of λ. Such effects are discussed in Section 6.

[66] R. F. Pettifer, P. W. McMillan, and S. J. Gurman, *in* "The Structure of Non-Crystalline Materials" (P. H. Gaskell, ed.), p. 63. Taylor & Francis, London, 1977.

phase and amplitude of t introduced by the "small-atom" approximation increase with the size of the scattering atom, and therefore with atomic number. Pettifer[56] concluded that the calculated backscattering amplitudes and phases of Teo and Lee,[40] which necessarily incorporate that approximation, should not be used for $Z > 52$ and $E < 200$ eV (see Section 5 for further discussion). It should be stressed that the small-atom approximation need not be made in comparisons between data and calculated spectra. In addition to the work of Gurman and Pendry[67] and of Pettifer and co-workers[55,56,66] in this regard, sophisticated treatments have been suggested by Grosso and Pastori-Parravicini[68] and by Müller and Schaich.[69]

The curvature of the outgoing wave and the r dependence of $v_{l'}^{+}(kr)$ will also affect comparisons among experimental spectra, resulting in subtle changes in the backscattering signature with near-neighbor distance. This is a strong argument for restricting such comparisons to neighbors at similar distances.

5. Calculation of EXAFS Spectra

An EXAFS spectrum has been shown to depend not only on near-neighbor structural information, but also on certain electron scattering properties of the atoms involved. A knowledge of these nonstructural elements is a prerequisite of any extraction of the structural information. The gaining of this knowledge could involve the detailed analysis of EXAFS spectra from structurally known systems, called "standards," as will be discussed in Section 10. Alternatively, the analysis of EXAFS data could be based on calculated values of the nonstructural components of $k\chi$, as will be discussed here. In the absence of multiple-atom scattering and in the limit where asymptotic forms for the outgoing plane wave are adequate, such a calculation would include the components in Λ: the phase shift in the outgoing electron wave due to the excited atom $\eta_{l'}$, the complex t matrix of the backscattering neighboring atoms, and the attenuation of the electron wave due to inelastic scattering. There have been many recent attempts to evaluate these quantities, each characterized by particular models for the electron–atom and electron–electron interactions and for the relaxation of the excited atom. The results of these studies are indicated below, but the reader is referred to the original articles for the detailed differences in approach.

The phase shifts, the t matrix, and the attenuation of the electron wave are various aspects of the underlying interactions between the photoelectron

[67] S. J. Gurman and J. B. Pendry, *Solid State Commun.* **20,** 287 (1976).
[68] G. Grosso and G. Pastori-Parravicini, *J. Phys. C* **13,** L919 (1980).
[69] J. Müller and W. Schaich, *Bull. Am. Phys. Soc.* [2] **26,** 320 (1981).

and the atoms and other electrons of the system. Phase shifts are a convenient representation of the electron scattering properties of atom potentials. In particular, the atomic t matrix may be expressed as follows:

$$(2\pi)^3\langle \mathbf{k}'|t(k)|\mathbf{k}\rangle = -\frac{2\pi\hbar^2}{mk}\sum_l (2l+1)\exp(i\eta_l)\sin(\eta_l)\,P_l\,(\cos\theta), \quad (5.1)$$

where $|\mathbf{k}'| = |\mathbf{k}| = k$, the phase shift η_l is an implicit function of k, P_l is a Legendre polynomial, and $\cos\theta = \mathbf{k}'\cdot\mathbf{k}/k^2$. For a localized spherical potential at the origin, $\eta_l(k)$ is the increase in the phase of the regular solution of energy $\hbar^2k^2/2m$ and angular momentum quantum number l at some large r due to the presence of that potential. It may be calculated straightforwardly, given the appropriate potential.[70] The phase shift will be complex if that potential incorporates the effects of inelastic electron scattering, leading to an attenuation of the electron wave. Accordingly, calculations of the phase shift of the excited atom potential, of the t matrix, and even of the effects of inelastic scattering reduce to the problem of constructing suitable potentials. These potentials should in principle reflect the constraints which were placed on them in the course of deriving the EXAFS equations. Recall that the full particle potential was divided into two parts, V_0 and V, such that V (and hence T) will be as nearly vanishing as possible in the vicinity of the excited atom core. The intent of this constraint is to enable one to neglect the spatial overlap between T and V_0–E_0, the spatially varying portion of V_0. To this end, V_0 includes the potential of the excited atom plus that of some effective medium representing the effect of the other atoms (and electrons) in the system. V_0 is constrained to be spherically symmetric, but may well be energy dependent and/or complex. The calculated phase shift of the excited atom and t matrix of the scattering atoms ought to reflect the resulting V_0 and V, respectively. In practice, however, the calculations are more typically based on model potentials derived from free atoms.

A comparison between calculations and EXAFS data requires that a one-to-one correspondence be established between the two independent variables: k of the phase-shift calculations and the X-ray photon energy $\hbar\omega$ of the experiment. As derived, these two variables are related by two expressions for the final-state electron energy, $E = E_{\alpha;nl} + \hbar\omega = E_0 + \hbar^2k^2/2m$. The needed correspondence depends accordingly on an explicit and accurate evaluation of $E_{\alpha;nl} - E_0$. This would be somewhat difficult in practice, even for two atomic levels, owing to the simple models often used for V_0. The present situation is complicated further by the difficulty in properly describing the interactions between a high-energy photoelectron and the other

[70] For example, see A. Messiah, "Quantum Mechanics," Vol. I, Chapter 10. North-Holland, Amsterdam, 1961.

electrons in the sample. As a consequence, E_0 is often treated as an adjustable parameter in comparisons with EXAFS data, as will be seen in the following discussion.[71] With this background, let us now consider representative calculations of EXAFS spectra.

In an early calculation, Ashley and Doniach[28] found only qualitative agreement with the EXAFS spectra of Cu metal and crystalline Ge. Kincaid and Eisenberger[72] made a more detailed comparison and found that their calculations for Br_2 and $GeCl_4$ led to errors in interatomic separation $\approx$ 5%. Lee and Pendry[29] found similar errors in a calculation for Cu metal. Lagarde[73] obtained somewhat less satisfactory results in calculations for NaBr and KBr. In a substantial improvement over previous attempts, Gurman and Pendry[67] analyzed data on Cu metal using a subtraction technique and found errors of $\leq$1% for the nearest-neighbor distance and $\leq$2% for the second- and third-neighbor distances. This is the first study in which calculated EXAFS spectra were used to determine interatomic distances with acceptable error levels.

A few years later, Gurman and Pettifer[74] compared their calculations with an experimental EXAFS spectrum from crystalline As_2O_3, an insulating molecular solid. They examined the phase shifts and backscattering amplitudes which result from alternative models for the electron–atom interaction potentials: nonoverlapping Hartree–Fock potentials and overlapping Hartree–Fock–Slater potentials with values of the Slater parameter α between 1 and ⅔. They found that the phase is substantially more sensitive to the potential model than is the backscattering amplitude, leading to expected uncertainties in the experimentally determined interatomic separations of approximately 0.05 Å, and that the effect of the core hole should not be neglected. They also concluded that similar phases result from two particular approximations for the electronic configuration of the excited atom: the "relaxed $Z + 1$ atom," where the extra nuclear charge is intended to approximate the effect of the core hole and all the other core electrons have relaxed to this potential; and the "unrelaxed Z ion," where the potential is that of the neutral atom except for the absence of a contribution from the core electron which has been photoexcited. The latter conclusion is particularly noteworthy, since these two alternatives correspond to the models put forth by Lee and Beni[46] for the low-k and high-k limits, respectively, of an EXAFS spectrum. It appears to be at variance, however, with the calculations of Teo and Lee,[40] in which significant differences in phase were found to result from comparable changes in potential.

[71] The assignment of a value to E_0 is discussed further in Section 10.

[72] B. M. Kincaid and P. Eisenberger, *Phys. Rev. Lett.* **34,** 1361 (1975).

[73] P. Lagarde, *Phys. Rev. B: Solid State* [3] **14,** 741 (1975).

[74] S. J. Gurman and R. F. Pettifer, *Philos. Mag.* [*Part*] *B* **40,** 345 (1979).

Beni *et al.*[64] proposed a more sophisticated treatment of electron–atom scattering that takes into account the polarization of the atom by the electron which is being scattered. They found that polarization of the 3d electrons in Cu leads to substantially enhanced forward scattering. A simpler approach with the same objective was formulated by Lee and Beni,[46] who incorporated the plasmon pole approximation for exchange and correlation into a Thomas–Fermi model. In this approximation due to Lundqvist and Hedin,[75,76] the elementary excitations of the electron gas are modeled with a single pole in the self-energy. It yields the necessary reduction in the exchange and correlation potential as the kinetic energy of the electron increases, and leads to quite good agreement with EXAFS measurements on Br_2, $GeCl_4$, Ge, and Cu (i.e., approximately 0.5% error in the nearest-neighbor distance, and 1% for further neighbors). It should be noted, however, that this agreement can be obtained only through an essentially arbitrary adjustment of E_0, as discussed earlier. Such adjustments do not address directly the deficiencies and uncertainties in the calculations and are, accordingly, not a wholly satisfactory means to bring the calculations into agreement with experiment.

The earliest attempts to provide extensive tables of the relevant quantities resulted in a pair of articles by Teo, Lee, and coworkers.[77,78] These works were succeeded 2 years later by an extensive presentation by Teo and Lee[40] of the results of such calculations in a form suitable for the analysis of EXAFS data. We summarize below the model used in this effort and their results.

Teo and Lee[40] calculated their phase shifts from potentials constructed using the approach of Lee and Beni[46] discussed above. In this theory, the effects of exchange and correlation are incorporated together with the electrostatic potential into an effective *complex* scattering potential. The resulting phase shifts are complex since they take into account, within the limits of the model, some of the inelastic scattering for which we introduced an effective electron mean-free-path length λ in Section 3. More precisely, the inelastic scattering due to the excited and backscattering atoms is included, but not that due to any intervening electron density. Therefore, in applying these tabulated results, one should include an additional attenuation factor *only* when analyzing the signal from neighbors more distant than the first. This factor should have a form similar to $\exp[-2(r - d)/\lambda]$, where d accounts for the attenuation already included through the phase

[75] B. I. Lundqvist, *Phys. Kondens. Mater.* **6,** 206 (1967).

[76] L. Hedin and B. I. Lundqvist, *Solid State Phys.* **23,** 1 (1969).

[77] B-K. Teo, P. A. Lee, A. L. Simons, P. Eisenberger, and B. M. Kincaid, *J. Am. Chem. Soc.* **99,** 3854 (1977).

[78] P. A. Lee, B.-K. Teo, and A. L. Simons, *J. Am. Chem. Soc.* **99,** 3856 (1977).

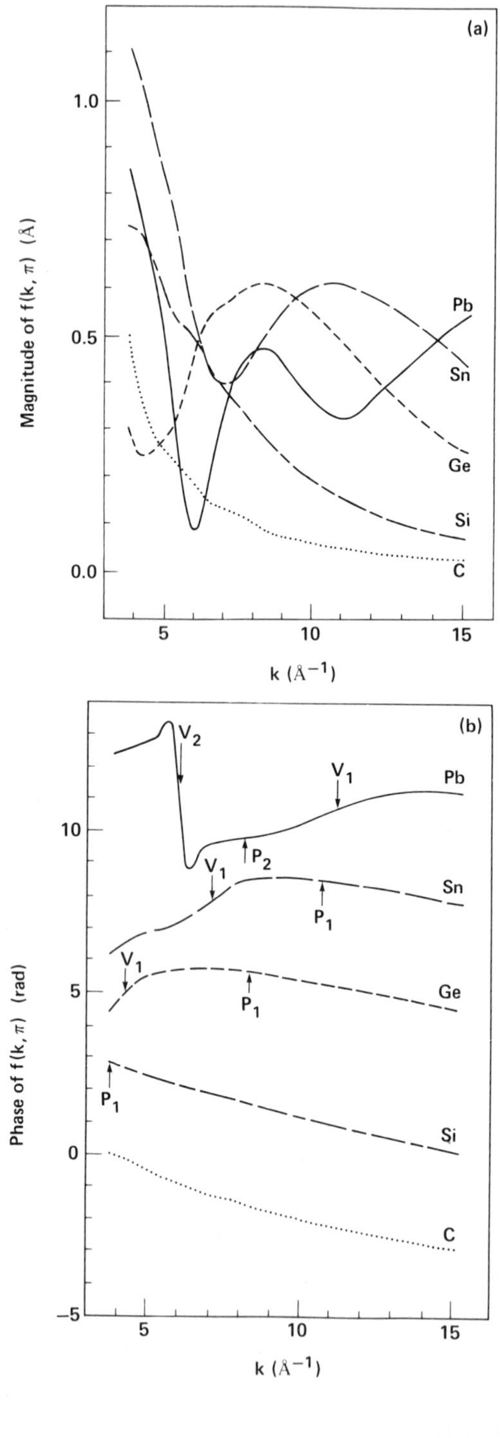
(a)
Magnitude of f(k, π) (Å)
1.0
0.5
0.0
5
10
15
k (Å^{-1})
Pb
Sn
Ge
Si
C
(b)
Phase of f(k, π) (rad)
10
5
0
−5
V$_2$
V$_1$
P$_2$
P$_1$
Pb
Sn
Ge
Si
C
5
10
15
k (Å^{-1})

shifts (i.e., d is approximately equal to the nearest-neighbor distance). As noted in Section 4,c, Stern *et al.*[54] have verified this procedure.

The potential of the excited atom will reflect the state of its relaxation to the developing core hole. The underlying issues were discussed at some length in Section 4,a. In constructing the potential of the excited atom, Teo, Lee, and co-workers followed initially[77,78] the precedent established by Lee and Pendry[29]; that is, they used the wave functions of the $Z + 1$ atom, but omitted the potential contributed by one of the valence electrons. For their newer set of calculations of the heavier elements,[40] Teo and Lee used a different approximation for the excited atom potential—constructing it from the wave functions of the Z atom, but omitting the potential contributed by the core electron that has been photoexcited. This represents a substantial change in the *ansatz* for the relaxation of the final-state atom to the developing core hole. The atom is totally unrelaxed to the presence of the core hole in the latter case, and fully relaxed except for conduction electron screening in the former case. As one might expect, this change in potential results in relatively significant changes in the phase shift.[40] The proper treatment of core relaxation is discussed in detail in Section 4,a.

A calculation of the phase shifts also requires assumptions about the charge on and the electronic configuration of the atom in question. We refer the reader to Teo and Lee[40] for an extensive discussion of these issues as well as the overall trends in the calculated quantities as a function of position in the periodic table. In particular, these authors discussed at some length the dependence of the calculated quantities on the state of ionization, on the electronic configuration, on charge, on relativistic effects, and so on, as well as the increasingly resonant behavior in backscattering amplitude and phase with increasing atomic number. In each instance, Teo and Lee noted that one should compensate for uncertainties in the calculated values through adjustment of E_0. As was discussed above, this procedure is not wholly satisfactory.

Using the potentials described above, Teo and Lee have calculated and tabulated[40] the excited atom phase shifts $\eta_l(k)$ and the phase and magnitude of the complex electron scattering amplitude $f(k, \pi)$ for a selection of elements between C and Pb ($Z = 6$–82), at 17 values of electron momentum k between 3.7 and 15.2 Å^{-1}. $f(k, \pi)$ is proportional to the backscattering matrix element of t between plane waves, $t(-\mathbf{k}, \mathbf{k})$ [see Eq. (3.29)]. The overall trends in these quantities are indicated in the following figures taken

FIG. 8. The magnitude (a) and phase (b) of the complex electron scattering amplitude $f(k, \pi)$ as a function of k for C, Si, Ge, Sn, and Pb. $f(k, \pi)$ is proportional to the backscattering matrix element of t between plane waves, $t(-\mathbf{k}, \mathbf{k})$ [see Eq.(3.29)]. The positions of the peaks and valleys in the amplitude of $f(k, \pi)$ are marked "P" and "V," respectively, in (b). [Taken from Figs. 2b and 4b, respectively, in B.-K. Teo and P. A. Lee, *J. Am. Chem. Soc.* **101**, 2815 (1979). Copyright 1979 American Chemical Society.]

from that paper. Figure 8a shows the magnitude of $f(k, \pi)$ for five Group 4A elements: C, Si, Ge, Sn, and Pb. Note that one or more sharp dips occur in $|f|$ for the heavier elements. These are associated with resonances in the electron–atom scattering. Each moves to progressively higher values of k with increasing Z. The locations of these local minima and maxima in $|f|$ are indicated in Fig. 8b, in which the phases of f are shown for the same five elements. (The positions of the peaks and valleys of $|f|$ are marked "P" and "V," respectively, in Fig. 8b.) As one might expect for resonant behavior, the sharp dips in $|f|$ coincide with relatively rapid changes in the phase of f. Each dip corresponds to a resonance in one of the individual phase shifts $\eta_l(k)$, which contribute to t and f [see Eq. (5.1)]. The positions of these resonances, and the resulting shape of t and f, are sufficiently characteristic of the backscattering atom species that its identity can often be deduced from a detailed analysis of the EXAFS data. This is one of the significant advantages of EXAFS spectroscopy as a structural probe. Finally, the excited atom phase shift for $l = 1$ is shown in Fig. 9 for the same five Group 4A elements. These phase shifts are comparatively featureless. This poses no problem in EXAFS data analysis, however, since the identity of the excited atom is already known from the energy of the X-ray absorption edge associated with the EXAFS spectrum. The resulting EXAFS spectra

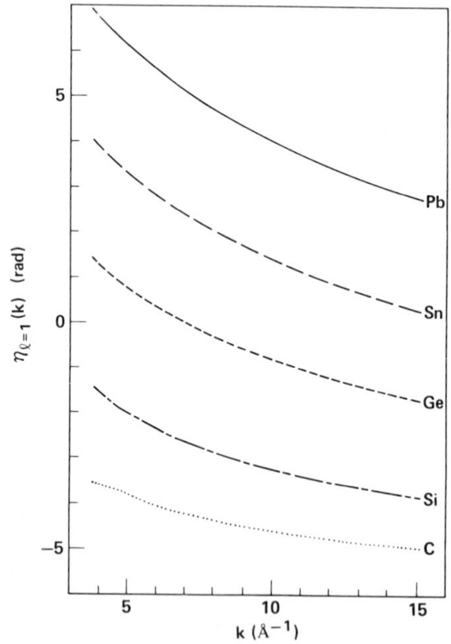

FIG. 9. The excited atom phase shift $\eta_l(k)$ for $l = 1$, as a function of k for C, Si, Ge, Sn, and Pb. [Taken from Fig. 6b in B.-K. Teo and P. A. Lee, *J. Am. Chem. Soc.* **101,** 2815 (1979). Copyright 1979 American Chemical Society.]

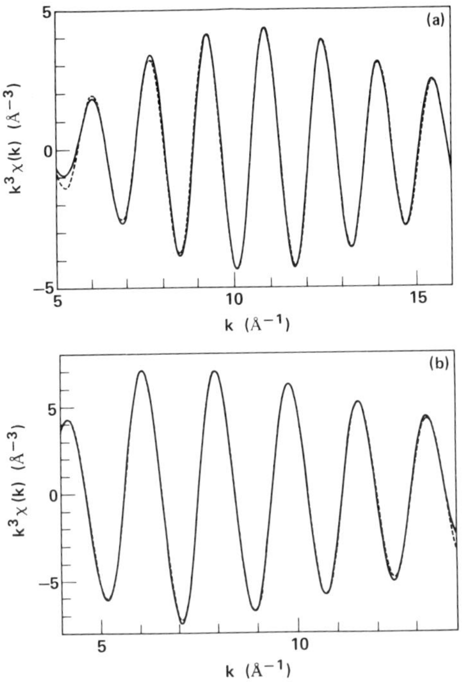

FIG. 10. Experimental K-shell absorption spectra $k^3\chi(k)$ from (a) Br_2 and (b) the Ge atoms in $GeCl_4$ shown as functions of k. The calculated spectra (dashed lines) are based on the phase shifts and scattering amplitudes of Teo *et al.* [B.-K. Teo, P. A. Lee, A. L. Simons, P. Eisenberger, and B. M. Kincaid, *J. Am. Chem. Soc.* **99,** 3854 (1977)]. The fit illustrated here resulted from adjustments in E_0, in the overall scattering amplitude (the calculated value was scaled by 0.66 in the case of Br_2 and by 0.41 in the case of $GeCl_4$), and in the width of an assumed Gaussian nearest-neighbor distribution function. [Taken from Fig. 2 in B.-K. Teo, P. A. Lee, A. L. Simons, P. Eisenberger, and B. M. Kincaid, *J. Am. Chem. Soc.* **99,** 3854 (1977). Copyright 1977 American Chemical Society.]

compare quite well in some instances with measured spectra, provided that the overall vertical scale and E_0 are treated as adjustable parameters (to compensate for some effects of "shake" transitions and for energy scale uncertainties, respectively, as discussed earlier). The quality of fit possible is illustrated for two molecular systems in Fig. 10.

The tables of Teo and Lee[40] furnish values for those quantities which enter into Λ [see Eq. (3.28)]. It should be stressed that Λ is part of an approximate expression for the EXAFS. In particular, the t matrix appears in Λ *only* as a matrix element between plane waves, $t(-\mathbf{k}, \mathbf{k})$, which neglects the curvature of the actual outgoing electron wave and the r dependence of $v_l(kr)$ over the spatial extent of the backscattering atom potential. The underlying "small-atom" approximation is discussed in Section 4,c. If this approximation is not adequate, the matrix element of t is intrinsically dependent on the distance r from the excited atom to the backscattering atom and will require a more sophisticated treatment than that of Teo and Lee.[40] Just such a treatment has been used by Pettifer and co-workers.[55,56,66] In particular, Pettifer[56] has calculated the t matrix for nearest-neighbor atoms in Se and Te, and has shown that the "small-atom" approximation introduces errors in the phase and amplitude of t which are as large as those resulting from a neglect of inelastic processes. These errors increase with the size of the backscattering atom, and therefore with atomic number. Pettifer concluded that the tabulations of Teo and Lee[40] should not be used for atomic numbers greater than 52 or for energies less than 200 eV. From their detailed study of Cu and Pt, Stern *et al.*[54] concluded similarly that the tables of Teo and Lee[40] are applicable only for $k > 10$ Å^{-1}.

Finally, it should be noted that Holland *et al.*[79] have calculated the "smooth, atomlike" falloff in absorption above an absorption edge, represented in this treatment by σ^0, and found evidence for low-frequency oscillations due to scattering resonances of the *excited atom* potential. They regarded the prominent peak at the As K edge as an unusually strong manifestation of this effect, a p-wave resonance in that specific case. These oscillations may well provide interesting information about the excited atom potential. They are unlikely to complicate the EXAFS analysis appreciably, however, since the small effective radius of the excited atom potential should lead to oscillations which are quite distinct in frequency from those due to substantially more distant near neighbors.

6. Structural Information Content of EXAFS

In the absence of significant effects due to multiple-atom scattering of the final-state electron, the EXAFS on the absorption cross section has been shown to depend upon the arrangement of the atoms only through atomic pair correlation functions. Furthermore, the EXAFS arising from the photoexcitation of an electron from an atom of species α depends upon only $\rho_{\alpha\beta}(\mathbf{r})$ for all species β, a subset of all possible pair correlation functions. The full angular information contained in $\rho_{\alpha\beta}(\mathbf{r})$ has been reduced in the EXAFS in even the most favorable instances by a weighted average over angles such as is shown in Eqs. (3.30) and (3.34). Finally, the structural information is reduced further by the limited range of k over which the EXAFS oscillations may be extracted from experiment. Each of these aspects of the information content in EXAFS will be discussed in this section, with emphasis on implications for the extraction of structural information from EXAFS data. Comparison with diffraction studies will be made where appropriate.

[79] B. W. Holland, J. B. Pendry, R. B. Pettifer, and J. Bordas, *J. Phys. C* **11**, 633 (1978).

a. Separation of Multiple Pair Correlation Functions

In the limit that multiple-atom scattering can be neglected, the EXAFS on the absorption spectrum of an excited atom of species α depends upon only those atom pair correlation functions $\rho_{\alpha\beta}(\mathbf{r})$ which involve that species, as is expressed clearly in Eq. (3.30). For a sample containing N atom species, this will include N out of a total of $N(N+1)/2$ independent pair correlation functions. On the other hand, a diffraction spectrum is a function of *all* of these pair correlation functions. For $N \geq 2$, this represents a significant complication in the interpretation of the diffraction data and is perhaps the only *fundamental* advantage of EXAFS over diffraction studies. Structural information enters into diffraction and EXAFS data in different ways, however, leading to other important differences in the kind and quality of the information which can be extracted. Let us compare these alternative probes of structure in some detail.

X-Ray[80] and neutron[81] diffraction are the structural probes of choice for many solids. In the ideal case of a single-crystal sample, they have the potential of yielding the complete distribution of atoms. In many instances, however, the wealth of information contained in a diffraction spectrum complicates its interpretation unnecessarily. Of particular interest in this context are those instances in which determining the local atomic order in a system is especially important, for only then can EXAFS be an attractive alternative.[82]

It is possible to derive an expression for the diffracted intensity which is analogous to Eq. (1.2). This emphasizes the similarities in the structural information contained in the two measurements and facilitates focusing on the differences. The basic quantity determined in an X-ray or neutron diffraction study is the diffracted intensity as a function of momentum transfer $i(q)$, where $q = (4\pi/\lambda_X) \sin\theta$, λ_X is the X-ray wavelength, and θ is the diffracted angle. When properly normalized and corrected for self-scattering, $i(q)$ is analogous to the EXAFS $\chi(k)$. In a general case, the diffracted intensity may be written as

$$i(\mathbf{q}) \propto \sum_{\alpha} c_\alpha \sum_{\beta} \int d\mathbf{r}\, \rho_{\alpha\beta}(\mathbf{r}) \exp(i\mathbf{q}\cdot\mathbf{r}) \frac{f_\alpha(q) f_\beta^*(q)}{\langle |f(q)|^2 \rangle}, \tag{6.1}$$

where c_α is the concentration of atom species α and $\rho_{\alpha\beta}(\mathbf{r})$ is the angle-dependent distribution of atom species β about atom species α, as defined in Section 3. The $p_{\alpha\beta}(r)$ appearing in Eq. (1.2) is the angular average of $4\pi r^2 \rho_{\alpha\beta}(\mathbf{r})$ [see Eq. (3.31)]. It is related to the "usual" pair distribution

[80] For example, see A. Guinier, "X-Ray Diffraction." Freeman, San Francisco, California, 1963.

[81] For example, see G. E. Bacon, "Neutron Diffraction." Oxford Univ. Press (Clarendon), London and New York, 1975.

[82] Diffraction has no competition from EXAFS as a source of information about long-range correlations.

function of diffraction studies, g, by

$$p_{\alpha\beta}(r) = 4\pi r^2 n c_\beta g_{\alpha\beta}(r), \tag{6.2}$$

where n is the density of atoms. $f_\alpha(q)$ is the scattering factor for atom species α and is effectively independent of q in the case of neutron scattering. Unlike Λ, this scattering factor is a well-known quantity. $\langle |f(q)|^2 \rangle$ is $\Sigma_\alpha c_\alpha |f_\alpha(q)|^2$, the average of $|f|^2$ over all atoms in the sample. In the limiting cases of a liquid, an amorphous solid, or a randomly oriented polycrystalline sample, all of the angular information in the diffraction data is averaged out, leaving only radial distributions such as those contained in EXAFS data. In that limit, Eq. (6.1) becomes

$$qi(q) \propto \sum_\alpha c_\alpha \sum_\beta \int_0^\infty dr\, r^{-1} p_{\alpha\beta}(r) \sin(qr) \frac{f_\alpha(q) f_\beta^*(q)}{\langle |f(q)|^2 \rangle}. \tag{6.3}$$

When the integral over $\mathbf{r}$ in Eq. (6.1) is evaluated, the long-range order in a single-crystal sample contributes only to Kronecker δ functions at the reciprocal lattice vectors, the usual Bragg peaks. For a randomly oriented polycrystalline sample, these become the Debye lines, independent of the angle of $\mathbf{q}$. If there is additional short-range order in the sample, then it will contribute to a diffuse background, oscillating slowly with q.

Apart from the added angular information which is contained in Eq. (6.1) for a single-crystal sample, the most obvious difference between Eqs. (1.2) and (6.1) is the additional sum over atom species α present in Eq. (6.1). As a consequence, each diffraction spectrum yields the value of a single linear combination of the Fourier transforms of *all* the independent pair correlation functions in the sample, totaling $N(N + 1)/2$ for a sample with N different atom species. In principle, the unique determination of these correlation functions requires the generation of $N(N + 1)/2$ independent equations, each obtained from a separate diffraction spectrum in which the coefficients of the correlation functions have been varied appropriately. The coefficient of the α–β pair correlation function is proportional to the real part of $c_\alpha c_\beta f_\alpha(q) f_\beta^*(q)$. The requirement that the structure be unchanged usually eliminates the possibility of varying significantly the concentrations, so that the most promising route to generating the required set of equations involves variations in the scattering factors f_α. There are several possibilities in this regard. X-Ray and neutron scattering factors are generally quite distinct from one another. Furthermore, X-ray scattering factors are sensitive to X-ray photon energy if that energy is in the vicinity of an absorption edge. The tunability of synchrotron radiation sources can be quite helpful in this regard. Neutron scattering factors are sensitive to nuclear isotope, which is convenient for a few elements. Both of these approaches are sufficiently difficult in practice, however, that rigorous separation of the pair correlation functions has been accomplished only occasionally, even in the limiting case of a two-component system, in which there are only three independent pair correlation functions.[83] It is more common to analyze a single diffraction spectrum with some simple model.[84] This procedure is most successful when applied to diffraction data from a single-crystal sample. The uniqueness of the result is difficult to establish, given the similar shapes of the scattering factors of the various elements involved. The neutron scattering factors are, in fact, identical in shape. This situation worsens rapidly as the number of atom species increases, since the number of independent $\rho_{\alpha\beta}$'s contributing to $i(q)$ is quadratic in N. For these reasons, the amount of local structural information which has been extracted from diffraction studies declines strongly as the chemical and structural complexity of the system increases. In contrast, as emphasized above, EXAFS data is a function of only N pair correlations. On a more subtle level, the scattering factors for different atoms tend to be more distinct in EXAFS than in diffraction studies. This is often a substantial aid in separating the $p_{\alpha\beta}(r)$ from one another. Taken in conjunction, these two differences greatly simplify the structural analysis of EXAFS data for a multicomponent system.

The contrast is even more dramatic in the case of an atom species present only in small quantities. Although the EXAFS measurement might well require a special technique as discussed in Section 9, the resulting data will involve only the environment of that rare species and is as easily interpreted as if that species were the predominant one. Diffraction data, on the other hand, will be determined largely by the predominant species to the exclusion of meaningful information about the rare species. This is apparent from Eqs. (1.2) and (6.1), given that $\rho_{\alpha\beta}$ is proportional to c_β. It may be concluded that the EXAFS technique can be a very desirable probe of the nearest-neighbor environment, even in the case of a single-crystal sample, if the species of interest is rare or only one of many.

b. *Angular Component of Structure*

As suggested by Eqs. (3.30) and (3.33)–(3.35), an EXAFS measurement probes principally the radial part of the pair correlation functions $\rho_{\alpha\beta}(\mathbf{r})$. There are, however, some interesting exceptions to this general rule based on the macroscopic directional selectivity of the photoexcitation event: The intensity of the beam of photoelectrons is largest in directions parallel to the polarization vector $\hat{\epsilon}$ of the incident X-ray photons.[85] For K-shell ex-

[83] For an application to Cu–Sn liquids, see J. E. Enderby, D. M. North, and P. A. Egelstaff, *Philos. Mag.* [8] **14,** 961 (1966).

[84] For an application to a-SiO_2, see R. L. Mozzi and B. E. Warren, *J. Appl. Crystallogr.* **2,** 164 (1969).

[85] The polarized nature of synchrotron radiation, discussed in Section 8,a, is essential in this property.

citation, the local environment is sampled selectively along the $\hat{\epsilon}$ direction with a probability $\propto \cos^2 \theta$, where θ is the angle between $\hat{\epsilon}$ and **r**. Similarly, for $L_{II,III}$-shell excitation (which is dominated by $l = 2$ final states), the directional weighting factor is proportional to $3 \cos^2 \theta + 1$. In the case of a single-crystal sample (or of an oriented polycrystalline sample), this macroscopic directionality allows one to probe crystalline directions selectively by orienting the crystal appropriately relative to $\hat{\epsilon}$. For example, Brown *et al.*[86] used this property in an EXAFS study of single-crystal hexagonal close-packed (hcp) Zn to determine the pair correlation functions perpendicular and parallel to the *c* axis. Similarly, Rabe *et al.*[87] measured substantial anisotropy in the EXAFS from single-crystal GeS. Such studies are discussed in detail in Section 11,h.

The polarization of synchrotron radiation can occasionally be turned to advantage in samples with a more macroscopic directionality. Consider, for example, the application of EXAFS spectroscopy to probe surface phenomena. By definition, such applications require that essentially all the detected signal originate from excited atoms residing on the surface in question.[88] By orienting this surface appropriately, one can selectively probe the environment of the excited atom parallel to and perpendicular to that surface. Such information is often extremely valuable, as is discussed at length in Section 14.

c. *Experimental Range of k and Its Effect on Information Content*

In general, the sample is such that the EXAFS measurement probes only $p_{\alpha\beta}(r)$, the angular average of the pair correlation functions. The information about $p_{\alpha\beta}$ which can be extracted from the EXAFS depends importantly on the range of k over which it can be measured. The factors which determine that range are discussed in this subsection, as well as the k-space distribution of information about $p_{\alpha\beta}$. This leads to an examination of the phenomena which underlie the shape of a peak in $p_{\alpha\beta}$, including static and thermal "disorder." The consequences for data analysis of peak asymmetry are discussed in detail. In order to be explicit, the following discussion will be based primarily on Eq. (3.33), appropriate for K-shell excitation in amorphous or randomly oriented polycrystalline samples. Its generalization to more complicated situations is straightforward.

There is an important distinction to be drawn between the results of EXAFS and diffraction, based on the respective regions of momentum space from which structural information can be extracted. The range of k over

[86] G. S. Brown, P. Eisenberger, and P. Schmidt, *Solid State Commun.* **24,** 201 (1977).
[87] P. Rabe, G. Tolkiehn, and A. Werner, *J. Phys. C* **13,** 1857 (1980).
[88] The means to accomplish this are discussed in Section 9.

which information can be obtained from a structural technique is an important descriptor of the nature of that information. It is possible to recast Eq. (3.33) so as to make explicit the one-to-one correspondence between the k range over which $k\chi$ can be extracted from experiment and the k range over which structural information can be obtained. Assume for simplicity that all the nearest neighbors of α are of only one atom species β, and that $p_{\alpha\beta}$ consists of one narrow peak with a mean nearest-neighbor distance r_i. The EXAFS due to these nearest neighbors alone can be represented approximately by

$$k\chi_{\alpha\beta}(k) \simeq (2\pi)^{1/2} r_i^{-2} 2 \operatorname{Re}[P_{\alpha\beta}(k)\Lambda_{\alpha\beta}(k, r_i)], \tag{6.4}$$

where

$$P_{\alpha\beta}(k) = (2\pi)^{-1/2} \int_0^\infty dr\, e^{2ikr} p_{\alpha\beta}(r). \tag{6.5}$$

$P(k)$, the Fourier transform of $p(r)$, is related to the $S(q)$ of diffraction studies. The salient difference from the viewpoint of this discussion is that the variable complementary to r is $2k$ in the EXAFS case, not q as in the diffraction case. This must be taken into account in any proper comparison of the nature of the structural information to be obtained from the different techniques.

One frequently used expression for the EXAFS is obtained only for the idealized case in which the peak in $p(r)$ is a Gaussian of half-width σ_i representing a single shell of N_i atoms at r_i. Under these circumstances,

$$P(k) = (2\pi)^{-1/2} N_i \exp[-2(\sigma_i k)^2] \exp(2ikr_i) \tag{6.6}$$

and Eq. (6 4) reduces to Eq. (1.4). It is unfortunate that Eq. (1.4), or its equivalent, appears commonly in the literature, since it incorporates a potentially unwarranted assumption about peak shape. As will be discussed later, the peak shape may well not be Gaussian, and this equation not appropriate.

The ultimate goal of most analyses of EXAFS data is to determine the nearest-neighbor peak in the pair correlation function p (i.e., the position, type, and number of the nearest neighbors). The procedure for accomplishing this is complicated by the limited range of k over which Eq. (6.4) [or, more properly, Eq. (3.33)] can be used to extract $P(k)$ from the EXAFS data. Implicit in Eqs. (3.33) and (6.4) is the assumption that the final-state electron wave function is determined completely by single-atom backscattering events. This neglects effects which are important for low k. Firstly, the final-state electrons have long mean free paths at low energies.[62] This leads to the emergence of multiple-atom scattering as a significant factor in determining the EXAFS for k less than 2 or 3 Å^{-1}, as discussed earlier. In other words, band structure effects become important. Moreover, the cross section near the absorption edge is often dominated by localized resonance phenomena.[79,89] This is the probable origin of the peaks below 2.5 Å^{-1} in Fig. 4. Owing to both of these omissions, Eq. (3.33) cannot describe the oscillations in the absorption cross section for k below approximately 2.5 Å^{-1} (a final-state electron energy of 24 eV). As a result, the usual analysis, based on Eq. (3.33), cannot be used to obtain structural information [i.e., $P(k)$] from this region of an EXAFS measurement. Finally, there is an additional problem associated with backscattering atoms coming from the vicinity of the fourth period of the periodic table. In these cases, the backscattering strength $t_\beta^+(-\mathbf{k}, \mathbf{k})$ has a local minimum at low k.[90] The resulting small magnitude of Λ makes it very difficult to get satisfactory information about P for low values of k [see Eq. (6.4)]. For several reasons, then, the practical lower limit for obtaining $P(k)$ ranges from 2 to 3 Å^{-1} in typical situations. An effective upper limit occurs because of the decrease in both the scattering factor (t matrix) and $P(k)$ for high k. The value at which noise becomes a significant factor might vary from $k = 9$ Å^{-1} for a light element backscatterer such as carbon to 20 Å^{-1} or more for a heavy element such as iodine.

As discussed above, the range of EXAFS data to be compared with a diffraction experiment is that of $2k$: approximately 5–30 Å^{-1}. In contrast, the information available from X-ray diffraction starts at $q \simeq 0$ and begins to deteriorate substantially in quality at 10–15 Å^{-1}. This important difference between EXAFS and diffraction data has been known for some time[31] and was the subject of a paper by Eisenberger and Brown.[91] The inability of EXAFS to yield structural information below $k \simeq 2.5$ Å^{-1} is a serious problem, because it is $P(k)$ in precisely that k-space region (i.e., as k approaches zero) which is capable of yielding definitively the coordination number N and the mean nearest-neighbor distance r_{nn}. Specifically, $P(0)$ is $(2\pi)^{-1/2} N$ and the wavelength of $P(k)$ as k approaches zero is π/r_{nn}. Since diffraction data approach closely to $k = 0$, N and r_{nn} may be determined in a straightforward manner. For example, a simple integral in r space leads to the proper coordination number. On the other hand, N and r_{nn} can be obtained from EXAFS data only through an extrapolation to $k = 0$ from $k > 2.5$ Å^{-1}. Any extrapolation requires some assumption about functional form—in this instance, the proper shape for the peak in the pair correlation function. The Gaussian shape which led to Eq. (1.4) is commonly assumed in EXAFS analysis, but is often not the proper choice. In particular, some materials have highly asymmetric near-neighbor peak shapes. In order to

[89] M. Brown, R. E. Peierls, and E. A. Stern, *Phys. Rev. B: Solid State* [3] **15,** 738 (1977).

[90] There is substantial experimental and theoretical evidence for this phenomenon; see, for example, the calculations of t_β^+ in Teo and Lee.[40]

[91] P. Eisenberger and G. S. Brown, *Solid State Commun.* **29,** 481 (1979).

distinguish among alternative shapes in an EXAFS analysis, one must understand the data at high k more completely than is required in the analysis of diffraction data.

Let us examine how the shape of a peak in p is determined. The photoexcitation event takes place very quickly on the time scale of atomic thermal motions.[92] The neighbors of each excited atom are frozen at specific distances. Each neighbor contributes $\delta(\mathbf{r} - \mathbf{r}_i)$ to the appropriate $\rho_{\alpha\beta}(\mathbf{r})$, where $\mathbf{r}_i$ is the position vector from the excited atom α to the neighbor β. The EXAFS experiment measures the average of this quantity over all excited atoms, including contributions from both thermal and static disorder. In an amorphous or randomly oriented polycrystalline sample, the average over excited atoms will correspond to an average over the directions of $\mathbf{r}$, motivating our definition of $p_{\alpha\beta}(r) \equiv 4\pi r^2 \langle \rho_{\alpha\beta}(\mathbf{r}) \rangle_{\Omega_\mathbf{r}}$ in Section 3. Even in cases where it is recognized that static disorder may contribute to the peak shape, the resulting peak in p is frequently approximated with a narrow Gaussian in the spirit of the harmonic approximation for lattice vibrations. With that *ansatz*, Stern *et al.*[93] were able to extract an estimate of the change in the width of the nearest-neighbor peak in Cu metal as the temperature is increased from 77 to 300°K. Similarly, Gurman and Pendry[67] were able to extract the width of the peaks in p for Cu metal at 77°K as a function of peak position. They found that the Gaussian half-width increases steadily from approximately 0.084 Å for the nearest neighbor to 0.097 Å for the fifth neighbor. They pointed out that the limiting value of this width at large r should correspond to the width observed in a diffraction experiment, as it does in this instance. They observed, furthermore, that the near-neighbor values should be reduced owing to increased correlation in the thermal vibrations, as observed.

The thermal contribution to the shape of peaks in $\rho_{\alpha\beta}(\mathbf{r})$ has been examined in some detail for a harmonic solid by Beni and Platzman.[94] In a harmonic solid, the time-averaged distribution of each atom about its own equilibrium position $\mathbf{r}_0$ will be proportional to $\exp(-\frac{1}{2}|\mathbf{r} - \mathbf{r}_0|^2/\Sigma^2)$, where Σ is a function of temperature (and possibly also of direction $\Omega_\mathbf{r}$). A proper calculation of the time-averaged near-neighbor separation which enters into the EXAFS requires that the correlated motions of the near neighbors be taken into account. Unlike more distant neighbors, near-neighbor atoms tend to vibrate in unison. This lowers the width of the nearest-neighbor

[92] The lifetime broadening observed experimentally for a threshold near 10 keV is ≈ 1 eV, suggesting a lifetime of $<10^{-15}$ sec. Atoms cannot respond in a much shorter time than the reciprocal of a phonon frequency, $\approx 10^{-13}$ sec.

[93] E. A. Stern, D. E. Sayers, and F. W. Lytle, *Phys. Rev. B: Solid State* [3] **11,** 4836 (1975).

[94] G. Beni and P. M. Platzman, *Phys. Rev. B: Solid State* [3] **14,** 1514 (1976).

peak from that of more distant peaks, this latter width being simply related to the Debye–Waller width of the peaks in diffraction.[95] It is straightforward to calculate the shape of the near-neighbor peaks in $\rho_{\alpha\beta}(\mathbf{r})$ from the phonon spectrum if it is known completely. For the more common situation in which we have only limited information about the phonon spectrum, Beni and Platzman derived an explicit expression for the effective Gaussian half-width σ_1 of the nearest-neighbor peak in the limiting case of an anisotropic Debye model. They were able to calculate the observed anisotropy in the Zn nearest-neighbor vibrations[86] with good agreement.

There have been many other comparisons of their approach with experiment, and in each case the agreement has been satisfactory. It is of particular interest that Sevillano *et al.*[96] concluded from a study of Cu, Fe, and Pt that correlations typically lessen the width of the nearest-neighbor peak at 77°K by 15%. Other work includes studies of Co, Cu, RbCl, and SrS,[97] of Cu,[98] of Ge,[99] and of Cu, Ge, and Pt.[54] In a detailed study of crystalline and amorphous Ge, Rabe *et al.*[100] showed that the Debye model of Beni and Platzman overestimates significantly the temperature dependence of the peak width when covalent bonds are present. These studies are discussed further in Section 11,a.

Note that these comparisons between theory and experiment use an expression for the EXAFS appropriate only in the limit of narrow peaks [equivalent to Eq. (1.4) rather than the more accurate Eq. (1.2)]. These approximations, taken all together, constitute only one of many possible models for $\rho(\mathbf{r})$. It may not be the proper model. For example, asymmetry is introduced by the factor $r^{-2}e^{-2r/\lambda}$ in Eq. (1.2) if the peak in $\rho(\mathbf{r})$ is not "very" narrow. More fundamental asymmetry would arise in the case of an anharmonic solid. The implications of such asymmetry are examined below.

The width of a peak in $p(r)$ determines the upper limit to the range of $q = 2k$ over which there will be detectable structural information about that peak. In either diffraction or EXAFS, the scattering signal falls off with increasing q as $\exp(-\frac{1}{2}q^2\sigma_r^2)$, where σ_r is the effective Gaussian half-width of the structural peak in r space and q is replaced by $2k$ in the EXAFS case. As has been discussed, both static and thermal disorder will contribute to this peak width. Both of these contributions are expected to be larger for

[95] In particular, the Debye–Waller width includes significant contributions from long-wavelength acoustic phonons that have little influence on the nearest-neighbor peak width.

[96] E. Sevillano, H. Meuth, and J. J. Rehr, *Phys. Rev. B: Condens. Matter* [3] **20,** 4908 (1979).

[97] W. Böhmer and P. Rabe, *J. Phys. C* **12,** 2465 (1979).

[98] R. B. Greegor and F. W. Lytle, *Phys. Rev. B: Condens. Matter* [3] **20,** 4902 (1979).

[99] O. H. Nielsen and W. Weber, *J. Phys. C* **13,** 2449 (1980).

[100] P. Rabe, G. Tolkiehn, and A. Werner, *J. Phys. C* **12,** L545 (1979).

more distant (and therefore less correlated) peaks in $p(r)$, leading to a smaller signal at large q. In addition, the angular average implicit in the scattering from a disordered or randomly oriented polycrystalline sample acts to smooth out the contributions of peaks at large r, further reducing the signal at large q. For these reasons, distant peaks in $p(r)$ often make little or no contribution for $q = 2k > 5$–10 Å^{-1}. The scattering at large values of q is expected instead to reveal the details of the nearest-neighbor peak in $p(r)$ (i.e., the sharpest feature). For low values of q, on the other hand, near and distant neighbors contribute on an equal footing, so that the scattering signal will be dominated by long-range correlations.

Specifically, the Bragg scattering portion of a measured diffraction spectrum is dominated by long-range interatomic correlations for low values of the scattering vector. The atom density pattern which is obtained from an analysis of this scattering is properly interpreted as revealing the spatial relationship between an atom and the lattice, rather than that atom's nearest neighbors. As such, it may contain valuable information about diffusion paths in a material such as a superionic conductor. This atom density pattern does not, however, yield as much information about the short-range pair correlations. *If* it were to be analyzed as an effective nearest-neighbor distribution, it would yield a peak which is artificially broad, corresponding to an apparent distance of closest approach which is unphysically short. The total scattering for *large* scattering vectors ($q > 10$–15 Å^{-1}) could be analyzed to yield the desired short-range information, but this is seldom done in diffraction studies.

In contrast, dependent upon information for $q = 2k > 5$Å^{-1}, the EXAFS data bear directly on the nearest-neighbor contribution to the interatomic correlation functions. Moreover, the extensive range of scattering vector available in an EXAFS experiment (q up to 20 or even 40 Å^{-1}) yields data which are sensitive to the details of the nearest-neighbor pair distribution. This was used to good advantage in a detailed study of the short-range portion of atom–atom pair interaction potentials.[101] EXAFS spectroscopy yields less information about more distant neighbors, since their positions are generally less well correlated (i.e., their peaks in p are broader).[102] The implicit extrapolation to $2k = 0$ to obtain the distance and coordination number (as discussed earlier) becomes more difficult as the scattering signal slowly disappears from the accessible range of k. Information about more distant neighbors is limited further by the short mean-free-path length $\lambda(k)$ of the electrons, which gives rise to the factor of $\exp(-2r/\lambda)$ in Λ [see Eq.

[101] T. M. Hayes and J. B. Boyce, *J. Phys. C* **13**, L731 (1980).

[102] This presumably explains the absence of any signal from atoms beyond the nearest neighbors in the amorphous Ge EXAFS data shown in Section 10 and can be a particular disadvantage in amorphous materials.

(1.3)]. Both of these factors contribute to the absence of direct information about long-range order from an EXAFS spectrum. It should be noted, however, that the nearest-neighbor peak is usually narrow enough to yield a strong EXAFS signal, even in the extreme case of a liquid (e.g., see the measurements on liquid Zn by Crozier and Seary[103] and the discussion in Section 13). Furthermore, the ability to identify atom species from the EXAFS peak functions $\xi_{\alpha\beta}$ results in part from the slow buildup of the EXAFS signal at low k, which varies significantly with atom species. In a sense, some sensitivity to broadened peaks has been exchanged for the ability to determine the neighboring atom species.

d. Accommodating Non-Gaussian Peak Shapes

Not every peak in a pair correlation function will exhibit the Gaussian profile for which the simplest model might be appropriate. On the contrary, we expect that many interesting candidates for EXAFS studies will exhibit nearest-neighbor peak shapes which are non-Gaussian. In some cases, including liquids and glasses, the peak shape is dominated by substantial disorder which does not lend itself to expression as thermal deviations about a mean position. In most cases, however, non-Gaussian peak shapes might be expected from purely thermal considerations, as in studies of the surfaces of solids, superionic conductors, anharmonic crystals, and, as a general class, materials at temperatures near or above their characteristic Debye temperatures (e.g., zinc, even at room temperature[91]). The complications of EXAFS data analysis due to non-Gaussian peak shapes have been discussed extensively in the literature.[91,103–107] In the following paragraphs, we will illustrate these complications and an approach with which to deal with them. A superionic conductor, CuI, has been chosen for this illustration because detailed analysis has demonstrated convincingly that only a strongly asymmetric peak shape can explain the EXAFS data on several of these materials at a wide variety of temperatures (i.e., AgI, CuI, CuCl, and CuBr, as discussed in detail in Section 11,b). Furthermore, their unusual ionic conductivity has been shown to arise naturally from the same excluded volume model which yields this peak shape.[108]

The EXAFS due to the nearest neighbors only of a Cu ion in superionic

[103] E. D. Crozier and A. J. Seary, *Can. J. Phys.* **58,** 1388 (1980).

[104] B. Lengeler and P. Eisenberger, *Phys. Rev. B: Condens. Matter* [3] **21,** 4507 (1980).

[105] P. Eisenberger and B. Lengeler, *Phys. Rev. B: Condens. Matter* [3] **22,** 3551 (1980).

[106] T. M. Hayes and J. B. Boyce, *in* "EXAFS Spectroscopy: Techniques and Applications" (B. K. Teo and D. C. Joy, eds.), p. 81. Plenum, New York, 1981.

[107] R. Haensel, P. Rabe, G. Tolkiehn, and A. Werner, *Proc. NATO Adv. Study Inst. Liq. Amorphous Met.* (in press).

[108] T. M. Hayes and J. B. Boyce, *Phys. Rev. B: Condens. Matter* [3] **21,** 2513 (1980).

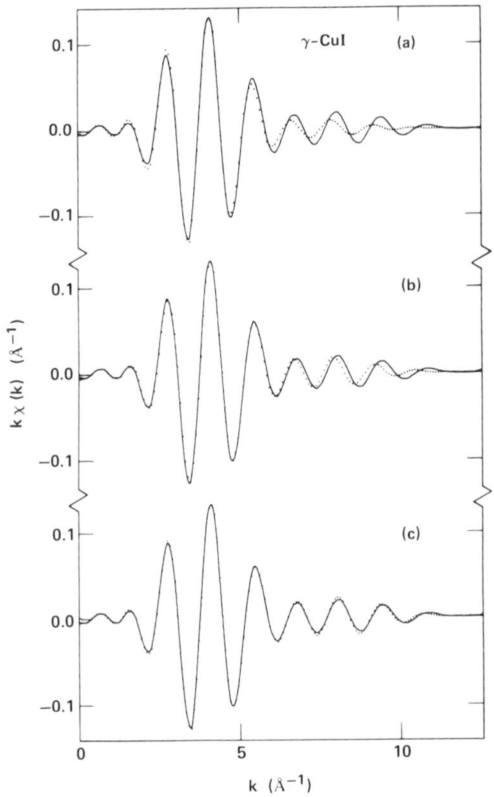

FIG. 11. The EXAFS oscillations $k\chi(k)$ from the nearest neighbors of Cu in γ-CuI measured at 573°K (solid line), plotted together with three alternative simulations (dotted): (a) a single Gaussian with N fixed at 4, (b) a single Gaussian with N optimized at 3.2, (c) an excluded volume model with N fixed at 4. The last of these corresponds most closely with the data. [Taken from Fig. 2 in T. M. Hayes and J. B. Boyce, *in* "EXAFS Spectroscopy: Techniques and Applications" (B. K. Teo and D. C. Joy, eds.), p. 81. Plenum, New York, 1981.

γ-CuI at 573°K is reproduced as the solid curve in all three parts of Fig. 11. This curve is obtained by Fourier transforming the data to r space, and then transforming *only* the nearest-neighbor peak back in k space. The nearest neighbors are iodine ions in each case. These data can be understood in the context of an excluded volume model, as described in detail in Section 11,b. For the present purposes, it is sufficient to note that this model leads to a nearest-neighbor peak in the radial distribution which is strongly asymmetric, as shown in Fig. 12. Two symmetric models for the peak shape have also been considered: a single Gaussian, where the amplitude has been constrained to $N = 4$, as expected for a tetrahedral site in this material; and

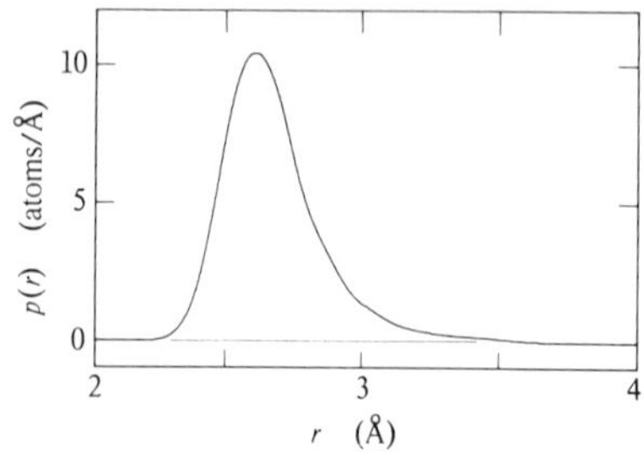

FIG. 12. The pair correlation function $p(r)$ for the excluded volume model with N fixed at 4. The model parameters have been chosen to give an excellent fit to the nearest-neighbor contribution to the EXAFS data from the Cu edge in γ-CuI measured at 573°K (as shown in Fig. 11c). [Taken from Fig. 3 in T. M. Hayes and J. B. Boyce, *in* "EXAFS Spectroscopy: Techniques and Applications" (B. K. Teo and D. C. Joy, eds.), p. 81. Plenum, New York, 1981.]

a single Gaussian, where the amplitude has been adjusted to fit the data. The analysis of the EXAFS data at 573°K took place in r space using Eq. (1.5), as described in the section on data analysis. For each model, the EXAFS spectrum at 573°K is compared numerically with a simulated spectrum based on EXAFS data from CuI at 77°K, where the structure is known.

The sensitivity of the simulated $k\chi$ to the shape of $p(r)$ is illustrated clearly in Fig. 11. Figure 11a shows the nearest-neighbor contribution to $k\chi$ for CuI measured at 573°K together with the Fourier transform of the simulated φ corresponding to the best Gaussian fit with $N = 4$, the physically reasonable value ($r_1 = 2.62$ Å, $\sigma_1 = 0.14$ Å, $R = 0.040$). R is a measure of the residual error in the simulation. The amplitude of the fit is reasonable for $k < 6$ Å^{-1}, but much too small for higher k. In addition, the phase of the simulation is far from that of the data at high k. It is obvious that this Gaussian peak is unable to reproduce the k dependence of either the amplitude or the wavelength of the data.

Figure 11b shows the simulated $k\chi$ for a Gaussian peak where N is allowed to vary from 4 to its favored value of 3.2 ($r_1 = 2.60$ Å, $\sigma_1 = 0.12$ Å). In this process, the value of R declines from 0.040 to 0.034, a significant improvement. By adopting a value of N which is physically unreasonable, this Gaussian model has achieved a better fit to the amplitude of the data at high k. If the data analysis had stopped at this point, the erroneous conclusion would have been reached that the Cu ions had found a new site where there are three iodine nearest neighbors. Notice, however, that the fit to the phase of the oscillations in the data has not improved. It is clear that further improvement in the model $p(r)$ is necessary. Part of the difficulty is that a $p(r)$ with a single Gaussian peak *cannot* yield a variable wavelength in k space such as is found in the data. The fitting procedure will then lead to a simulation which fits the data only at low k, where the signal is large. To the extent that the resulting simulation does not extrapolate properly to $k = 0$, the coordination number and the mean nearest-neighbor distance obtained in the fit will be misleading.

Figure 11c shows the simulated $k\chi$ obtained from the excluded volume model, where $N = 4$. Since the R value has dropped substantially, to 0.009, a high confidence level can be attached to the structural information deduced from this fit. Not only is the amplitude of the data reproduced more closely for all k, but so is the phase. The mean nearest-neighbor distance r_{nn} is properly deduced to be the distance appropriate to the tetrahedral Cu site, 2.64 Å. Inspection of the excluded volume $p(r)$ shown in Fig. 12 reveals how this model is able to fit the variable wavelength observed in the data. Recall that the wavelength of $P(k)$ as k approaches zero is just π/r_{nn}. In contrast, the high-k portion of $P(k)$, and therefore of $k\chi$, is dominated by the sharpest feature in $p(r)$. In this case, that is the sharp rise at $r \simeq 2.43$ Å (which is less than r_{nn}). Since a shorter r corresponds to a longer wavelength in k space, the simulated $k\chi$ will have a slightly longer wavelength at high k than at low k, just as is seen in the data. The phase information in the k-space data has specified directly the principal features of $p(r)$: The wavelength at low k determined the mean nearest-neighbor distance r_{nn} and the slightly longer wavelength at high k demanded a sharp rise in p at a value of r somewhat less than r_{nn}. In other words, $p(r)$ is unavoidably asymmetric in precisely the manner shown in Fig. 12. Had the analysis been confined to Gaussian peak shapes, not only would the coordination number have been underestimated, but also the mean nearest-neighbor distance. In this instance, the absence of structural information for $k < 2.5$ Å^{-1} was overcome by demanding a truly good fit to the data over a wide range of k.

From the above discussion, it is clear why the values of r_1 which were deduced from the Gaussian fits are close to the proper r_{nn}, and why the Gaussian with N held at 4 atoms is the closer of the two. Both Gaussians fit the data better at low k, where the wavelength yields the proper r_{nn}, than at high k, where the wavelength would yield a value of r which is too small. Additionally, the greater width σ_1 of the Gaussian with $N = 4$ further confined the fit to low k, yielding a value of r_1 even closer to r_{nn}. From this perspective, there is a clear disadvantage to the common practice of multiplying $k\chi$ by k or k^2 prior to analyzing the EXAFS data. By weighting preferentially the high-k portion of the spectrum, this procedure aggravates all of the problems discussed here, making it still more difficult to obtain reliable values of N and r_{nn}. If such additional factors of k had been introduced into the above analysis, the Gaussian fits would have yielded substantially lower, and more misleading, values of both N and r_1. It should also be noted that the dramatic effect of the examples of Eisenberger and Brown[91] was enhanced by such additional factors of k.

From this example, it is clear that determining the proper shape for a peak in the pair correlation function is an essential aspect of EXAFS data analysis. Its neglect can lead to errors in determining the coordination number and the nearest-neighbor distance. Eisenberger and Brown[91] were led by their analysis to conclude similarly that caution is appropriate in interpreting EXAFS data on systems with broad nearest-neighbor peaks (as in the above example). On the other hand, *if* the quality and the range of the data at high k are adequate, then it is possible to recognize those cases in which a Gaussian is a poor approximation to the actual peak in $p(r)$. The proper functional form for the peak (i.e., the appropriate structural model) can be determined by fitting the data so that the extrapolation to $k = 0$ to obtain the coordination number and the mean nearest-neighbor distance will be meaningful.

A similar conclusion was reached by Crozier and Seary[103] in their study of polycrystalline and liquid Zn. They investigated several alternative forms for the peak shape in the liquid state and concluded that the added complications due to non-Gaussian peak shapes are indeed surmountable, as discussed in Section 13.

It is important, of course, to exercise some judgment, limiting the investigation to those peak shapes which are consistent with the known properties of the material in question. The study of superionic conductors was facilitated by prior knowledge of the number and symmetry of the nearest neighbors to be expected in the possible alternative sites for the cations. Even in cases where prior knowledge is minimal, however, the reasonableness of a Gaussian peak shape can be tested straightforwardly using simple functional alternatives. One possible set of functions might include $\exp[-A(r - r_1) - B(r - r_1)^2]$, $\exp[+A(r - r_1)^2 - B(r - r_1)^4]$, and so on. Eisenberger and Brown[91] have suggested a slightly different approach, which is convenient when the data is to be fit in k space: The $p(r)$ for an anharmonic potential is expanded in moments and Fourier transformed analytically to obtain a model expression for fitting in k space. Crozier and Seary[103] have used yet other functional forms. The range of possible approaches is wide, but all would have the same purpose: to test for *deviations* from a simple Gaussian shape. Once deviations are found, however, the analysis of those deviations will necessarily depend upon the nature of the system being studied; that is, a structural model consistent with the known properties of the system must be created and tested.

7. Absorption Edge Features

The absorption spectrum of a solid within approximately 20 eV of an edge is dominated by features which do not lend themselves to analysis on the basis of single-atom scattering. One type of complication arises from the increased importance of multiple-atom scattering at low final-state electron energies, as discussed in Section 4,b. This leads to features which are related more to the long-range structure of the sample than they are to the short-range structure, and which are therefore of little interest to most practitioners of EXAFS spectroscopy. Other features can arise from the presence of complicated many-electron effects such as were discussed in Section 4,a. These may include transitions to exciton-like states "pulled down" out of the continuum by the core hole potential. Finally, peaks can arise from resonances or large densities of states in the continuum of one-electron states owing to residual atomic or molecular effects. These latter effects all reflect to some extent the local environment of the excited atom, but our understanding of them is not such as to allow an interpretation which is as straightforward as that of the EXAFS. We summarize below some of the studies of this region of the absorption spectrum.

Since the number of features in the near edge region is limited, it is conceivable that one could identify systematic variations in these features as a function of various known environments of the excited atom. Determination of the excited atom environment in unknown systems might then be possible even in the absence of a microscopic understanding of the near edge region. This idea has been the basis for many studies of the near edge structure.[109–112] Rather than examine these studies in detail, let us turn to those studies of which the goal was an understanding of the edge region.

It is not uncommon to observe a prominent peak at the onset of an absorption edge. These peaks are referred to as "white lines," since they were first observed as relatively unexposed lines on X-ray plates. Brown *et al.*[89] studied these large peaks in several materials and concluded that the L_I white lines (s $\rightarrow$ p) in Te, Sn, Sb, and In are due to transitions to a large density of p-symmetry final states, the $L_{II,III}$ white lines (p $\rightarrow$ s,d) in Ta and Pt are due to transitions to a large density of d-symmetry final states, while

[109] R. A. van Nordstrand, *Roentgenspektren Chem. Bindung, Vortr. Int. Symp., 1965* p. 255 (1966).

[110] R. G. Shulman, Y. Yafet, P. Eisenberger, and W. E. Blumberg, *Proc. Natl. Acad. Sci. U.S.A.* **73,** 1384 (1976).

[111] S. P. Cramer, K. O. Hodgson, W. O. Gillium, and L. E. Mortenson, *J. Am. Chem. Soc.* **100,** 3398 (1978).

[112] H. W. Huang, S. H. Hunter, W. K. Warburton, and S. C. Moss, *Science* **204,** 191 (1979).

the K-edge white lines (s → p) in Ge and Se are excitonic in origin. Holland *et al.*[79] attributed the white line at the As K edge to an s → p resonance.

Heald and Stern[43] have observed a strong anisotropy of the Se K-edge white line in a single crystal of 2H–WSe_2, which correlates well with anisotropy in the EXAFS proper.[113]

Wei and Lytle[114] compared the shape of the Ta L_{III} white line with Lorentzian and Breit–Wigner–Fano lineshapes, with inconclusive results. Lytle *et al.*[115] studied the systematics of the effect of chemical environment on the L_{III} white line of Ir, Pt, and Au, and concluded that the area of the peak correlates well with the number of holes in the d band, as one might expect for this p → d transition.

Recently, the edge structure has been used to measure the fractional valence in the mixed valent systems TmSe[116] and SmS alloyed with Y.[117,118] These are discussed in Section 11,c.

In analogy with early theories of the EXAFS, the sparse theoretical work on this region of the absorption spectrum may be classified into two approaches: long-range order (LRO) and short-range order (SRO). Muller *et al.*[119] have calculated the 4d transition metal absorption edge structure from the band structure by ignoring the effect of the core hole (i.e., a LRO approach). Szmulowicz and Pease[120] have obtained good agreement with the first 20 eV of the Ni K- and $L_{II,III}$-edge absorption spectra using an augmented-plane-wave band structure calculation. An approach based on the short-range structure would be more consistent, however, with the EXAFS analysis. Dill and Dehmer[121–123] have extended the X_α method of Johnson[124] and coworkers to treat continuum states and have discussed both the near

[113] Note that the anisotropy in the Ta L_{III}-edge white line in 1T-TaS_2 previously reported by E. A. Stern, D. E. Sayers, and F. W. Lytle [*Phys. Rev. Lett.* **37,** 298 (1976)] is actually due to sample thickness variations.

[114] P. S. P. Wei and F. W. Lytle, *Phys. Rev. B: Condens. Matter* [3] **19,** 679 (1979).

[115] F. W. Lytle, P. S. P. Wei, R. B. Greegor, G. H. Via, and J. H. Sinfelt, *J. Chem. Phys.* **70,** 4849 (1979).

[116] H. Launois, M. Rawiso, E. Holland-Moritz, R. Pott, and D. Wohlleben, *Phys. Rev. Lett.* **44,** 1271 (1980).

[117] R. M. Martin, J. B. Boyce, J. W. Allen, and F. Holtzberg, *Phys. Rev. Lett.* **44,** 1275 (1980).

[118] J. B. Boyce, R. M. Martin, J. W. Allen, and F. Holtzberg, *in* "Valence Fluctuations in Solids" (L. M. Falicov, W. Hanke, and M. B. Maple, eds.), p. 427. North-Holland, Amsterdam, 1981.

[119] J. E. Muller, O. Jepsen, O. K. Anderson, and J. W. Wilkins, *Phys. Rev. Lett.* **40,** 720 (1978).

[120] F. Szmulowicz and D. M. Pease, *Phys. Rev. B: Solid State* [3] **17,** 3341 (1978).

[121] D. Dill and J. L. Dehmer, *J. Chem. Phys.* **61,** 692 (1974).

[122] J. L. Dehmer and D. Dill, in *Proc. Int. Conf. Inn. Shell Ioniz. Phenom., 2nd, 1976* p. 221, (1976).

[123] J. L. Dehmer and D. Dill, *J. Chem. Phys.* **65,** 5327 (1976).

[124] For a review, see K. H. Johnson, *Adv. Quantum Chem.* **7,** 143 (1972).

edge and the EXAFS regions in a unified context. Natoli *et al.*[125] have used the approach of Dill and Dehmer to treat the near edge structure in $GeCl_4$. They achieved reasonable agreement but cautioned that the results are quite sensitive to the model used for the molecular potential.

Having explored the physical origin of an EXAFS spectrum and gained some appreciation for its information content, it is now appropriate to consider its measurement.

III. Experimental Techniques

Strictly speaking, an EXAFS spectrum can be derived only from a measurement of the photoabsorption cross section over an appropriate range of X-ray photon energies. The more direct approaches to this involve measuring the normal incidence transmittance or grazing angle reflectance of the sample. The absorption cross section can also be inferred indirectly through measurement of some particular response of the sample to the absorption events. Such indirect schemes may involve detection of the photoemitted electrons, or of the fluorescent X rays or Auger electrons emitted when the hole in the core of the excited atom is ultimately filled, or of ions which desorb from the surface of the sample in response to photoabsorption. There are also techniques such as electron energy loss and appearance potential spectroscopy in which an EXAFS spectrum or something similar is inferred from the interaction of energetic electrons with the sample. These various techniques differ widely, not only in experimental configuration and detection technique, but also in their relative appropriateness for different types of systems. Some are especially suited to dilute systems, for example, whereas others are suited to studies of surface structure. In Sections 8 and 9 we discuss several experimental techniques for acquiring an EXAFS spectrum, with particular emphasis on their relative advantages in obtaining structural information on different types of systems. A brief description of the experimental equipment for each technique is followed by a discussion of the factors which determine the ratio of signal to noise, S/N. The strengths and weaknesses of the various measurement methods will be apparent from the expressions for S/N. We begin with a discussion of transmission experiments in the hard X-ray region, since this is the simplest and the most commonly used method for obtaining an EXAFS spectrum.

8. Transmission Experiments

The experimental configuration for a transmission EXAFS experiment is shown schematically in Fig. 13. The apparatus consists of five essential

[125] C. R. Natoli, D. K. Misemer, S. Doniach, and F. W. Kutzler, *Phys. Rev. A.* **22,** 1104 (1980).

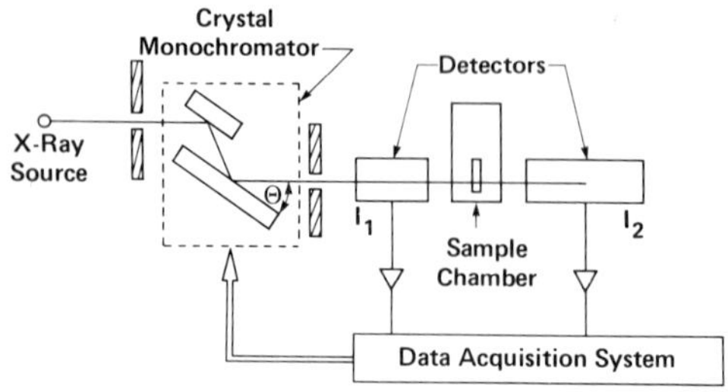

FIG. 13. Schematic representation of the experimental configuration for a transmission X-ray absorption experiment. A double-crystal monochromator, such as is shown here, is typically used with synchrotron radiation sources. The data acquisition system steps the monochromator to pass the desired photon energy, moves the sample to keep it in the X-ray beam, and monitors the resulting signals in I_1 and I_2.

components: the X-ray source, the monochromator, the detectors, the sample, and the data acquisition system. Put briefly, the X-ray source must provide a continuous spectrum of radiation. The X-ray monochromator is a crystal which selects a particular photon energy or wavelength from that spectrum through diffraction according to Bragg's law[126]:

$$n\lambda_X = 2d \sin \Theta, \qquad (8.1a)$$

where n is the order of the reflection, λ_X is the X-ray wavelength, d is the spacing of the planes in the crystal that diffract the radiation, and Θ is the angle that the incident and reflected beams make with these planes, as shown in Fig. 13. This equation can be rewritten in terms of the photon energy $\hbar\omega$ as

$$\begin{aligned} \hbar\omega &= 2\pi\hbar cn/2d \sin \Theta \\ &= (6199.3 \text{ eV})n/d \sin \Theta, \qquad (8.1b) \end{aligned}$$

where c is the speed of light and d is in angstroms.

After monochromatization, the X-radiation passes through a detector which monitors the total incident photon flux in the beam I_1. The X-ray beam then passes through a sample and the transmitted flux I_2 is measured in a second detector. For a sample of thickness z_s, the attenuation coefficient μ_s is given by the well-known expression[127]

$$\mu_s z_s = \ln(I_1/I_2). \qquad (8.2)$$

[126] For example, see L. V. Azároff, "Elements of X-Ray Crystallography." McGraw-Hill, New York, 1968.

[127] For example, see R. M. Eisberg, "Fundamentals of Modern Physics," Chapter 14. Wiley, New York, 1964.

In the X-ray region where most EXAFS experiments are performed (i.e., $\hbar\omega < 100$ keV), the attenuation is dominated by the photoelectric effect. As a result, μ_s is the photoelectric absorption coefficient of the sample. It is given by

$$\mu_s = \sum_{\alpha} n_\alpha \sigma_\alpha(\omega), \tag{8.3}$$

where n_α is the density of atoms of species α, σ_α is the atomic photoabsorption cross section discussed earlier, and the sum includes all atom species in the sample.

The apparatus is typically controlled by a computer data acquisition system. The computer steps Θ, and thus photon energy, through the region of interest, positions the sample to remain in the monochromatized X-ray beam, and acquires the digitized I_1 and I_2 readings. One thereby obtains $\mu(\omega)$ using Eqs. (8.1) and (8.2). We now discuss some of the pertinent features of each of the components of the apparatus, using as an example the specific parameters appropriate for the Stanford Synchrotron Radiation Laboratory (SSRL) and the storage ring SPEAR at the Stanford Linear Accelerator Center.[128]

a. X-Ray Sources

Two sources of continuous X radiation are available for EXAFS experiments: bremsstrahlung from a conventional X-ray source and synchrotron radiation. Before contrasting the relative advantages of these two types of sources, we first review some of their spectral characteristics.

X rays are produced in an X-ray tube when high-energy electrons (≈ 50 keV) strike the anode. The spectrum consists of both continuous bremsstrahlung radiation and characteristic line spectra. The continuous radiation is typically three orders of magnitude less intense than the characteristic lines. Since the continuous radiation is used in an EXAFS scan, a strong characteristic line can severely distort the EXAFS spectrum if it lies in the range of the scan. The intensity of the continuous radiation varies only slowly up to a cutoff given by the kinetic energy of the electrons striking the anode. The electric field vector $\hat{\epsilon}$ of the radiation is partially polarized in the plane containing both the velocity vector of the electron beam and the vector from the anode to the observation point. For the case of small-angle scattering of the electron beam at the anode, the expressions for the angular distribution of the radiation are simple.[129] The angular distribution of the two polarizations differ, with the perpendicular component being

[128] S. Doniach, I. Lindau, W. E. Spicer, and H. Winick, *J. Vac. Sci. Technol.* **12,** 1123 (1975).

[129] For example, see J. D. Jackson, "Classical Electrodynamics," Chapter 15. Wiley, New York, 1965.

isotropic and the parallel component varying as the square of the cosine of the angle between the two vectors identified above.

Synchrotron radiation is produced by electrons or any other charged particles traveling at relativistic velocities in curved orbits in a storage ring.[130] Its intensity increases slowly with increasing photon energy $\hbar\omega$ for photon energies well below a critical energy, beyond which it falls off exponentially with energy. The critical energy is given by

$$\hbar\omega_c = (3\hbar c/2R_s)\gamma_e^3 , \tag{8.4}$$

where c is the speed of light, R_s is the radius of curvature of the electron beam in the storage ring, and $\gamma_e \equiv E/mc^2$ is the relativistic enhancement factor for electrons of rest mass m and energy E. For the SPEAR radius of 12.7 m and an operating energy of 3 GeV, $\gamma_e = 5871$ and $\hbar\omega_c = 4.7$ keV. This radiation is well collimated in the direction perpendicular to the plane of the circulating beam (i.e., in the vertical direction), with a divergence angle of $\approx 1/\gamma_e$ for photon energies near the critical energy. For SPEAR operating at 3 GeV, the vertical divergence angle is $\approx$0.2 mrad at $\hbar\omega_c$. This small divergence is particularly convenient for experiments, since the radiation is concentrated in a small vertical distance on the order of a sample size. This 0.2-mrad divergence corresponds to a beam height of only 4 mm at an experimental station 20 m from SPEAR. In addition, synchrotron radiation is almost completely linearly polarized. For an observation point in the plane of the electron orbit, $\hat{\epsilon}$ is polarized 100% parallel to the plane. A small component perpendicular to the plane develops as the observation point moves out of this plane, increasing with distance from the plane. The total intensity integrated over all energies and angles contains seven times as much energy with parallel polarization as with perpendicular polarization. Synchrotron radiation also has a pulsed time structure, since the electrons travel in bunches in the storage ring. In the "one-bunch, one-bucket" mode of operation at SPEAR, the pulses are 0.3 ns wide and occur once every 780 ns.

As concerns a typical EXAFS experiment, the most relevant differences between these two sources are intensity and polarization. For photon-stimulated desorption experiments, the pulsed time structure of the radiation from the synchrotron is unusually important and will be discussed in Section 9,c. We consider the intensity first, as it is of great importance for all types of EXAFS experiments.

The EXAFS signal typically contributes only a few percent to the total

[130] For a thorough discussion of the properties of synchrotron radiation see H. Winick, *in* "Synchrotron Radiation Research" (H. Winick and S. Doniach, eds.), Chapters 2 and 3, and references contained therein. Plenum, New York, 1980.

absorption and must often be measured with a signal-to-noise ratio S/N greater than 100 in order to determine the structure accurately. This implies that the absorption coefficient must be determined to better than 1 part in 10^4, placing stringent requirements on all components of the apparatus, particularly on the X-ray source. Since S/N is proportional to the square root of the X-ray intensity, as discussed below, the X-ray source must be very intense. The early EXAFS experiments were performed using a 1-kW sealed X-ray tube and a flat crystal monochromator.[131] This arrangement produced intensities of 10^3–10^4 photons/s in a 4-eV bandwidth. This intensity is to be contrasted with that at SSRL of 10^8–10^{10} photons/s in a 2-eV bandwidth. This means that experiments which required weeks to perform on the conventional X-ray source could be completed in a half hour at SSRL.[72]

Improvements have been made on both types of sources since these first experiments. The intensity from conventional X-ray tubes can be increased by collecting a larger fraction of the radiation diverging from the tube and focusing it on the sample. This is done with a curved crystal which simultaneously focuses and monochromatizes the X-ray beam. Such an arrangement can increase the intensity by a factor of 100.[132] Further increases to 10^7 photons/s have been obtained using powerful rotating anode X-ray sources. By additionally degrading the resolution to ≈ 10 eV at the Cu K edge (i.e., at 8.98 keV), intensities approaching the lower end of the range for SSRL can be achieved (i.e., 10^8 photons/s).[133]

The X-ray flux available from a synchrotron can be increased in analogous ways. Focusing mirrors can be utilized to collect photons diverging in the horizontal and vertical directions, increasing the flux with some loss in energy resolution.[134] Such a mirror at SSRL increases the flux by a factor of 30.[135] The brightness of the synchrotron source can be increased dramatically by running at higher stored currents under dedicated operation and by employing special electron-accelerating devices like wigglers and undulators. With these modifications the photon flux can be increased greatly and the spectral characteristics of the synchrotron radiation modified substantially.[136,137] In the end, despite recent advances in conventional sources, the synchrotron remains the most intense continuous X-ray source

[131] F. W. Lytle, D. E. Sayers, and E. A. Stern, *Phys. Rev. B: Solid State* [3] **11,** 4825 (1975).

[132] G. S. Knapp, H. Chen, and T. E. Klippert, *Rev. Sci. Instrum.* **49,** 1658 (1978).

[133] J. A. del Cueto and N. J. Shevchik, *J. Phys. F* **7,** L215 (1977).

[134] J. A. Howell and P. Horowitz, *Nucl. Instrum. Methods* **125,** 225 (1975).

[135] J. B. Hastings, B. M. Kincaid, and P. Eisenberger, *Nucl. Instrum. Methods* **152,** 167 (1978).

[136] J. E. Spencer and H. Winick, *in* "Synchrotron Radiation Research" (H. Winick and S. Doniach, eds.), Chapter 21. Plenum, New York, 1980.

[137] B. M. Kincaid, *J. Appl. Phys.* **48,** 2684 (1977).

and thus the one with which the best S/N can be obtained. For many EXAFS experiments on concentrated systems, however, adequate S/N can be obtained using conventional sources, which may then be preferable in these instances owing to their obvious relative convenience.

A second important parameter of the X-radiation is polarization. The polarized photons can be used to probe the local structure in the various directions of an anisotropic material, since the EXAFS from a neighbor is proportional to $(\hat{\epsilon} \cdot \hat{\mathbf{r}})^2$ for K edges, where $\hat{\epsilon}$ is the direction of electric field polarization and $\hat{\mathbf{r}}$ is the direction vector from the excited atom to that neighbor in the sample. Synchrotron radiation is ideally suited to exploit the directionality of the EXAFS, since it is linearly polarized in the plane of the electron orbit of the storage ring. This polarization is complete for an observation point in the plane, but decreases as one moves out of the plane, as discussed above. The degree of polarization is enhanced further by Bragg diffraction from the monochromator and is adequate to obtain EXAFS information as a function of direction in an oriented sample.[86] The X rays from a conventional source cannot be used directly to obtain such information, since they are only partially polarized. A high degree of polarization can, however, be obtained upon diffraction from the monochromator.[138] If the monochromator crystal scatters as an ideal mosaic, the resulting polarization is given by[139]

$$P = (1 - \cos^2 2\Theta)/(1 + \cos^2 2\Theta). \tag{8.5}$$

For a Bragg angle Θ near 45°, P is near unity. By a judicious choice of the d-spacing of the crystal, one can arrange that Θ be near 45° at the energy of interest. During an EXAFS scan, however, Θ is varied and so the degree of polarization changes. For example, $P = 99\%$ if $\Theta = 43°$ at the Cu K edge. In a scan to $E = 1500$ eV above the edge [i.e., to $k \simeq (2mE/\hbar^2)^{1/2} \approx 20$ Å^{-1}], Θ changes to 36° and P is reduced to 82%. This disadvantage of using the monochromator to polarize the X rays makes synchrotron radiation the obvious choice for studies of anisotropic structures.

b. Monochromators

The preferred way to monochromatize an X-ray beam is by diffraction from a crystal. The emerging beam then has a photon energy given by Eq. (8.1). Both flat and bent crystal monochromators have been used. A bent crystal monochromator is often preferable for conventional X-ray sources, since it can focus a large fraction of that source's diverging radiation on the

[138] J. A. del Cueto and N. J. Shevchik, *J. Phys. E* **11,** 1 (1978).

[139] For example, see B. E. Warren, "X-ray Diffraction." Addison-Wesley, Reading, Massachusetts, 1969.

sample, with a resulting increase in intensity (but with some loss in energy resolution).[140]

For synchrotron sources, a double crystal or a channel cut crystal with a horizontal axis of rotation is usually employed (see Fig. 13). This arrangement has the advantage that the direction of the naturally collimated beam is unchanged. It is merely deflected in the vertical direction by a small distance $2D\cos\Theta$, where Θ is the Bragg angle and D is the spacing between the two crystals. As Θ is scanned, the detectors and sample have to be moved slightly to remain in register with the beam. For example, using a Si(220) double-crystal monochromator with $D = 1$ cm, the vertical position of the beam changes by about 0.4 mm in a scan from the Cu K edge to 1500 eV above the edge ($k \simeq 20$ Å^{-1}). Double-crystal monochromators which provide a constant offset of the exit beam have also been employed.[141]

The energy resolution of a monochromator is of particular interest. In general, the resolution requirements for EXAFS are not too severe. The most rapid variation of the EXAFS with k (or $\hbar\omega$) comes from the $\exp(2ikr_j)$ factor in Eq. (1.2), where r_j is the distance to a neighboring atom. The phase of this factor changes by 2π when k changes by $\approx\pi/r_j$. To resolve the EXAFS, one requires that $\delta k \ll \pi/r_j$ or

$$\delta(\hbar\omega) \ll (2\pi\hbar^2/m)k/r_j\,. \tag{8.6}$$

The information at low k and about more distant neighbors is thereby more severely affected by poor resolution. The most stringent requirement corresponds to $k \approx 2.5$ Å^{-1} and $r_j \approx 5$ Å, yielding $\delta(\hbar\omega) \ll 23$ eV. A resolution of $\delta(\hbar\omega) \approx 1$ eV is therefore perfectly adequate, whereas $\delta(\hbar\omega) \approx 10$ eV would cause distortions of the EXAFS which could affect the determination of distances, coordination numbers, and peak shapes.[104]

The resolution of the monochromator can be estimated by differentiating Eq. (8.1) to get

$$\delta(\hbar\omega)/\hbar\omega = \cot\Theta\,\delta\Theta. \tag{8.7}$$

Here $\delta\Theta$ is the total spread in Bragg angle due to all sources. A discussion of the contributions to $\delta\Theta$ for a focusing monochromator and a conventional X-ray source is given by Knapp *et al.*[132] and Cohen *et al.*[142] Typical values are $\delta(\hbar\omega) \approx 10$ eV at the Cu K edge ($\hbar\omega = 8.98$ keV). As discussed above, such resolution can cause some distortion of the EXAFS.

For a synchrotron source with a double-crystal monochromator, the

[140] E. A. Stern, ed., "Laboratory EXAFS Facilities-1980," Chapters 1, 3, and 9. Am. Inst. Phys., New York, 1980.

[141] J. Cerino, J. Stöhr, N. Hower, and R. Z. Bachrach, *Nucl. Instrum. Methods* **172,** 227 (1980); J. A. Golovchenko, R. A. Levesque, and P. L. Cowan, *Rev. Sci. Instrum.* **52,** 509 (1981).

[142] G. G. Cohen, D. A. Fisher, J. Colbert, and N. J. Shevchik, *Rev. Sci. Instrum.* **51,** 273 (1980).

spread in Bragg angles is due to the vertical divergence $\delta\Theta_V$ of the beam, the intrinsic angular width $\delta\Theta_D$ of the diffraction (i.e., the Darwin width), and the horizontal divergence $\delta\Theta_H$ of the beam. Since these are independent, one can write the total width as $\delta\Theta \approx (\delta\Theta_V^2 + \delta\Theta_D^2 + \delta\Theta_H^2)^{1/2}$. $\delta\Theta_V$ is affected both by the finite vertical dimension of the source h_s and the height of the monochromator entrance slits h_m. For the entrance slit at a distance z_0 from the source, one has $\delta\Theta_V \approx (h_m + h_s)/z_0$. Typical values at SSRL are $h_m \approx$ 1 mm, $h_s \approx 0.3$ mm, and $z_0 = 20$ m, yielding $\delta\Theta_V \approx 10^{-4}$ rad. $\delta\Theta_D$ is typically smaller than this value and depends on the monochromator crystal, the particular reflection, and the photon energy.[139] As an example, a Si(220) crystal at 9 keV has $\delta\Theta_D \approx 5 \times 10^{-5}$ rad. $\delta\Theta_H$ is given by $\frac{1}{2}(w/z_0)^2 \tan \Theta$, where w is the horizontal width of the beam at the monochromator. This contribution is usually negligible.[143] For a w of 2 cm, $\delta\Theta_H \approx 10^{-6}$ rad, much smaller than $\delta\Theta_V$. The foregoing values give $\delta\Theta \approx 10^{-4}$ rad, which yields a resolution of about 2 eV at a 9-keV photon energy for a Si(220) monochromator, according to Eq. (8.7). As discussed above, such a width will cause negligible distortion of the EXAFS. If a focusing mirror is used, $\delta\Theta_V$ can be increased to the entire angular divergence of the synchrotron radiation, which at SSRL is $\approx 3 \times 10^{-4}$ rad at 2.5 GeV stored energy and 8-keV photon energy. This increases the photon flux but degrades the resolution to about 7 eV, which can distort the EXAFS, as discussed above.

Another consideration pertinent to the monochromator is higher-order reflections. According to Bragg's law, if a photon energy $\hbar\omega$ is an allowed reflection, then the higher harmonics of this energy are also allowed reflections at the same Bragg angle, provided that their geometrical structure factor is not zero. These higher harmonics contaminate the monochromatic photon beam and can distort the EXAFS spectra. The predominant distortion is a reduction in the EXAFS amplitude, as discussed below. Two techniques exist to reduce the harmonic content of the beam using the monochromator. The first is to use a crystal for which the geometrical structure factor vanishes for specific harmonics which are troublesome. For Si(111) and Ge(111), for example, the structure factor for the second harmonic (222) beam is zero. The first allowed beam is the third harmonic (333) beam. A second technique is to detune slightly the parallelism of a double monochromator.[144] Since the Darwin width is narrower for the higher-energy reflections than for the fundamental, the monochromator may be adjusted to pass nearly all the first harmonic, while essentially excluding the higher orders. Other methods exist to reduce the harmonic

[143] M. Lemonnier, O. Collet, C. Depautex, J. M. Esteva, and D. Raoux, *Nucl. Instrum. Methods* **152,** 109 (1978).

[144] G. S. Brown and S. Doniach, *in* "Synchrotron Radiation Research" (H. Winick and S. Doniach, eds.), Chapter 10. Plenum, New York, 1980.

contamination of the beam, such as using a mirror with a high-energy cutoff which falls between the fundamental and the harmonic energies, or using detectors with some energy selectivity. Further discussion of the problem of the higher-order Bragg reflections is given below in the section on signal-to-noise considerations.

c. Detectors

The standard X-ray detectors are photographic films, gas ionization detectors, and solid-state detectors.[145] Film is seldom used to record EXAFS scans. Two types of gas ionization detectors are commonly used in EXAFS experiments: the ionization chamber and the proportional counter. Both consist of an inert gas between two electrodes. Electron–ion pairs are produced when an X-ray photon is absorbed. In argon, for example, it takes 30 eV to produce an electron–ion pair. Approximately 300 electron–ion pairs are produced for each 9-keV photon absorbed. If the voltage across the electrodes produces an electric field which exceeds $\approx$100 V/cm, the electrons and ions do not recombine but are swept apart toward the electrodes. This produces an electric current in the external circuit which is proportional to the number of X-ray photons absorbed. If the detector is operated in the region between about 100 and 25,000 V/cm, the gain is unity and independent of the applied electric field. In this case the detector is referred to as an ionization chamber. If the electric field in the gas exceeds about 25,000 V/cm, then the electrons produced by the X rays are accelerated sufficiently to produce secondary electrons. This gives the detector a gain which is proportional to the applied electric field, up to the field where an electron avalanche or discharge occurs in the gas. The detector operated in this mode is referred to as a proportional counter, since the current pulse produced is proportional to the energy of the X-ray photon absorbed. As a result, the proportional counter can be used to discriminate among various photon energies. Its energy resolution is $\approx$1300 eV for 10-keV X rays. Standard photon counting techniques can be used with the proportional counter, provided that the mean time between the arrival of X-ray photons is greater than the dead time of the detector, typically 300 ns. The proportional counter saturates for a flux greater than $\approx 10^6$ photons/s.

Two types of solid-state detectors are commonly used: scintillation counters and semiconductor detectors. In a scintillation detector, pulses of visible light are produced by fluorescence resulting from the absorption of X rays in the scintillating material. The visible radiation is detected with a pho-

[145] See L. V. Azároff, "Elements of X-Ray Crystallography," Chapter 12. McGraw-Hill, New York, 1968.

tomultiplier tube. These detectors have high gain, reasonably short dead times, and relatively poor energy resolution. For standard NaI(Tl) scintillation detectors, as an example, the dead time and energy resolution are on the order of 1 μs and 4000 eV at a 10-keV photon energy. Semiconductor detectors are Si or Ge crystals compensated with Li. They have an intrinsic gain of unity but short dead times ($\approx$10 ns) and good energy resolution ($\approx$200 eV) at a 10-keV photon energy. A discussion of detectors and the relevant operating characteristics of each is contained in the articles by Brown and Doniach[144] and Stern.[146]

For synchrotron sources where the photon flux is large (i.e., $\geq 10^8$ photons/s), digital photon counting techniques are not feasible and gas ionization chambers with analog counting techniques are employed. The current from the chamber is amplified, digitized, and stored by a computer. For conventional X-ray sources, the flux is often low enough so that digital pulse counting techniques *can* be used. Then a scintillation detector with fast counting electronics is most often employed. This detection scheme has the added advantage of X-ray energy resolution, so that higher harmonics from the X-ray source can be excluded. It has the disadvantage that nonlinearities occur for count rates exceeding $10^5\ s^{-1}$, so that care must be taken with the more intense laboratory sources to avoid distortion of the EXAFS.

Another consideration relevant to the detectors is their response as a function of photon energy. As the X-ray energy is changed during an EXAFS scan, the efficiency of the detectors will also be changing. Consider the experimental setup of Fig. 13, where the incident intensity I_1 and transmitted intensity I_2 are monitored by gas ionization chambers. The current in the chambers is proportional to the product of the number of electron–ion pairs produced by each photon and the number of photons absorbed in the gas; that is, it is proportional to $\hbar\omega[1 - \exp(-\mu_i z_i)]$, where μ_i and z_i are the absorption coefficient and length, respectively, for each gas ionization chamber. The incident intensity is attenuated in the first counter and also in the sample, so that the measured signal, $\ln(I_1/I_2)$, is not simply $\mu_s(\omega)z_s$ as given by Eq. (8.2) but, rather,

$$\ln(I_1/I_2) = \mu_s(\omega)z_s + \ln(\{\exp[\mu_1(\omega)z_1] - 1\}/\{1 - \exp[-\mu_2(\omega)z_2]\}). \quad (8.8)$$

The second term on the right-hand side is a smoothly varying function of energy, provided that $\hbar\omega$ does not cross an absorption edge in μ_1 or μ_2 during a scan. It simply adds to the measured slowly varying background absorption of the sample, causing no distortion of the EXAFS, and is removed in the data reduction together with the prethreshold absorption, as discussed in Section 10.

[146] E. A. Stern, ed., "Laboratory EXAFS Facilities-1980," Chapter 4. Am. Inst. Phys., New York, 1980.

d. Samples

EXAFS has been used to study a variety of gases, liquids, and solids over a wide range of temperatures and pressures. While different sample considerations apply in each case, there are also some general criteria which should be satisfied for a transmission experiment concerning the sample container and the sample thickness and homogeneity. Firstly, the sample container should not be too absorbing in the photon energy region of interest. It adds to the background absorption and so is simply removed in data reduction, causing no distortion of the EXAFS, provided that it is a smoothly varying function of photon energy. It does deteriorate the S/N ratio, however, since it absorbs photons which contribute nothing to the EXAFS. This is discussed further in the next section. As a result, the sample containers should be kept as thin as possible and constructed of low-Z elements to minimize the absorption. (Note that atomic absorption is proportional to Z^4/ω^3.) Substrates and sample holders are therefore made of materials like graphite and boron nitride for high-temperature work.[147,148] At moderate and low temperatures, the holders, substrates, and windows are constructed of light metal foils like Be and Al or hydrocarbon films like Kapton. Special considerations of sample-holder strength apply for high-pressure work, but sample cells consisting of boron and lithium hydride[149] and of boron carbide and diamond[150,151] have proven successful.

Furthermore, the sample itself should be homogeneous and approximately two absorption lengths thick. This thickness condition, that $\mu_s z_s \approx 2$ at the absorption edge of interest, gives the optimum S/N ratio, as discussed in the next section. It corresponds to $\approx$90% of the photons being absorbed in the sample. As examples, the optimum thickness for Cu K-edge EXAFS is $\approx 10\ \mu$m for Cu metal and 25 μm for CuCl, while that for the Br K edge in Br_2 gas at STP is $\approx$2 cm.

The sample should also be homogeneous, that is, of uniform thickness with no pinholes. Any variation in the sample uniformity will lead to a distortion of the EXAFS, which takes the form of an apparent decrease in the EXAFS amplitude. This will affect determinations of the coordination number and the peak width, but will not seriously affect near-neighbor spacings. These points are discussed further in the next section.

[147] E. D. Crozier, F. W. Lytle, D. E. Sayers, and E. A. Stern, *Can. J. Chem.* **55,** 1968 (1977).

[148] J. C. Mikkelsen, Jr., J. B. Boyce, and R. Allen, *Rev. Sci. Instrum.* **51,** 388 (1980).

[149] R. Ingalls, G. A. Garcia, and E. A. Stern, *Phys. Rev. Lett.* **40,** 334 (1978).

[150] O. Shimomura, T. Fukamachi, T. Kawamura, S. Hosoya, S. Hunter, and A. Bienenstock, *Jpn. J. Appl. Phys.* **17,** Suppl. 17-2, 221 (1978).

[151] R. Ingalls, E. D. Crozier, J. E. Whitmore, A. J. Seary, and J. M. Tranquada, *J. Appl. Phys.* **51,** 3158 (1980).

e. Signal-to-Noise Considerations

There are many sources of signals which can modify or add structure to an absorption spectrum, thereby reducing the information which can be extracted from the measured EXAFS. These can be classified into two categories: sources of statistical noise and sources of systematic distortions. Included in the former category are all fluctuations of a statistical nature, such as the intrinsic measurement error in detecting a finite number of photons and the noise figure of the amplifiers. Sources of systematic distortions include X-ray source instabilities and variations, competition among diffraction processes within the monochromator, harmonic contamination of the beam, detector nonlinearities, and sample inhomogeneities. Such systematic distortions are often the limiting factor in determining the quality of the EXAFS data. We first consider statistical noise, however, since it is possible to derive general expressions for it and to draw conclusions from them about various detection schemes. Also, it is the statistical noise that determines the best S/N ratio obtainable for a perfect system.

(i) Statistical Noise. Generally the amplifier noise can be made smaller than the $N_{\text{ph}}^{1/2}$ fluctuations in detecting N_{ph} photons. This is the case for a synchrotron source and gas ionization chambers. If, on the other hand, the detected flux is lowered below 10^5 photons/s, as with laboratory sources, then detectors with some intrinsic gain (i.e., proportional counters or scintillation counters) can be used together with digital pulse counting techniques to minimize amplifier noise. In either case, the $N_{\text{ph}}^{1/2}$ in the photon statistics most often dominates the noise, while the amplifier merely degrades S/N by a small factor.

Consider for simplicity the case where the first detector contributes no noise and passes N_1 photons. The number of photons detected in the transmitted intensity monitor is $N_2 = N_1 \exp(-\mu_s z_s)$. Recalling that μ_s is related simply to the atomic absorption cross sections [see Eq. (8.3)], we can separate the various contributions to μ_s in analogy with Eq. (1.1). Specifically, divide the total absorption coefficient μ_s into a contribution μ_α due to the absorption edge of interest and a background contribution μ_{bg} due to the other absorption of the sample:

$$\mu_s = \mu_{\text{bg}} + \mu_\alpha = \mu_{\text{bg}} + \mu_\alpha^0[1 + \chi_\alpha(k)]. \tag{8.9}$$

The EXAFS oscillations χ give rise to small oscillations in I_2 as a function of photon energy,

$$S \propto N_2 \mu_\alpha^0 \chi_\alpha(k) z_s\,. \tag{8.10}$$

The noise is proportional to the square root of the total number of photons detected, $N_2^{1/2}$, yielding

$$\text{S/N} \approx N_1^{1/2} \mu_\alpha^0 \chi_\alpha(k) z_s \exp(-\mu_s z_s/2). \tag{8.11}$$

Maximizing this with respect to sample thickness gives an optimum sample thickness of two absorption lengths:

$$(\mu_s z_s)_{opt} = 2, \tag{8.12}$$

where μ_s is evaluated just above the edge of interest. For this thickness,

$$(S/N)_{opt} = (2/e)N_1^{1/2}\chi_\alpha(k)(\mu_\alpha^0/\mu_s), \tag{8.13}$$

where $e \simeq 2.718$. If the finite absorption of the beam monitor is taken into account, then the $2/e$ factor in Eq. (8.13) is modified slightly to 0.56, $(\mu_s z_s)_{opt} = 2.55$ instead of 2, and the optimum beam monitor absorption corresponds to $(\mu_1 z_1)_{opt} = 0.24$.[152,153] This corresponds to the absorption of about 20% of the incident photons by the I_1 beam monitor. The optimum sample thickness yields a transmission of about 10% of the beam. Equation (8.13) shows that the S/N increases with the square root of the incident photon flux, as expected. It also decreases as the background absorption increases according to $\mu_\alpha/(\mu_{bg} + \mu_\alpha)$, so that for dilute samples, where $\mu_{bg} \gg \mu_\alpha$, the transmitted EXAFS signal can be quite small. In this case, other techniques such as fluorescence are more advantageous, as discussed below.

(ii) Systematic Distortions. The S/N determined by considering photon statistics alone can be quite large for an intense X-ray source. This S/N is often not realized, owing to other sources of error, given the stringent requirement that the total absorption of the sample be measured to roughly 1 part in 10^4. For example, fluctuations in the X-ray beam intensity due to source instabilities may not cancel out in I_1/I_2 to this accuracy, owing to nonlinearities in the detectors. Also, the incident intensity may decrease, owing to the Bragg condition being satisfied simultaneously for some additional diffracted beam. The appearance of such spurious beams can abruptly reduce the primary beam intensity by as much as 50%. Other signals present in the detectors, such as higher harmonics or background radiation from the sample, may be affected differently by these unwanted beams. In this case, the reduction in beam intensity will not appear in the same proportion in the two detectors, I_1 and I_2, and will not "ratio out" precisely. This gives rise to the well-known "glitches." Ways to minimize their effect have been discussed by G. S. Brown.[154]

Since harmonic contamination of the X-ray beam is a prevalent problem that causes significant distortions and aggravates the "glitch" problem, we consider it further. If the primary and second harmonics are present in the beam, leaving the monochromator with intensities $I_0(\omega)$ and $I_0(2\omega)$, respectively, one has for the first and second detectors, respectively,

[152] M. E. Rose and M. M. Shapiro, *Phys. Rev.* [2] **74,** 1853 (1948).
[153] B. M. Kincaid, SSRL Rep. No. 75/03 (1975).
[154] G. S. Brown, "Some Notes on X-ray Absorption Spectroscopy" (unpublished).

$$I_1 = G_1(\omega)I_0(\omega) + G_1(2\omega)I_0(2\omega)$$

and

$$I_2 = G_2(\omega)I_0(\omega)\exp[-\mu_s(\omega)z_s] + G_2(2\omega)I_0(2\omega)\exp[-\mu_s(2\omega)z_s], \quad (8.14)$$

where $\mu_s(\omega)$ and $\mu_s(2\omega)$ are the sample absorption coefficients for the first and second harmonics, respectively, and the G's are the appropriate sensitivities of the I_1 and I_2 detectors to the first and second harmonics. $G_2(\omega)$ and $G_2(2\omega)$ include the attenuation of $I_0(\omega)$ and $I_0(2\omega)$ due to the I_1 detector, so that for gas ionization counters

$$G_1(\eta\omega) \propto \eta\hbar\omega\{1 - \exp[-\mu_1(\eta\omega)z_1]\}$$

and

$$G_2(\eta\omega) \propto \eta\hbar\omega\exp[-\mu_1(\eta\omega)z_1]\{1 - \exp[-\mu_2(\eta\omega)z_2]\}. \quad (8.15)$$

In these expressions, the subscripts 1 and 2 refer to the first and second detectors, respectively, and $\eta = 1$ or 2 for the first or second harmonic. The EXAFS is obtained from the natural logarithm of I_1/I_2, which is given by

$$R_I(\omega) \equiv \ln(I_1/I_2) = \mu_s(\omega)z_s - \ln(1 + A\{\exp[\mu_s(\omega)z_s] - 1\}) - \ln B, \quad (8.16)$$

where

$$A = C/(1 + C),$$

$$C = G_2(2\omega)I_0(2\omega)\exp[-\mu_s(2\omega)z_s]/[G_2(\omega)I_0(\omega)],$$

and

$$B = [G_2(\omega)I_0(\omega) + G_2(2\omega)I_0(2\omega)\exp(-\mu_s(2\omega)z_s)]$$

$$\times [G_1(\omega)I_0(\omega) + G_1(2\omega)I_0(2\omega)]^{-1}. \quad (8.17)$$

C is proportional to the ratio of the detected intensity of the second harmonic to that of the first harmonic. It is zero for a truly monochromatic incident beam. $\ln B$ is a smoothly varying background term similar to the gain term in Eq. (8.8). It will cause distortions of the EXAFS only if there are large changes in the first harmonic intensity which do not occur in the same proportion in the second harmonic intensity, as is the case for the "glitch" problem mentioned above. The second term on the right-hand side of Eq. (8.16) can, however, cause severe systematic distortions. Due to the decrease in absorption with increasing $\hbar\omega$, one usually has $\mu(\omega) > \mu(2\omega)$ and $G(\omega) > G(2\omega)$. Since $0 \leq A \leq 1$, $\exp[\mu_s(\omega)z_s] \geq 1$, and the second term on the right-hand side of Eq. (8.16) appears with a negative sign, the effect of any change in $\mu_s(\omega)z_s$ will be reduced by the contribution of that second term. Consequently, the measured edge height and the amplitude of the EXAFS oscillations will appear to be reduced in the presence of harmonic radiation. This reduction is greatest when the transmitted signal is least, so that the EXAFS portion of the spectrum is reduced more than the edge height. As a result, when the EXAFS is normalized to the edge height in the usual data analysis procedure, it will appear with a reduced amplitude. The oscillations are proportional to the derivative of Eq. (8.16) with respect to changes in $\mu_s(\omega)z_s$:

$$\delta R_I(\omega) = \delta[\mu_s(\omega)z_s]/\{1 + C\exp[\mu_s(\omega)z_s]\}. \tag{8.18}$$

The measured step height at the edge, $\Delta R = R_I(\omega^+) - R_I(\omega^-)$, can also be obtained from Eq. (8.16). The + and − denote energies above and below the edge, respectively. Since the EXAFS χ is approximately $\propto \delta R_I(\omega)/\Delta R_I$, one has the following for the ratio of the measured to the actual χ's:

$$\chi_M/\chi = [\mu_s(\omega^+) - \mu_s(\omega^-)]z_s(D^+\{[\mu_s(\omega^+) - \mu_s(\omega^-)]z_s - \ln(D^+/D^-)\})^{-1}, \tag{8.19}$$

where $D^\pm = \{1 + C\exp[\mu_s(\omega^\pm)z_s]\}$. The term on the right-hand side of Eq. (8.19) is always less than or equal to one, so that the EXAFS amplitude is always reduced by higher harmonics. This expression also shows that the reduction of the EXAFS amplitude is larger for thicker samples. We treat an example for specificity.

Consider the K edge of Cu (at 8.98 keV), for which $\mu_s(\omega^+)/\mu_s(\omega^-) = 7.6$, and a sample with the optimum thickness, $\mu_s(\omega^+)z_s = 2.55$. Then $[\mu_s(\omega^+) - \mu_s(\omega^-)]z_s = 2.21$, $D^+ = 1 + 12.8\ C$, and $D^- = 1 + 1.4\ C$. χ_M/χ decreases monotonically from 1.0 for $C = 0$ to 0.35 for $C = 1$. A reasonable estimate of C is needed to calculate the amplitude reduction. For this thickness $\exp[-\mu_s(2\omega)z_s] = 0.72$ and for SSRL operating at 3 GeV the ratio of the intensity of the first harmonic to the second harmonic is ≈ 5. For a perfect Si(220) monochromator, the total intensity of the second harmonic is reduced over that of the first harmonic at 9 keV by a factor ≈ 6. This yields $I_0(2\omega)/I_0(\omega) \approx 0.03$, so that $C \approx 0.02G_2(2\omega)/G_2(\omega)$. Consider a 30-cm-long gas ionization counter for I_2 and ignore the attenuation of the beam in the I_1 counter. If Ar gas is used in the I_2 counter, then from Eq. (8.15), $G_2(2\omega)/G_2(\omega) \approx 1$ and $C \approx 0.02$, yielding $\chi_M/\chi \approx 0.88$. For Ne gas in the second detector, $G_2(2\omega)/G_2(\omega) \approx 0.3$, giving C ≈ 0.01 and $\chi_M/\chi \approx 0.93$. This would be reflected as an EXAFS amplitude of 10 neighbors for the Ar gas detector and 11 neighbors for Ne gas, instead of the correct value of 12. These are relatively small corrections due to the small values of C for the parameters considered in this case. The value of C can be significantly larger for edges at lower energy, causing a more drastic reduction in the EXAFS amplitude. Thick samples are more severely affected. For example, if $\mu_s(\omega^+)z_s = 4$ instead of 2.55, as considered above, then $\chi_M/\chi \approx 0.6$ for the Ar gas detector instead of 0.88. This example also illustrates the fact that the amplitude reduction is larger for the heavier detector gas [for which $G_2(2\omega)/G_2(\omega)$ is larger].

Similar considerations apply in the case of an inhomogeneous sample. For a uniform sample with pinholes, Eqs. (8.16)–(8.19) apply, where A is the fraction of the illuminated area of the sample taken up by the pinholes. For 10% pinholes in Cu, $A = 0.1$ and $C = 0.11$, for which $\chi_M/\chi = 0.62$. Again the amplitude reduction is more severe for thicker samples for a given area of pinholes.[155]

f. Reflectivity Experiments

A technique that is complementary to the transmission experiments is that of measuring the reflectivity.[156–159] The X-ray beam strikes the sample at a small angle from the surface ($\approx$10 mrad), and the specularly reflected beam is detected. This technique makes use of the fact that the reflectivity is high (>50%) and the penetration depth is small (<20 Å) below a critical angle given by

$$\sin \Theta_c = (2\delta)^{1/2}, \tag{8.20}$$

where $1 - \delta$ is the real part of the index of refraction. (This is also the principle of operation for X-ray mirrors.) The reflectivity falls rapidly above this critical angle to almost zero, and the penetration depth increases substantially to the reciprocal of the absorption coefficient ($\approx 10^5$ Å). Near the critical angle, fine structure is observed in the reflectivity above the edge which is very similar to the EXAFS obtained from an absorption measurement.[156]

One advantage of this technique over transmission is that it is surface sensitive. The penetration depth for angles of incidence less than Θ_c is of order 20 Å, comparable to that of electron yield techniques. But, unlike electron yield, for which the count rates are relatively low, the signals in reflectivity are large, comparable with those from bulk absorption. This technique is suitable even with thick samples, which is particularly useful in the soft X-ray region, where the samples for transmission experiments would have to be too thin to be practical.

The reflectivity technique has some disadvantages that may limit its usefulness in some applications. Firstly, the sample lengths are quite large at the glancing angles required. For example, at an angle of incidence of 5 mrad for a study of the K edge in Cu metal,[160] a sample length of 10 cm would be needed for an incident beam of height 0.5 mm. The sample must

[155] P. Rabe, G. Tolkiehn, and A. Werner, *Nucl. Instrum. Methods* **171,** 329 (1980).

[156] R. Barchewitz, M. Cremonese-Visicato, and G. Onori, *J. Phys. C* **11,** 4439 (1978).

[157] G. Martens and P. Rabe, *Phys. Status Solidi A* **57,** K31 (1980).

[158] R. Fox and S. J. Gurman, *J. Phys. C* **13,** L249 (1980).

[159] G. Martens and P. Rabe, *J. Phys. C* **14,** 1523 (1981).

[160] G. Martens and P. Rabe, *Phys. Status Solidi A* **58,** 415 (1980).

be flat and smooth over this entire area, since surface roughness can severely reduce the reflectivity. A second difficulty arises if one wants to extract detailed structural information from the reflectivity. If distances only are needed, a simplified analysis of the oscillatory structure may be adequate provided that proper account is taken of the phase shift, which is different from that in absorption. A more elaborate analysis is required, however, if quantitative information on the number of near neighbors and the near-neighbor peak width is desired. One approach is to relate the reflectivity to the absorption coefficient, for which the expression for the EXAFS is well known. A Kramers–Kronig analysis is required in this case.[158] In fact, a series of energy scans at different angles of incidence is needed to obtain the real and imaginary parts of the index of refraction. From this data the absorption can be obtained. Even then, however, a reliable determination of the EXAFS amplitude may be difficult.[159,161]

9. Indirect Methods

The indirect methods monitor some characteristic by-product of the X-ray absorption process. In some cases the intensity of this by-product is proportional to the number of absorption events that occur (i.e., the absorption coefficient), and so it may also exhibit EXAFS. The indirect methods for which fine structure has been observed include fluorescence, electron yield, ion desorption, and luminescence. There are also methods in which the initial excitation is accomplished with electrons instead of X rays, such as electron energy loss and appearance potential spectroscopy. Each of these will be discussed in turn, with more emphasis given to fluorescence and electron yield spectroscopy, since these two are the more widely used of the indirect methods. Many of the experimental considerations for the transmission technique discussed in Section 8 apply as well to the indirect techniques. In the following, we will discuss principally the differences. Particular emphasis will be placed on the advantages and disadvantages of the various indirect techniques relative to transmission.

a. Fluorescence Experiments

The core hole that is created by the photoabsorption process can decay by either a radiative or a nonradiative transition, as illustrated in Fig. 14. The nonradiative decay is a two-or-more-electron transition and is discussed in the next section. In radiative decay, a higher-lying electron fills the core hole with the emission of a photon of energy $\hbar\omega_f$ equal to the difference

[161] G. Martens and P. Rabe, *J. Phys. C* **13,** L913 (1980).

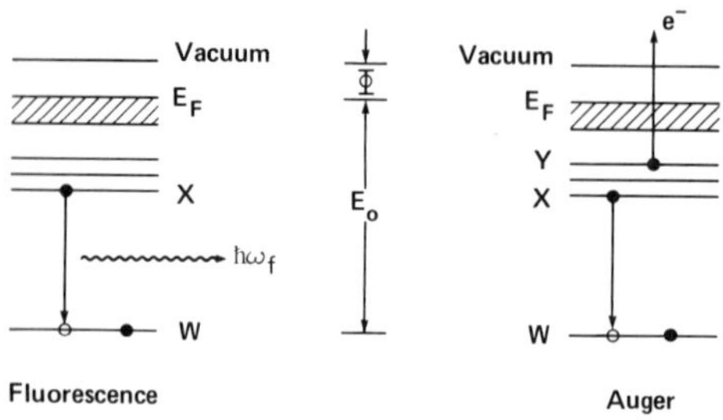

FIG. 14. The two processes by which the core hole created in the photoexcitation process can decay. In radiative decay, a fluorescent photon is emitted with an energy equal to the difference in the two energy levels, $\hbar\omega_f = E_X - E_W$. In the nonradiative decay process, an Auger electron is emitted to take up the energy from the X → W transition. The transitions giving rise to the WXY Auger peak are shown. Radiative decay is more probable for heavy atoms, and nonradiative decay more probable for light atoms, with the equal probability point for a 1s core hole occurring for $Z \approx 30$ (Zn).

in the two energy levels. This fluorescent photon energy is characteristic of the atomic transition that fills the core hole and is therefore independent of the incident photon energy. The intensity of the fluorescent yield is proportional to the probability that the incident photon has created a core hole; that is, it is proportional to the absorption coefficient. Thus the fluorescence will exhibit EXAFS.

In these EXAFS experiments, the fluorescent radiation at $\hbar\omega_f$ is monitored as a function of the energy of the incident radiation absorbed in the sample.[162] Monochromatized incident radiation of intensity $I_0(\omega)$ impinges on the sample of thickness z_s at an angle Θ_i relative to the front plane of the sample. The resulting fluorescent radiation is monitored by a detector that subtends a solid angle Ω at a mean angle Θ_f relative to the front plane of the sample. We assume that Ω is small, so that the angle between the sample surface and the detector is essentially constant over the area of the detector and equal to Θ_f. If z_n measures the depth into the sample normal to the front plane, then the intensity of the fluorescence coming from the region of the sample between z_n and $z_n + dz$ is

$$I_f(z_n)dz = (\Omega/4\pi)I_0 \exp[-\mu_s(\omega)z_n/\sin\Theta_i] \times (\epsilon_f \mu_\alpha\, dz/\sin\Theta_i) \exp[-\mu_s(\omega_f)z_n/\sin\Theta_f]. \quad (9.1)$$

In this expression, $\mu_s(\omega)$ is the total absorption coefficient at the incident photon energy $\hbar\omega$ due to both the absorption edge of interest μ_α and the other edges and atoms in the sample, μ_{bg}. ϵ_f is the fluorscent yield—the probability that the core hole created in the photoabsorption process decays

[162] J. Jaklevic, J. A. Kirby, M. P. Klein, A. S. Robertson, G. S. Brown, and P. Eisenberger, *Solid State Commun.* **23**, 679 (1977).

by means of a radiative transition at energy $\hbar\omega_f$. It increases monotonically with atomic number Z and is larger for K-line emissions than for L-line emissions. For K-shell fluorescence, for example, ϵ_f varies from 0.006 at $Z = 8$ (oxygen) to 0.5 at $Z = 30$ (Zn) to 0.96 at $Z = 78$ (Pt). For $Z > 50$, ϵ_f for K lines is essentially unity. For L-shell fluorescence, ϵ_f reaches 0.5 at $Z = 90$ (Th).[163] Integrating Eq. (9.1) over z_n from 0 to the sample thickness z_s yields the total fluorescence intensity

$$I_f(\omega_f) = I_0(\omega)\epsilon_f(\Omega/4\pi)[\mu_\alpha(\omega)/(\mu_s(\omega) + \mu_s(\omega_f)\sin\Theta_i/\sin\Theta_f)]$$
$$\times (1 - \exp\{-[\mu_s(\omega) + \mu_s(\omega_f)\sin\Theta_i/\sin\Theta_f]z_s/\sin\Theta_i\}). \quad (9.2)$$

The total detected photon flux I_t is then the sum of the fluorescence signal of interest $I_f(\omega_f)$ and a background signal I_b due to elastic scattering, Compton scattering, and other fluorescences of the sample and sample holder. The advantage of fluorescence over transmission is the much larger background rejection possible in the former. Before examining these expressions further and quantifying the relative advantages, we discuss some of the experimental methods used to minimize the background signal and maximize I_f.

For simplicity one often chooses $\Theta_i = \Theta_f$. In addition, it is often advantageous to have $\Theta_i = 45°$ so that the angle between the incident and fluorescent beams, $\Delta\Theta = \pi - (\Theta_i + \Theta_f)$, is 90°. This minimizes the contribution of both the elastic and inelastic scattered radiation to the background. For example, Thomson scattering due to an unpolarized source varies as $2 - \sin^2\Delta\Theta$, which is at a minimum at 90°. Likewise, Compton scattering from a free electron has a broad minimum at 90°.[164] In addition, with a polarized source, the fluorescence can be observed along the direction of the polarization to minimize the scattered background.

Furthermore, detectors with energy discrimination are utilized. The fluorescence signal of interest remains at $\hbar\omega_f$ as the incident photon energy is varied, while I_b is at $\hbar\omega$ and other energies. A detector that is sensitive only to a small band of energies about $\hbar\omega_f$ can exclude much of the quasi-elastic and matrix-fluorescence background. The detectors used range from scintillation counters with a resolution $\approx$ 5 keV to focusing, curved-crystal analyzers with a resolution $\approx$ 20 eV.[165] The latter involve an array of Bragg

[163] For a discussion of fluorescence yield, see W. Bambynek, B. Crasemann, R. W. Fink, H.-U. Freund, H. Mark, C. D. Swift, R. E. Price, and P. V. Rao, *Rev. Mod. Phys.* **44,** 716 (1972).

[164] See J. D. Jackson, "Classical Electrodynamics," p. 490. Wiley, New York, 1965.

[165] A discussion of detectors is given by G. S. Brown and S. Doniach, *in* "Synchrotron Radiation Research" (H. H. Winick and S. Doniach, eds.), Chapter 10. Plenum, New York, 1980; E. A. Stern, ed., "Laboratory EXAFS Facilities—1980," Chapter 4. Am. Inst. Phys., New York, 1980.

diffracting crystals with a point source and point focus,[166,167] which limits somewhat the total solid angle available but provides excellent energy discrimination. In addition, X-ray filters having an absorption edge between $\hbar\omega_f$ and $\hbar\omega$ can be used to remove much of the elastic and Compton scattered radiation.[168] Fluorescence from the filter can, of course, add to the background and thereby lessen the effect of the filter. Soller-slit assemblies can be used to reduce the filter fluorescence background.

Another consideration is the total solid angle subtended by the detector. This can be maximized by increasing the detector area using, for example, an array of many scintillation counters or placing the detector as close as possible to the sample.[169] In this way $\Omega/4\pi \approx 0.1$ may be achieved. Detector saturation at large subtended solid angles is very possible and must be avoided. Also, as mentioned above, crystal monochromator analyzers severely limit the subtended angle. As a result, other fluorescence X-ray detection systems have been employed, using detector arrays to increase Ω and filters to reduce I_b.[170]

We now return to the expression for the intensity of the fluorescence, Eq. (9.2). For simplicity we take $\Theta_i = \Theta_f$. We also treat the case of most interest for fluorescence, that of a thick sample, that is, $z_s \gg [\mu_s(\omega) + \mu_s(\omega_f)]^{-1}$. Then Eq. (9.2) reduces to

$$I_f(\omega_f) = I_0(\omega)\epsilon_f(\Omega/4\pi)\{\mu_\alpha(\omega)/[\mu_s(\omega)+ \mu_s(\omega_f)]\}. \tag{9.3}$$

It is interesting to note that this expression shows that I_f is nearly constant for a thick, concentrated sample, and therefore the EXAFS signal is essentially zero. In this case one has $\mu_s = \mu_{bg} + \mu_\alpha \simeq \mu_\alpha$ and, typically, $\mu_s(\omega) \gg \mu_s(\omega_f)$, so that the factor in curly brackets in Eq. (9.3) is approximately unity and independent of $\mu_\alpha = \mu_\alpha^0[1 + \chi_\alpha(k)]$. This disappearance of the EXAFS oscillations from I_f for thick, concentrated samples can be explained as follows. When μ_α increases or decreases owing to the EXAFS oscillations, the incident photon beam merely penetrates less far or further into the sample. All the incident photons continue to be absorbed and all contribute to the fluorescent signal. So the number of fluorescent photons produced is essentially independent of μ_α, provided that $\mu_s(\omega_f) \ll \mu_s(\omega)$. There must be a significant background absorption (i.e., $\mu_{bg} > \mu_\alpha$) in order for I_f to vary

[166] J. B. Hastings, P. Eisenberger, B. Lengeler, and M. L. Perlman, *Phys. Rev. Lett.* **43,** 1807 (1979).

[167] M. Marcus, L. S. Powers, A. R. Storm, and B. M. Kincaid, *Rev. Sci. Instrum.* **51,** 1023 (1980).

[168] E. A. Stern and S. M. Heald, *Rev. Sci. Instrum.* **50,** 1579 (1979); *Nucl. Instrum. Methods* **172,** 397 (1980).

[169] J. A. del Cueto and N. J. Shevchik, *J. Phy. C* **11,** L833 (1978).

[170] S. P. Cramer and R. A. Scott, *Rev. Sci. Instrum.* **53,** 395 (1981).

and thereby reflect changes in μ_α. Then the penetration depth $[\approx\mu_s^{-1}(\omega)]$ will respond only slowly to the EXAFS oscillations. The number of α atoms encountered by the incident photons will then vary slowly and thus the number of fluorescent photons produced will reflect the more rapid changes in μ_α, (i.e., the EXAFS).

We now derive the expression for the S/N ratio when statistical noise is the only consideration. The EXAFS signal is proportional to the changes in I_f, whereas the noise is proportional to $(I_f + I_b)^{1/2}$. Using Eq. (9.3), one has for thick samples

$$S/N = \chi_\alpha(k)(\epsilon_f\,\Omega/4\pi)^{1/2}I_0(\omega)^{1/2}\{\mu_\alpha^0/[\mu_s(\omega) + \mu_s(\omega_f)]\}^{1/2} \times \{[\mu_{bg} + \mu_s(\omega_f)]/[\mu_s(\omega) + \mu_s(\omega_f)]\}[1 + I_b/I_f]^{-1/2}. \quad (9.4)$$

The second from the last factor is approximately unity for dilute samples. The last factor gives the deterioration in S/N due to the background radiation. Comparing Eq. (9.4) with the analogous expression for transmission, Eq. (8.13), we see that the transmission S/N is proportional to $\mu_\alpha^0/\mu_s(\omega)$, while the fluorescence S/N is proportional to $[\mu_\alpha^0/\mu_s(\omega)]^{1/2}$. As a result, fluorescence is the method of choice for very dilute systems, for which $\mu_\alpha^0/\mu_s(\omega) \ll 1$. Using Eqs. (8.13) and (9.4), we see that fluorescence is preferred over transmission when

$$\mu_\alpha^0/\mu_s(\omega) \approx c_\alpha\sigma_\alpha^0/\sigma_{bg} < \epsilon_f(\Omega/4\pi). \quad (9.5)$$

In this expression, we have assumed that the background radiation is small $(I_b/I_f \ll 1)$ and that we are in the dilute limit where $[\mu_{bg} + \mu_s(\omega_f)]/[\mu_s(\omega) + \mu_s(\omega_f)] \approx 1$ and $\mu_s(\omega) \approx \mu_s(\omega_f)$. c_α is the concentration of species α and the σ's are the atomic absorption coefficients. If the background radiation is not small, a factor of $1 + I_b/I_f$ should divide the right-hand side of Eq. (9.5) and reduce the region where fluorescence is preferred over absorption. Consider Fe in Cu as a specific example. In this case, $\epsilon_f = 0.2$ and $\mu_\alpha^0/\mu_s(\omega) \simeq 5c_{Fe}$. Assuming a large acceptance solid angle of $\Omega/4\pi = 0.1$, we see that fluorescence is best for c_{Fe} less than 0.5%. If the background radiation is taken into account, it will modify this condition somewhat. The major consideration in this regard, however, is the feasibility of the experiment, that is, whether or not the S/N is adequate. In this example of Fe in Cu, the fluorescence intensity is only 10^{-4} times that of the scattered intensity for $c_{Fe} = 75$ ppm.[166] In this case, a focusing crystal analyzer is required to reduce I_b/I_f to a reasonable value to achieve an adequate signal-to-noise ratio. Another example of a dilute system is the small protein rubredoxin. Figure 15 shows a comparison of the transmission and fluorescence EXAFS on the Fe K edge in this system.[171] The fluorescence tech-

[171] R. G. Shulman, P. Eisenberger, B. K. Teo, B. M. Kincaid, and G. S. Brown, *J. Mol. Biol.* **124,** 305 (1978).

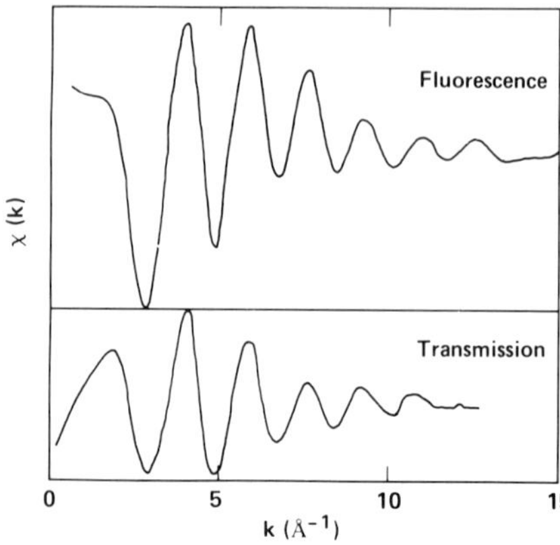

FIG. 15. A comparison of the EXAFS signals above the Fe K edge of powdered samples of oxidized *Peptococcus aerogenes* rubredoxin measured in fluorescence and in transmission. The fluorescence technique yields a better signal-to-noise ratio in this dilute system (one Fe atom in a molecule with molecular weight ≈6000). [Taken from Fig. 2 in R. G. Shulman, P. Eisenberger, B. K. Teo, B. M. Kincaid, and G. S. Brown, *J. Mol. Biol.* **124,** 305 (1978). Copyright: Academic Press Inc. (London) Ltd.]

nique yields better S/N, allowing data to be obtained out to a larger value of k. This system is discussed further in Section 16.

The criterion of Eq. (9.5) included only statistical noise. Some of the systematic distortions discussed in Section 8 are more easily overcome using the fluorescence technique than transmission. This will tend to make fluorescence more attractive for concentrations higher than those given by Eq. (9.5). Two such troublesome distortions are higher harmonics and sample inhomogeneity. The higher harmonics in fluorescence will merely add to the background I_b in Eq. (9.4). This will deteriorate the signal-to-noise ratio somewhat, but will not distort the spectrum as it does for transmission. Similarly, sample inhomogeneity is generally not a problem in fluorescence, since thick samples can often be used.

In summary, the fluorescence technique is useful for dilute systems, the critical concentration being given by Eq. (9.5). Energy-discriminating detectors can be utilized to enhance the S/N if background radiation is substantial. In addition, a fluorescence experiment is less severely affected than a transmission experiment by problems such as harmonic contamination of the beam and sample inhomogeneities.

b. Electron Yield Experiments

In these types of experiments, the electrons ejected from the sample due to the photon absorption event are monitored as a function of incident photon energy. Thresholds and fine structure above the edge are observed in the electron yield as $\hbar\omega$ is scanned through a characteristic absorption edge.[172–175] These experiments are analogous to the fluorescence experiments, in that the detected signal corresponds to a by-product of the absorption event, so that similar considerations apply. Since electrons rather than X-ray photons are detected, however, noteworthy differences do exist in the experimental setup, in the spectra of the by-products, and in considerations of the sensitivity of the technique. We consider first the electron yield spectrum itself.

The emitted electrons can be classified as elastic and inelastic. The elastically emitted electrons are ejected with no energy loss and can be either direct photoelectrons or Auger electrons. The direct photoelectrons are created in the photoabsorption process. They can be ejected from the sample, provided that $\hbar\omega$ is sufficient to excite them above the vacuum level. As $\hbar\omega$ is increased, the direct photoemission peak moves to higher kinetic energies. This is to be contrasted with the Auger electrons, which are ejected at a fixed kinetic energy independent of $\hbar\omega$. These electrons are created in a nonradiative two-electron decay of the core hole as shown in Fig. 14. An electron in a level X drops to fill the core hole in a lower-lying level W, with energy conservation being provided by the promotion of a second electron, in level Y, up above the vacuum level. This gives rise to the WXY Auger peak, whose kinetic energy depends on the levels W, X, and Y and not on the incident photon energy $\hbar\omega$. The Auger process is complementary to fluorescence for the decay of a core hole. The probability that the core hole will be filled by an Auger process is large for low-Z elements and small for the heavy elements. The two processes have equal probability for K-edge excitations at Zn ($Z = 30$) and for L-edge excitations at Th ($Z = 90$).[176]

In addition to the electrons that are elastically emitted from the sample, a much larger number of inelastically scattered secondary electrons are emitted. These are created when the "hot" Auger electrons and direct (or primary) photoelectrons give up energy in collisions with the atoms of the material. The total electron yield spectrum then consists of relatively small Auger and photoemission peaks together with a large inelastic tail at lower energies. A schematic spectrum and energy level diagram are shown in Fig. 16 for a hypothetical sample consisting of O and Ni atoms excited by photons with energy just above the oxygen K-edge threshold. [177] In addition

[172] A. P. Lukirskii and I. A. Brytov, *Sov. Phys.—Solid State (Engl. Transl.)* **6,** 33 (1964).

[173] A. P. Lukirskii, O. A. Ershov, T. M. Zimkina, and E. P. Savinov, *Sov. Phys.—Solid State (Engl. Transl.)* **8,** 1422 (1966).

[174] W. Gudat and C. Kunz, *Phys. Rev. Lett.* **29,** 169 (1972).

[175] H. Petersen and C. Kunz, *Phys. Rev. Lett.* **35,** 863 (1975).

[176] V. O. Kostroun, M. H. Chen, and B. Crasemann, *Phys. Rev. A* **3,** 533 (1971); M. H. Chen, B. Crasemann, and V. O. Kostroun, *ibid.* **4,** 1 (1971).

[177] J. Stöhr, *Jpn. J. Appl. Phys.* **17,** Suppl. 17-2, 217 (1978).

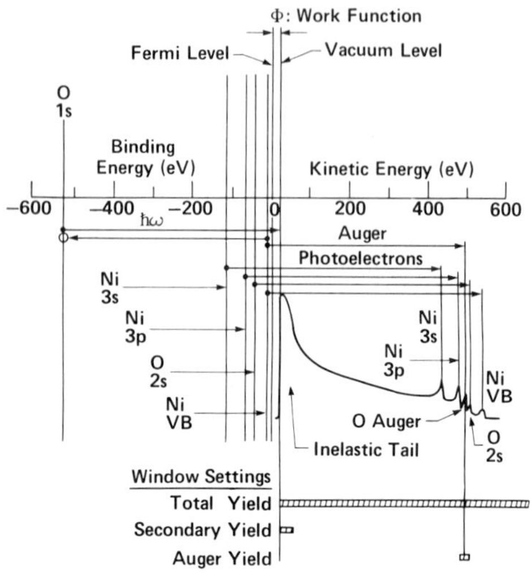

FIG. 16. A spectrum and energy level diagram for a sample consisting of O and Ni atoms excited by photons with energy just above the oxygen K-edge threshold. Shown are various direct photoemission peaks from the O and Ni levels as well as the oxygen *KVV* Auger peak and the large inelastic tail. The electron energy window settings for the three different electron yield EXAFS techniques are also indicated. [Taken from Fig. 1 in J. Stöhr, *Jpn. J. Appl. Phys.* **17**, Suppl. 17-2, 217 (1978).]

to the photoemission peaks from the various O and Ni atomic levels, there is a peak due to the oxygen *KVV* Auger transition, where *V* indicates the valence band. Also shown is the large inelastic tail, as well as the energy "windows" for the various yield techniques used to acquire EXAFS data. These include total yield, where all the electrons are detected; Auger yield, where only a specific Auger peak is monitored; and secondary yield, where electrons in the inelastic tail are detected, typically with kinetic energies < 10 eV. Before discussing the EXAFS signals obtained from these measurements, we mention briefly some of the experimental considerations.

Since electrons are detected in this technique, both the sample and the electron detector must be in a high-vacuum environment. For experiments in the hard X-ray region ($\hbar\omega > 3$ keV), this vacuum chamber can be separated from the photon source. In this case an ionization chamber can be used to measure the incident photon flux, as for transmission and fluorescence experiments, and the photons can enter the vacuum chamber through a thin Be window. For lower photon energies ($\hbar\omega < 3$ keV), the photon source, monochromator, and I_1 counter must also be in a high-vacuum environment. In this case the electron yield from a high-transmission ($\approx$80%) metal grid can be used as the beam monitor.[178] The grid is typically coated with a material which does not exhibit any absorption edges in the photon energy range of interest. The electron detectors used to monitor the photoemitted signal are either electron multipliers or electron energy analyzers (e.g., a cylindrical mirror analyzer). The former are typically used to detect all the electrons emitted from the sample (i.e., for total yield), while the latter are needed when only a portion of the photoemission spectrum is to be monitored. This is the case when only the elastically emitted Auger electrons are detected (i.e., for Auger yield) or when only the low-kinetic-energy portion of the electron spectrum is detected (i.e., for partial yield). The energy analyzer detection scheme also discriminates against background emission and can thus enhance the signal-to-noise ratio. We now consider the EXAFS from these electron yield techniques.

Of the various electron yield EXAFS techniques, it is the Auger yield that is most obviously proportional to the X-ray absorption coefficient.[179,180] The number of elastically emitted Auger electrons is proportional to the probability that the incident photon has created a core hole; that is, it is proportional to the absorption coefficient. These Auger electrons leave the sample with a fixed kinetic energy as the incident photon energy is scanned. As a result, any diffraction effects, which depend on the wave vector of the outgoing electrons, remain constant as a function of $\hbar\omega$. The direct proportionality between the Auger yield and the absorption coefficient is preserved as $\hbar\omega$ is varied, so that the Auger intensity rigorously exhibits EXAFS.[181,182] It is the nonradiative analog of the fluorescence yield.

This is not necessarily the case for the direct photoemitted peak. Lee[180] has shown that the oscillatory term in the emission intensity detected in any particular direction $\mathbf{k}$ is not that of the EXAFS oscillations. The phase factor $2kr_j$ of EXAFS is replaced by $kr_j - \mathbf{k}\cdot\mathbf{r}_j$, where $\mathbf{r}_j$ is the vector to an atom neighboring the photoexcited atom. The period of the oscillations in the direct peak will then be different for the various possible angles of scattering from the near neighbors. An angular average of the photoemission yield over all 4π sr yields the EXAFS expression, however, with a phase of $2kr_j$. This is as it must be, since such an integration yields all the photoemitted electrons, and this total intensity should be proportional to the

[178] J. Stöhr, R. Jaeger, J. Feldhaus, S. Brennan, D. Norman, and G. Apai, *Appl. Opt.* **19,** 3911 (1980).

[179] U. Landman and D. L. Adams, *Proc. Natl. Acad. Sci. U.S.A. Solid State* [3] **73,** 2550 (1976).

[180] P. A. Lee, *Phys. Rev. B: Solid State* [3] **13,** 5261 (1976).

[181] A Bianconi and R. Z. Bachrach, *Phys. Rev. Lett.* **42,** 104 (1979).

[182] P. H. Citrin, P. Eisenberger, and R. C. Hewitt, *Phys. Rev. Lett.* **41,** 309 (1978).

total absorption. The fact that the angular average of the photoemission is properly EXAFS may explain why quantum yield measurements on bulk materials compare well with EXAFS.[175,183] In these cases the electrons can originate from anywhere within one escape depth of the surface ($\approx$50 Å). The various scattering angles that the electrons can undergo in leaving these tens of atomic layers and still reach the detector may well provide an effective 4π sr average. This is not, however, the case for a single adsorbed surface layer. The electrons emitted from the top layer can provide an average over no more than 2π sr, which is not sufficient for the photoelectron yield to be proportional to EXAFS. Accordingly, caution must be exercised in treating the electron yield from a surface layer, which, as discussed below, is the type of study for which electron yield spectroscopy is most advantageous.

The situation for the total yield is complicated in theory by the many inelastic collisions that the "hot" Auger electrons and photoelectrons must undergo in giving rise to the large inelastic tail (see Fig. 16). As a result, a rigorous relationship between the total yield and the absorption coefficient has not been established. Nonetheless, if the total yield is dominated by the cascades of secondary electrons from the primary Auger electrons and from reabsorbed X-ray fluorescence photons, then the total electron yield will exhibit EXAFS, since both the fluorescence and Auger emission do. If the total yield has a significant component derived from the "hot" photoemitted electrons, then the difficulties of the direct peak discussed above may also affect the total yield. There is some evidence that the total yield is dominated by the former processes,[184] so that the oscillations in the total yield should be EXAFS. An experimental comparison of the X-ray absorption and total yield fine structure above the Cu K edge has also been made.[185] They compare favorably, yielding the same near-neighbor spacing, within experimental error (i.e., $\approx$0.02 Å), and amplitudes that differ by about 30%. The amplitude difference may arise from an L-edge contribution to the yield.

The situation for the partial yield is very similar to that for the total yield. This is due to the fact that the number of elastically emitted Auger electrons and photoelectrons is typically much smaller than the number of inelastically emitted secondary electrons[186] (see Fig. 16). As a result, the partial yield is approximately equal to the total yield and the same considerations discussed above apply. We now discuss the expressions for the signal-to-noise ratio for the electron yield techniques, keeping in mind the possible differences for the different yield spectroscopies.

[183] G. Margaritondo and N. G. Stoffel, *Phys. Rev. Lett.* **42,** 1567 (1979).
[184] G. Martens, P. Rabe, G. Tolkiehn, and A. Werner, *Phys. Status Solidi* A **55,** 105 (1979).
[185] G. Martens, P. Rabe, N. Schwentner, and A. Werner, *J. Phys. C* **11,** 3125 (1978).
[186] J. Stöhr, *J. Vac. Sci. Technol.* **16,** 37 (1979).

The expressions for the electron yield signals and their S/N ratios can be derived in the same way as they were for fluorescence. Equations (9.1)–(9.5), which were derived for fluorescence, apply as well to the Auger yield and to the partial and total yield (with the above-mentioned provisos). There are only two changes. Firstly, the fluorescence efficiency ϵ_f is replaced with the electron emission efficiency ϵ_e. For Auger emission ϵ_e is complementary to ϵ_f (i.e., $\epsilon_e + \epsilon_f \approx 1$). Secondly, the fluorescence sampling depth of $[\mu_s(\omega) + \mu_s(\omega_f)]^{-1}$ is replaced with the much smaller electron yield sampling depth of $[\mu_s(\omega) + \lambda^{-1}(k)]^{-1}$. Here $\lambda(k)$ is the electron mean-free-path length for electrons of wave vector k. It varies approximately as k, changing from ≈ 10 Å at 100 eV to ≈ 100 Å at 10^4 eV.[62] Since $\mu_s(\omega) \approx 10^3\ \mathrm{cm}^{-1}$, one typically has that $\lambda^{-1}(k) \gg \mu_s(\omega)$. In fact, the major differences between fluorescence yield and electron yield arise from the much smaller electron mean-free-path length. Equation (9.4) for the signal-to-noise ratio for the electron yield then becomes

$$S/N = \chi_\alpha(k)(\epsilon_e\Omega/4\pi)^{1/2}I_0(\omega)^{1/2}[\mu_\alpha^0\lambda(k)]^{1/2}(1 + I_b/I_e)^{-1/2}. \tag{9.6}$$

I_e is the intensity of the electron yield signal of interest and I_b is the background signal. For simplicity we have taken the case of a symmetric geometry (i.e., $\Theta_i = \Theta_f$) and the case of thick samples (i.e., $z_s \gg \lambda \approx 50$ Å). We have also used the fact that typically $\lambda^{-1}(k) \gg \mu_s(\omega)$.

We now compare the electron yield S/N to that for transmission. For simplicity we consider the electron background signals to be small (i.e., $I_b/I_e \ll 1$). Then Eqs. (8.13) and (9.6) indicate that electron yield is to be preferred over transmission when

$$\mu_\alpha^0/\mu_s \approx c_\alpha\sigma_\alpha^0/\sigma_{bg} < \mu_s(\omega)\lambda(k)(\epsilon_e\Omega/4\pi). \tag{9.7}$$

Here c_α is the concentration of species α and the σ's are the atomic absorption coefficients. The equality is appropriate for the dilute limit (i.e., $\mu_\alpha \ll \mu_{bg}$). For typical values of $\sigma_\alpha \approx \sigma_{bg}$, $\mu_s\lambda \approx 10^{-4}$, and $\epsilon_e\Omega/4\pi \approx 0.1$, we see that electron yield is preferred over transmission only for very small concentrations (i.e., $c_\alpha < 10^{-5}$). As a result, electron yield is hardly ever to be preferred over transmission except for surface studies. In this case, the surface constitutes a very small fraction of the entire sample and provides the dilution whereby electron yield becomes the technique of preference.

Comparing the electron yield result with that for fluorescence, we see that the ratio of signal to noise is

$$(S/N)_f/(S/N)_e \approx (\epsilon_f/\epsilon_e)^{1/2}[\mu_s(\omega)\lambda(k)]^{-1/2}(1 + I_b^e/I_e)^{1/2}/(1 + I_b^f/I_f)^{1/2}. \tag{9.8}$$

If the background signals I_b^e and I_b^f are small, then electron yield is hardly ever preferred over fluorescence, since $\mu_s \ll \lambda^{-1}$. For typical values of $\mu_s\lambda \approx 10^{-3}$–$10^{-4}$, the fluorescence S/N is 10–100 times larger than that for electron yield for comparable emission efficiencies (i.e., for Zn). Only for the very light elements, where $\epsilon_e \gg \epsilon_f$, can this factor be overcome. This corresponds to $Z \approx 6$ (carbon) and below. If the background signals are nonnegligible, however, as is the case for surface studies, then this situation can be reversed.

For surface studies, electron yield techniques have a S/N advantage over both transmission and fluorescence. Comparing with transmission, the small number of surface atoms relative to the bulk provides the large dilutions required for electron yield to be preferred. Only for systems with an anomalously large surface area, such as Br on Grafoil, will transmission be preferred for surface studies.[187] Comparing with fluorescence, we see that surface studies provide a situation where the background signal can be very large owing to the large penetration depth of the incident photons. Atoms well within the sample are excited by the incident X rays and will produce background fluorescent X radiation that competes with the fluorescent signal from the surface atoms. For the electron yield techniques, on the other hand, only absorption events within λ of the surface region ($\approx$50 Å) are detected. So there is a natural background reduction for surface studies using the electron yield techniques. The 50-Å path length is still substantial, however, so that bulk structure rather than the structure of the surface layer itself will be probed, unless the surface layer is of a different atom species than the bulk. One then tunes the photon energy to a characteristic absorption edge of the surface species. This is typically the procedure for surface EXAFS studies using electron yield.

c. *Miscellaneous Techniques*

In addition to the above three EXAFS techniques, which have been thoroughly studied, there are a variety of other techniques that have been examined. They differ in the type of excitation source that is used and/or in the specific by-product of the absorption event that is monitored. They have not as yet been as thoroughly studied as the other techniques, either theoretically or experimentally, and have therefore not been put on as firm a footing. Nonetheless, their differences from the well-established EXAFS techniques may provide advantages in certain situations. As a result, we describe briefly each of these other techniques and point out their possible advantages. Before they can be used routinely to obtain structural information, however, further studies will be required to explore thoroughly their possible limitations.

(i) Electron Energy Loss Spectroscopy. In electron energy loss spectros-

[187] E. A. Stern, D. E. Sayers, J. G. Dash, H. Shechter, and B. Bunker, *Phys. Rev. Lett.* **38,** 767 (1977).

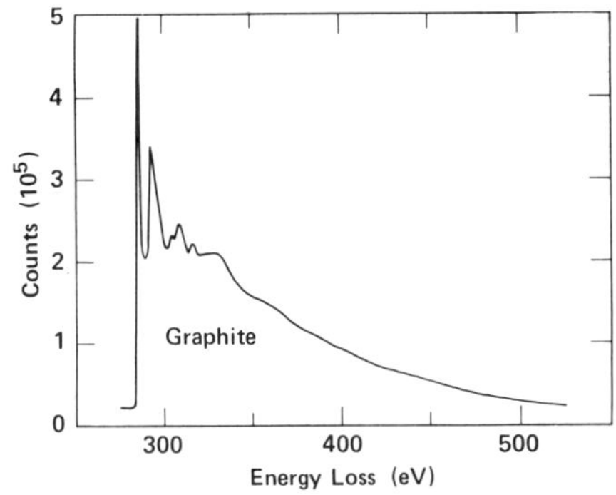

FIG. 17. Electron energy loss spectrum of crystalline graphite showing the white line at the K edge and the EXAFS oscillations. The loss signal falls off as $\approx E^{-4}$, more rapidly than the E^{-3} falloff of the photoabsorption cross section. [Taken from Fig. 1 in B. M. Kincaid, A. E. Meixner, and P. M. Platzman, *Phys. Rev. Lett.* **40,** 1296 (1978).]

copy (EELS), a monoenergetic, high-energy ($\approx$100 keV) electron beam is used as the probe rather than the X-ray photon beam of the previously discussed experiments. This electron beam is transmitted through a thin sample ($\approx$500 Å thick) and is detected and energy analyzed to determine the energy loss. The electron flux is recorded as a function of energy loss. Since the energy loss arises from electronic excitations in the sample, the electron flux will exhibit steps at the energies which characterize the X-ray absorption edges. The flux above those edges contains EXAFS-like fine structure.[188–191] In the limit of small momentum transfer from the incident electron to the sample, the cross section for the scattering of high-energy electrons is proportional to the photoabsorption coefficient; that is, it is proportional to the dipole matrix element between the initial and final electronic states of the sample. As a result, the energy loss spectrum for high-energy electrons is proportional to the photoabsorption spectrum, and the fine structure in the electron flux in this limit is EXAFS. An example of a loss spectrum is given in Fig. 17, showing both the white line at the edge and EXAFS above the edge. The momentum transfer dependence of the energy loss can also be measured. It is obtained by adjusting the electron scattering angle and is a quantity not obtainable from an X-ray absorption experiment. Only a few electron energy loss fine structure studies have been made to date, and their primary emphasis has been in demonstrating various

[188] C. Colliex and B. Jouffrey, *Philos. Mag.* [8] **25,** 491 (1972).

[189] J. J. Ritsko, S. E. Schnatterly, and P. C. Gibbons, *Phys. Rev. Lett.* **32,** 671 (1974).

[190] P. C. Gibbons, S. E. Schnatterly, and J. J. Ritsko, *Phys. Fenn.* **9,** 106 (1974).

[191] R. D. Leapman and V. E. Cosslett, *J. Phys. D* **9,** L29 (1976).

aspects of the technique. As a result, we will confine our discussion of this technique to a comparison with transmission studies.

Kincaid *et al.*[192] have shown that the ratio of the signal counting rate for electron energy loss experiments, S_e, to that of photons in an absorption experiment, S_{ph}, is given approximately by

$$S_e/S_{ph} \approx (\alpha_{fs}/2\pi)(mc^2/E_i)\omega^{-1}[N_e/N_{ph}(\omega)](z_e/z_{ph}) \ln(1 + q_0^2/q_{min}^2). \quad (9.9)$$

Here $\alpha_{fs} = 1/137$ is the fine structure constant, mc^2 is the electron rest mass energy, E_i is the incident energy of the electron beam, $\hbar\omega$ the photon energy, N_e the incident electron flux, $N_{ph}(\omega)$ the photon flux per unit bandwidth, and z_e and z_{ph} the sample thicknesses for electron and photon experiments, respectively. q_{min} is the minimum momentum transfer and q_0 is the transverse momentum resolution of the electron detector. The weak energy dependence of the last term in Eq. (9.9) will be ignored. From this expression it is evident that the electron energy loss method can compete with the transmission method only for the low-energy edges, such as the K edges of the first row elements and possibly the L, M and N edges of some heavier elements. The close spacing of the higher-lying L, M, and N edges is a difficulty, as it limits the accessible range of the EXAFS. In practice, then, only the light elements can be studied readily. This restriction to low-energy edges is due to two factors in Eq. (9.9). First, the ratio of the counting rates is roughly proportional to $1/\omega$, owing to the fact that electron energy loss falls off more rapidly with increasing energy loss than does absorption with increasing photon energy (see Fig. 17). Secondly, this ratio is proportional to z_e/z_{ph}, which is much smaller than unity, except for the light elements. z_e is very small, typically a few hundred angstroms. If z_e is too large, multiple scattering events will become significant and can cause additional oscillations which interfere with the EXAFS and which are difficult to remove. Also, a thick sample has more elastic scattering that removes electrons from the detection system, reducing thereby the counting rates. The optimal thickness for the photon absorption case is $z_{ph} \approx 2/\mu_s$, which is considerably larger than a few hundred angstroms, except for the very light elements. The result of these two factors is to reduce the sensitivity of the electron energy loss method at high energies.

Nonetheless, even for the low-energy K edge of carbon and a typical electron beam energy of $E_i = 100$ keV, $S_e/S_{ph} \approx 10^{-5}$ in comparing the electron flux of the spectrometer described by Kincaid *et al.*[192] with the photon flux of SPEAR at 3 GeV and 50 mA. In this comparison, the low transmission of the uv monochromator was not taken into account in estimating $N_{ph}(\omega)$. When this is done, the actual ratio may be closer to unity

[192] B. M. Kincaid, A. E. Meixner, and P. M. Platzman, *Phys. Rev. Lett.* **40**, 1296 (1978).

at the carbon K edge. By improvements in electron energy loss spectrometers, it is thought that this ratio can be improved further.

One area where electron energy loss EXAFS has a possible advantage over photon transmission experiments is in probing very small areas (radius $\approx$ 10 Å). The advantage lies in the availability of high-brightness field-emission electron sources which can produce an intense electron beam with a radius between 3 and 50 Å. Electron energy loss EXAFS of such small areas has been demonstrated for carbon, where as few as 10^4 atoms were detected in 4 min.[193] This is a bare minimum averaging time, since the counting statistics were such that the fine structure was barely visible above the noise. These types of studies have the potential of yielding information on the location of light atoms in small precipitates and on surfaces. A brief review of electron energy loss spectroscopy has been given by Joy and Maher.[194] Additional discussions of the relative advantages and disadvantages of electron energy loss relative to photon absorption EXAFS are contained in Ref. 8.

The question of radiation damage at such high electron intensities has not been thoroughly explored as yet. Radiation damage is not a problem for the photoabsorption case, even for the high photon fluxes currently available from synchrotron sources, owing to the small ratio of the number of photons absorbed to the number of atoms available to do the absorbing. One has at most 10^{11} photons/s absorbed in a sample volume of about 10^{-4} cm^3 (1 cm wide by 1 mm high by 10 μm thick). This corresponds to only 1 photon absorbed per second for every 10^8 atoms. The situation is rather different, however, for the intense electron sources.[195] Consider the experiment described by Batson and Craven.[193] In one case, an electron beam with a radius of 50 Å and a current of 10 nA impinged on an amorphous carbon sample 100 Å thick. This corresponds to about 6×10^{10} energetic (80 keV) electrons bombarding approximately 10^5 carbon atoms each second, or almost 10^6 electrons/s for each atom. The total cross section for the inelastic scattering of 100-keV electrons is approximately 0.03 $Å^2$ per atom for carbon.[194] This yields 10^3–10^4 scattered electrons per carbon atom per second for the above flux. It is clear that sample damage by the electron beam is not as easily ruled out as in the X-ray photon case. This striking contrast arises from a fundamental difference between these experiments. Since the photon beam is monochromatized before striking the sample, each absorbed or transmitted photon contributes to the EXAFS signal of interest. In EELS, however, energy analysis must take place after the beam

[193] P. E. Batson and A. J. Craven, *Phys. Rev. Lett.* **42,** 893 (1979).

[194] D. C. Joy and D. M. Maher, *Science* **206,** 162 (1979).

[195] D. E. Johnson, S. Csillag, and E. A. Stern, *in* "Laboratory EXAFS Facilities—1980" (E. A. Stern, ed.), Chapter 6. Am. Inst. Phys., New York, 1980.

passes through the sample. For each relevant electron energy loss event, there are a great many inelastic scattering events which are in the wrong energy range to contribute to the EXAFS signal. Low-energy losses are a particular problem. In fact, the electron energy loss situation is similar to placing the X-ray monochromator *after* the sample in a transmission experiment, allowing all the white radiation to fall on the sample.

(ii) Appearance Potential Spectroscopy. In appearance potential spectroscopy, an electron beam of varying but relatively low energy (≈ 1 keV) is used as the excitation source. The excitation probability of a core state is monitored as a function of the energy of these bombarding electrons through and above a core level excitation threshold known as the appearance potential. Fine structure is observed in this excitation probability above the threshold which has been attributed to the same interference phenomenon that gives rise to EXAFS.[196] The core-state excitation probability can be measured by monitoring the by-products resulting from the filling of the core hole created (i.e., the fluorescent X-ray yield or the Auger electron yield). Another detection procedure is to measure the elastic scattering yield, which decreases at threshold owing to the opening of the new inelastic channel.

This technique differs substantially from the X-ray absorption case and its theory has been discussed by Laramore and co-workers.[197–199] The low-energy electron excitation of a core electron is clearly a two-electron process, while photoabsorption is a one-electron problem, as illustrated in Fig. 18. In the X-ray case, a photon of energy $\hbar\omega$ promotes an electron with energy E_{nl} below the Fermi level E_F to a state with energy $E = (\hbar\omega + E_{nl})$ above the Fermi level. In the electron bombardment case, an electron with incident energy E_i loses energy to excite the bound core electron above the Fermi level, resulting in a final state with one electron having energy E_1, a second electron having energy $E_2 = E_i + E_{nl} - E_1$, and a core hole *nl*. All states that are empty and satisfy energy conservation are allowed (i.e., $E_F \leq E_1$, $E_2 \leq E_i + E_{nl}$). As a result, the final state does not have a single wave vector, $k \propto E^{1/2}$, as in the X-ray absorption case, but, rather, a wide range of wave vector values for both electrons. The transition density for each of these two final-state electrons must then be convoluted over the allowed energies.[200] Since the individual transition densities contain the EXAFS, this integration

[196] P. I. Cohen, T. L. Einstein, W. T. Elam, Y. Fukuda, and R. L. Park, *Appl. Surf. Sci.* **1,** 538 (1978).

[197] G. E. Laramore, *Phys. Rev. B: Condens. Matter* [3] **18,** 5254 (1978).

[198] G. E. Laramore, *Surf. Sci.* **81,** 43 (1979).

[199] G. E. Laramore, T. L. Einstein, L. D. Roelofs, and R. L. Park, *Phys. Rev. B: Condens. Matter* [3] **21,** 2108 (1980).

[200] R. L. Park, Surf. Sci. **86,** 504 (1979).

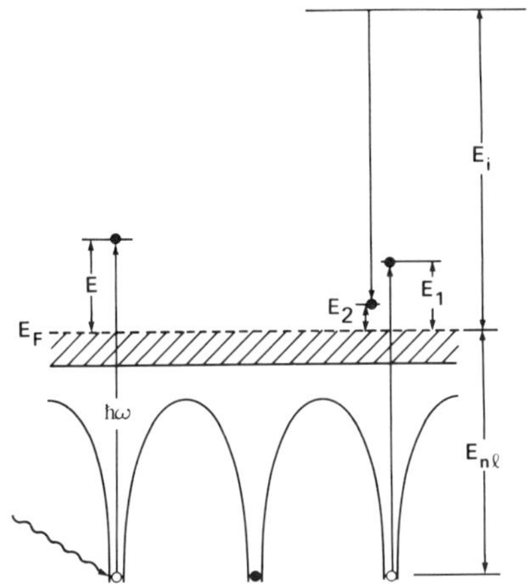

FIG. 18. Electron energy level diagram comparing the excitation process in a photoabsorption experiment (left) with that for low-energy electron bombardment. In the photoabsorption case, the final state consists of a single electron with kinetic energy E and a core hole. In the electron-excited case, the final state contains a core hole and two electrons: the incident electron that has lost energy from its incident kinetic energy E_i to its final kinetic energy E_2 and the ejected electron with kinetic energy E_1. Differentiation with respect to E_i has the effect of picking out the term in the response corresponding to one of these two electrons at E_F, thereby approximating the photoabsorption case. [Taken from Fig. 1 in M. L. den Boer, T. L. Einstein, W. T. Elam, R. L. Park, L. D. Roelofs, and G. E. Laramore, *Phys. Rev. Lett.* **44,** 496 (1980).]

tends to smear out the fine structure. The fine structure oscillations can, however, be "recovered" by differentiating the response.[196,198] This is done experimentally by adding a sinusoidally varying potential to the electron-accelerating voltage, thereby varying E_i, and detecting the first derivative of the response with a lock-in amplifier. The differentiation has the effect of picking out the term in the response that has one electron pinned at the Fermi level. Thus as E_i is swept, the derivative of the response is dominated by the term where the final-state kinetic energy of only one electron varies. The momentum of this one electron becomes the relevant variable for the EXAFS, $k \propto E_1^{1/2}$, with E_2 pinned at the Fermi level. This is similar to the X-ray absorption case, in that a known amount of energy is given up by exciting radiation.

A second difference is that the final-state angular momentum is not limited by the dipole selection rule. In the X-ray case, the final state for K-shell absorption is a p-symmetry wave, and for L-shell absorption, predominantly a d-symmetry wave. This greatly simplifies the analysis of EXAFS, since the outgoing electron then experiences a single phase shift, rather than a different phase shift for each partial wave. For the Coulomb matrix elements appropriate for the appearance potential technique, there are no simple angular momentum selection rules and an explicit calculation is required to determine the coupling to each partial wave of the higher-energy final-state electron. Such a calculation has been performed for the case of the oxygen K-edge excitations,[201] leading to the conclusion that coupling to the s-symmetry final state dominates. The expression for the fine structure in this case contains only one phase shift, as in the X-ray absorption case. This result cannot be generalized, however, so that an explicit calculation is required for each case.

A third difference is that since the probe is an electron beam rather than a photon beam, electron scattering processes in the initial state as well as in the final state have to be considered. The situation is further complicated for electron yield detection, where multiple scattering and diffraction effects of the detected electrons as well as the incident electrons must be taken into account.

Despite these differences which make the appearance potential technique substantially more complicated than standard absorption EXAFS, the technique may be competitive in certain situations owing to two advantages.[200] First, the technique is surface sensitive. This is due to the short mean free path for inelastic scattering of the relatively low-energy incident electrons ($\approx$1 keV).[202] Thus only atoms near the surface are excited and contribute to the signal. This is to be contrasted with the other electron yield techniques, where the incident photon beam, being more penetrating than an electron beam, excites the surface plus many layers below the surface. Secondly, the experimental apparatus is relatively simple and readily available in surface science laboratories. As a result, the technique has already been applied to studying V surfaces,[196,203] O on Al(100),[201] and O on Ni(100).[204] Nonetheless, further experimental and theoretical work will be required to put this technique on the same firm footing as standard absorption EXAFS and to deal with the question of possible sample damage from high incident electron fluxes.

(iii) Photon-Stimulated Desorption. This technique is analogous to electron yield spectroscopies, with the difference that ion yield rather than electron yield is monitored. A monochromatic photon beam impinges on a sample contained in a high vacuum. The photon absorption and subsequent Auger decay of the core hole results in adsorbed ions being desorbed from the surface, as discussed below. These desorbed ions are collected in a de-

[201] M. L. den Boer, T. L. Einstein, W. T. Elam, R. L. Park, L. D. Roelofs, and G. E. Laramore, *Phys. Rev. Lett.* **44,** 496 (1980).

[202] R. H. Ritchie, C. J. Tung, V. E. Anderson, and J. C. Ashley, *Radiat. Res.* **64,** 181 (1975).

[203] W. T. Elam, P. I. Cohen, L. Roelofs, and R. L. Park, *Appl. Surf. Sci.* **2,** 636 (1979).

[204] M. L. den Boer, T. L. Einstein, W. T. Elam, R. L. Park, L. D. Roelofs, and G. E. Laramore, *J. Vac. Sci. Technol.* **17,** 59 (1980).

tector, and their intensity recorded as a function of the incident photon energy. The ion yield is typically mass analyzed in a time-of-flight spectrometer in order to distinguish among the various possible desorbed species. This requires a pulsed photon source such as that provided by a synchrotron.[205] The ion yield obtained as a function of incident photon energy exhibits increases at the photoabsorption thresholds of both the adsorbate and substrate. In addition, EXAFS-like oscillations occur above these thresholds. It is argued that the ion yield is proportional to the photoabsorption cross section and that, as a result, these oscillations are properly EXAFS.[206–208] We now describe briefly a model for the desorption process and the EXAFS.

A photon is absorbed by either an adsorbate or substrate atom creating a core hole, as illustrated in Fig. 19. Some fraction of these core holes will be filled by an Auger process. If the Auger decay results in the adsorbate atom becoming positively charged, then the resulting Coulomb repulsion can cause desorption.[209,210] The desorbed species come off the surface as either neutral or positive ions. For an adsorbate core hole, this process could occur via an intra-atomic Auger decay; for a substrate core hole, on the other hand, it must occur via an inter-atomic Auger transition. In both cases the adsorbate atom loses electrons and thereby acquires a positive charge (see Fig. 19). The resulting Coulomb repulsion can cause desorption, provided that the recapture processes and neutralization processes are not very effective (only charged desorbed ions are detected). Accordingly, the ion yield $S_I(\omega)$ at photon energy $\hbar\omega$ is given by[208]

$$S_I(\omega) \approx P_d P_n \mu_n(\omega). \tag{9.10}$$

$\mu_n(\omega)$ is the partial photoabsorption coefficient due to photoexcitation of electrons from atomic level n. P_n is the probability for Auger transitions into the excited core hole in atomic level n which result in the repulsive ionic state which gives rise to desorption. P_d is the probability that a charged ion actually desorbs. It is argued that P_d and P_n are independent of photon energy, so that all the energy dependence is in $\mu_n(\omega)$. Since $\mu_n(\omega)$ contains EXAFS oscillations, the ion yield must also. It should be noted that

[205] At SPEAR in the "one-bunch, one-bucket" mode, for example, the synchrotron radiation comes in 0.3-ns pulses every 780 ns.

[206] M. L. Knotek, V. O. Jones, and V. Rehn, *Phys. Rev. Lett.* **43,** 300 (1979).

[207] R. Franchy and D. Menzel, *Phys. Rev. Lett.* **43,** 865 (1979).

[208] R. Jaeger, J. Feldhaus, J. Haase, J. Stöhr, Z. Hussain, D. Menzel, and D. Norman, *Phys. Rev. Lett.* **45,** 1870 (1980).

[209] M. L. Knotek and P. J. Feibelman, *Phys. Rev. Lett.* **40,** 964 (1978).

[210] P. J. Feibelman and M. L. Knotek, *Phys. Rev. B: Condens. Matter* [3] **18,** 6531 (1978).

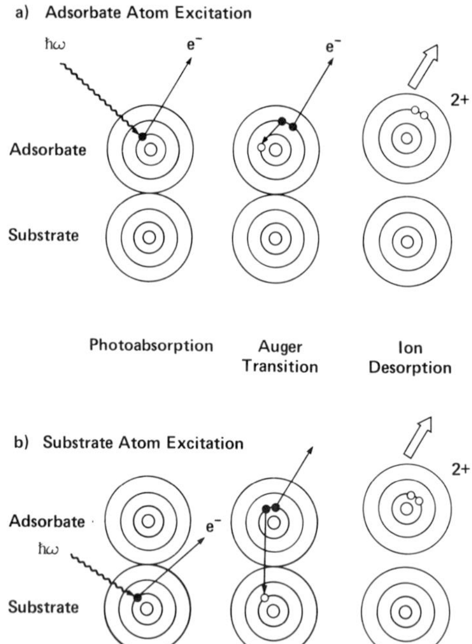

FIG. 19. Schematic representation of the ion desorption process proposed by Knotek and Feibelman [M. L. Knotek and P. J. Feibelman, *Phys. Rev. Lett.* **40,** 964 (1978)] for both (a) adsorbate atom and (b) substrate atom excitations. For the two types of excitations, an inter- or intra-atomic Auger transition results in the adsorbate atom becoming positively charged. The resulting Coulomb repulsion causes desorption of the positively charged adsorbate ion. [Taken from Fig. 9, in J. Stöhr, SSRL Rep. No. 80/07 (1980).]

$P_d \approx 0$ for atoms not in the outermost surface layer, so that the desorption signal arises exclusively from the surface atoms.

The photon-stimulated desorption (PSD) process itself has been studied on a variety of systems. These include H^+, OH^+, and F^+ from TiO_2,[206] O^+ and CO^+ from W(100),[207] and H^+ and OH^+ from H_2O-doped BeO, Al_2O_3, and SiO_2 surfaces.[211] The PSD EXAFS, however, has been studied only for O^+ desorbed from the α-oxygen phase on Mo(100).[208] In this study the PSD fine structure above the Mo L_I absorption edge has been analyzed to demonstrate the technique and to obtain information on the Mo–Mo distances for the surface atoms. Extensions of this work to the PSD signals due to excitation of the other Mo core levels and the O 1s level have been made.[212]

[211] M. L. Knotek, V. O. Jones, and V. Rehn, *Surf. Sci.* **102,** 566 (1981).

[212] R. Jaeger, J. Stöhr, J. Feldhaus, S. Brennan, and D. Menzel, *Phys. Rev. B: Condens. Matter* [3] **23,** 2102 (1981).

The advantage of this technique is that it is adsorbate specific and has extreme surface sensitivity. Only atoms on the surface can be desorbed with appreciable probability, so that structural information is probed only at the surface. In addition, structural information about an adsorption site only is obtained. This is to be contrasted with electron yield techniques, where electrons from many atomic layers below the surface are detected, as well as those from occupied adsorption sites. In addition, a specific bonding site may be studied, since the PSD yield efficiencies can differ strongly for different chemisorption sites. Furthermore, adsorbates present in different atomic or molecular forms have the possibility of being studied separately via PSD.

(iv). Luminescence. Oscillations have been observed in the luminescent yield as a function of X-ray excitation energy and have been correlated with EXAFS from absorption measurements.[213] The particular case that has been studied is that of CaF_2, where the X-ray excitation energy is in the vicinity of the Ca K edge at 4 keV and the intrinsic luminescence at 4.44 eV is detected. The process is thought to be the following.[213] The thick sample absorbs all the incident X-rays, producing photoelectrons. Electron–electron scattering then leads to the production of many electron–hole pairs. Self-trapped excitons are formed and these recombine to yield the 4.44-eV intrinsic luminescence line. All photoelectrons and Auger electrons and their secondaries contribute to the pair production and thus the luminescence. Any process that competes with the production of high-energy electrons will therefore lower the luminescent yield. This competing process has been proposed to be the radiative recombination of the core hole. The fluorescent X rays thus produced are below the Ca K-edge energy and are therefore less strongly absorbed than the incident X rays. As a result, variations in the luminescent yield are attributed to a decrease in the number of high-energy electrons produced due to the competing fluorescence process. Since the fluorescent yield contains EXAFS, it is argued that the oscillations in the reduction in the luminescent yield are also EXAFS. A difficulty with this explanation is that the fluorescence of thick, concentrated samples, as the CaF_2 in this case, shows no EXAFS, as discussed earlier. Another possibility is that the change in the luminescence is due to an opposite change in the Auger yield from atoms near the sample surface.[214] These electrons escape from the sample without creating any self-trapped excitons, thereby reducing (increasing) the luminescence in direct proportion to the increase (decrease) in the Auger yield. In any case, the inverse of the measured luminescent yield does approximately reflect the EXAFS from absorption in CaF_2. When the origin of these oscillations in the luminescent yield has

[213] A. Bianconi, D. Jackson, and K. Monahan, *Phys. Rev. B: Solid State* [3] **17,** 2021 (1978).
[214] K. Monahan, private communication.

been firmly established, the luminescence approach may well provide an alternative means of acquiring EXAFS data, being especially useful on thick samples.

(v) Laser EXAFS. These experiments yield transmission EXAFS spectra using a high-intensity, pulsed X-ray source.[215] The pulse of X-rays is generated by a laser-produced plasma. A high-intensity ($\approx$100-J) laser pulse $\approx$ 3.5 ns wide is focused onto a solid metal target, producing a surface plasma with a temperature in the kilovolt region. The radiation from this hot plasma extends into the X-ray region and has sufficient intensity to yield EXAFS spectra up to $\approx$3 keV. This radiation is dispersed by Bragg reflection from a crystal, passed through the sample, and recorded on photographic film. Film is used, since the entire spectrum is obtained in the few nanoseconds of the X-ray pulse. Both the incident and transmitted intensities are recorded in displaced positions on the same film to allow for normalization. In this way, an entire transmission EXAFS spectrum is obtained in a few nanoseconds.

This technique has been used to obtain EXAFS spectra of Al and Mg films in a few nanosecond.[216] The estimated photon flux falling on the sample in one pulse is $\approx 10^6$ photons/eV at the Al K edge (at 1.56 keV). This is to be contrasted with the photon flux at SSRL, which, at 10^{10} photons eV^{-1} s^{-1}, yields about 8000 photons/eV for the 0.3-ns pulses, which occur every 780 nsec. The two-orders of magnitude higher peak intensity from the laser-produced plasma enables these fast experiments to be performed. It should be observed, however, that obvious differences do exist between the Al EXAFS spectrum obtained in this manner and that obtained in standard absorption experiments.[217–219] If these differences can be eliminated or understood, this technique would offer the exciting potential of providing flash EXAFS spectra of transient species with short lifetimes, on the order of a few nanoseconds.

IV. Data Analysis

10. Data Analysis Techniques

In the following, we shall discuss various approaches to the analysis of EXAFS data. The rigor required at each step in the analysis depends some-

[215] P. J. Mallozzi, R. E. Schwerzel, H. M. Epstein, and B. E. Campbell, *Science* **206,** 353 (1979).
[216] P. J. Mallozzi, R. E. Schwerzel, H. M. Epstein, and B. E. Campbell, *Phys. Rev. A* **23,** 824 (1981).
[217] R. D. Deslattes, *Acta Crystallogr., Sect. A* **25,** 89 (1969).
[218] C. Senemaud and M. T. Costa Lima, *J. Phys. Chem. Solids* **37,** 83 (1976).
[219] A. Fontaine, P. Lagarde, D. Raoux, and J. M. Esteva, *J. Phys. F* **9,** 2143 (1979).

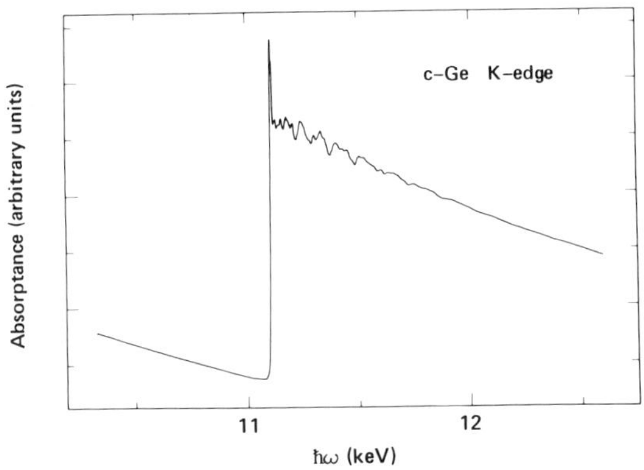

FIG. 20. The absorptance of c-Ge at 77°K as a function of X-ray photon energy, including the onset of Ge K-shell absorption at 11.1 keV.

what on the means by which the structural information will ultimately be extracted. Specifically, will it be deduced through comparisons among *experimental* EXAFS data sets, or through comparisons with *calculated* spectra? In the former case, it is essential that all the data sets be reduced according to the same detailed prescription. It is less important that each step be rigorous on an absolute scale, however, since small absolute errors which occur systematically will distort each data set in the same way without affecting the final comparison. On the other hand, comparison with a calculated spectrum demands that each data analysis step be precise on an absolute scale. We will comment on the relative rigor of each step where appropriate.

a. Removal of Background Absorption

The experimentally determined absorptance of crystalline Ge at 77°K is shown as a function of X-ray photon energy in Fig. 20.[220] The measured quantity is proportional to the product of the density of Ge atoms in the sample, the sample thickness, and $\sigma_{\mathrm{Ge;\,total}}$, the total absorption cross section for a Ge atom. The substantial increase in absorptance as the incident photon energy is increased past 11.1 keV marks the onset of absorption due to photoexcitation of electrons from the Ge K shell. The EXAFS is manifest as sharp oscillations in the absorption cross section extending for more than 1 keV above the K edge.[221,222]

[220] S. H. Hunter, unpublished (1976).

[221] Note that the prominent spike in the absorption at the K edge proper is not part of the EXAFS. It is more properly viewed as arising from either an unusually high density of final states or the substantial overlap of an unusually localized final-state wave function with the K-shell wave function.

[222] For a discussion of the near edge features, see Brown *et al.*[89] and Section 7.

It is convenient to regard the total experimental absorption cross section as the sum of two distinct contributions: the Ge K-shell absorption cross section σ, which contains the EXAFS information, and a slowly varying background absorption σ_{bg}. This background absorption is due to the photoexcitation of all those electrons in the system which are less tightly bound than the K electrons of interest in this example. As obtained experimentally, the prethreshold absorption also includes energy-dependent contributions from the X-ray photon counters, the sample holder, and so on, as discussed in Section 8. The first step in data reduction is often to fit a functional form to this quantity and to subtract the result from the measured spectrum. The sophisticated treatments in the literature of atomic absorption cross sections are of little use in this step for two reasons: The energy dependence of the prethreshold signal has been modified by the instrumental contributions noted above; and the functional forms in the literature are for $\omega \gg \omega_{\text{threshold}}$, which does not apply to the EXAFS region. Instead, σ_{bg} can be approximated as a polynomial of the form $A + B(\hbar\omega)^{-C}$. For a given C, the parameters A and B are determined through a least-squares fit to the absorptance over an energy range extending from $\approx$800 to $\approx$200 eV below the K edge. The resulting σ_{bg} is subtracted from the measured absorptance over its complete range, yielding a quantity proportional to the $\sigma_\alpha(\omega)$ defined in Eq. (1.1). The parameter C is determined by the following requirement: The resulting fractional falloff in $\sigma_\alpha^0(\omega)$ far above the K edge (between 0.5 and 1.5 keV in this example of Ge) must be identical to that measured for atomic K-shell absorption.[223] C is typically found to be $\approx$2.

b. *Extraction of the EXAFS* χ

The next step in data reduction is to divide the signal by $\sigma_\alpha^0(\omega)$ and thereby extract $\chi_\alpha(\omega)$ using Eq. (1.1). It would be possible in many cases to extract the needed σ_α^0 from a separate measurement of the X-ray absorption spectrum of atom species α in a monatomic gas phase. There are, however, several complications: (*a*) It would be necessary to shift the photon energy to account for changes in the initial- and final-state energies; (*b*) the low-energy portion of the continuum of final states in the solid is replaced with discrete bound states in the free atom (eliminating a portion of σ_α^0); and (*c*) it would be necessary to scale the overall strength of σ_α^0 to account for the differing number of absorbing atoms in the two experiments. Principally owing to these complications, the usual procedure for obtaining σ_α^0 is to extract it from the data set in question by fitting a slowly varying functional form to σ_α above the absorption edge. In one approach to this step, σ_α^0 is

[223] For example, see the tables in W. H. McMaster, N. K. Del Grande, J. H. Mallet, and J. H. Hubbell, *Lawrence Radiat. Lab.* [*Rep.*] **UCRL-50174,** Sect. II, Rev. 1.

approximated as a polynomial of fourth or sixth order in $(\hbar\omega - E_{th})^{1/2}$. E_{th} is that value of $\hbar\omega$ which corresponds to the center of the step in absorption at the K-shell threshold, determined by extrapolating the absorption from below and from above, but excluding the contribution of the threshold spike. The expansion variable is sublinear in $\hbar\omega - E_{th}$ so as to accommodate the relatively rapid variation of σ_α^0 in the region of the K edge. The parameters in the polynomial are chosen by a least-squares fit to the K-shell absorptance, weighted by $\hbar\omega - E_{th}$, over an energy range extending from just above the threshold spike to the upper limit of the data range. The resulting expression for $\sigma_\alpha^0(\omega)$ is used with Eq. (1.1) to extract $\chi_\alpha(\omega)$, the EXAFS associated with excited atom species α. The principle which underlies this approach is the following: The sinusoidal variations in χ as a function of $(\hbar\omega - E_{th})^{1/2}$ are of sufficiently high frequency relative to variations in σ^0 that fitting the absorptance to a polynomial of low order will ignore χ and approximate σ^0. This principle has been verified through extensive tests on data wherein the nearest-neighbor peak in the Fourier transform of $k\chi$ has been shown to be independent of the order of the polynomial for low orders (i.e., n less than 10 for an extensive data range such as in this example).

The net effect of first subtracting σ_{bg} and then dividing by σ_α^0 is to normalize the EXAFS properly, so that it will be possible to deduce the width (or, more generally, the shape) of a peak in $p(r)$ and the number of neighboring atoms from the data.[224] There are other ways to accomplish this. One alternative approach omits the separate extraction of σ_{bg}, fitting instead both σ_{bg} and σ_α^0 in the EXAFS region with a slowly varying functional form. The resulting curve is subtracted from the data to yield an unnormalized EXAFS spectrum. Proper normalization then follows from separate quantitative determinations of the step in absorption at the K edge and of the subsequent falloff in the measured spectrum.[225]

The remaining step in obtaining $k\chi_\alpha(k)$ as a function of electron momentum k is the identification of the zero of conduction electron energy E_0, that value of the final-state electron energy $E = E_{nl} + \hbar\omega$ which corresponds to $k = 0$ for the final-state electrons (i.e., $\hbar^2k^2/2m = E - E_0 = E_{nl} - E_0 + \hbar\omega$). This is very difficult to accomplish in an absolute sense, as discussed in Section 5. $E_0 - E_{nl}$ differs from the threshold for continuum excitation ($\hbar\omega \approx E_{th}$) by an undetermined energy-dependent self-energy correction and a Fermi energy (which is zero in the case of Ge). Furthermore, the possible presence of excitons and/or resonances in the spectrum

[224] The energy dependence of the EXAFS amplitude relates to the peak shape, while the absolute amplitude near threshold depends in part on the number of atoms.

[225] Note that cubic splines could be used to fit σ_α^0 in either approach, as an alternative to a polynomial expansion over the complete EXAFS range.

complicates the experimental identification of the threshold for continuum excitation. If the ultimate step in the analysis will involve comparisons among reduced data sets only, however, it is not necessary to determine E_0 in an absolute sense. It is important only that the difference between the assigned value of E_0 and its true value be the same in all data sets to be compared. The value of that difference can be chosen more or less arbitrarily. In reducing the Ge data, for example, we have chosen $E_0 - E_{nl}$ to be E_{th}, making k proportional to $(\hbar\omega - E_{th})^{1/2}$. The systematic k-scale distortion introduced in this way will not affect comparisons among data sets in most cases. Note, however, that this approach could lead to errors in comparing EXAFS spectra from substantially different systems, such as a metal and an insulator, where the self-energy correction could be significantly different. In fact, the correct determination of E_0 is often a significant and unavoidable problem in analyzing data from insulating solids and molecules.[226]

In principle, one cannot avoid assigning an absolute value to E_0 if analysis of the EXAFS data will involve comparisons with calculated spectra, as is discussed later in this section. In practice, however, it is usually necessary to adjust E_0 freely to correct for errors in assigning the "muffin-tin zero" and for numerous other uncertainties in the calculations, as is discussed in Section 5. A procedure for this adjustment is discussed in detail by Teo and Lee.[40]

When $E_0 - E_{nl}$ is set equal to E_{th}, the resulting $k\chi(k)$ for crystalline Ge[220] is shown in Fig. 21a. The procedure for extracting χ has normalized it properly to the number of Ge atoms in the sample, allowing it to be compared directly with Eq. (1.2). The broad feature at $k \approx 1$ Å^{-1}, corresponding to the threshold spike in Fig. 20, is seen to be a superposition of two peaks. The EXAFS starts above that feature and extends beyond the upper limit of the data at 20 Å^{-1}. The influence of many near-neighbor positions is clearly manifest in the presence of many frequencies in the EXAFS. In dramatic contrast, the $k\chi$ shown in Fig. 21b for amorphous (a-)Ge[220] exhibits only one distinct frequency, implying the dominance of a single near-neighbor distance in the EXAFS signal. This spectrum was obtained from a sample prepared by plasma decomposition of GeH_4, but is identical to one obtained from evaporated a-Ge.[220] The structural implications of the obvious differences between the c- and a-Ge EXAFS spectra will be discussed further below.

The ultimate extraction of structural information from $k\chi$ can be accomplished either in k space or in r space (where r is the variable complementary to $2k$ in a Fourier transform). As will be discussed, there are ad-

[226] One approach to this problem has been to treat E_0 as an adjustable parameter in analyzing the EXAFS data, even in those instances where two data sets are compared.

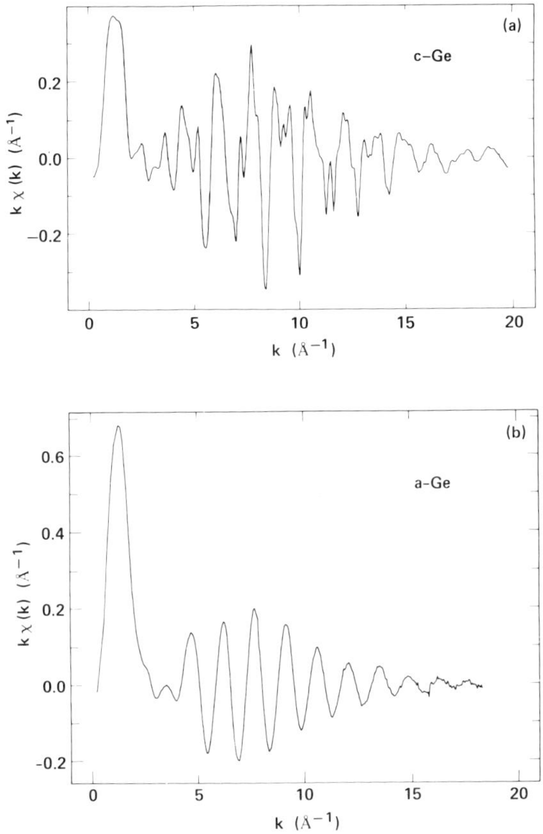

FIG. 21. The EXAFS oscillations $k\chi(k)$ on the Ge K-shell absorption cross section in (a) c-Ge and (b) a-Ge at 77°K as a function of photoelectron momentum k.

vantages associated with each approach. In most cases, several near-neighbor distances contribute their individual frequencies to $k\chi$, complicating substantially the analysis of it in its present form (i.e., as in Fig. 21a). Accordingly, it is common practice to separate these frequencies from one another through a Fourier transform, even in those cases where the ultimate analysis will be accomplished in k space. This procedure was suggested originally by Sayers, Stern, and Lytle.[26]

c. *Fourier Transform into r Space*

In transforming the EXAFS into r space, one immediately faces the problem of minimizing the undesirable effects of the limited data range available for the transform. The limits of the data at both low and high k were discussed in Section 6. The low-k cutoff is the more persistent problem in EXAFS analysis, arising from the necessity of avoiding the threshold spike. In the absence of a reliable method for removing that spike, the EXAFS data set must be terminated in the region of 2–3 Å^{-1}, where the EXAFS signal is usually strong. The large-k cutoff is much less of a problem, since it typically results from the EXAFS signal being reduced in strength to the noise level. In any case, the inaccessibility of a portion of the k-space range of $k\chi$ results in the shape and extent of the peak function ξ being quite sensitive to the manner in which the data is terminated in the transform. The problem of selecting an appropriate window function occurs analogously in the transformation of X-ray or neutron diffraction results and has been treated extensively.[227] The optimum window function is chosen as a compromise between a long-range or oscillating peak function on the one hand, and an intolerable broadening of the r space information on the other. The customary choice in diffraction has been to eliminate the oscillations in r space at the expense of structural resolution. In the following analysis, we will take a different approach and use a square window which has been broadened by convolution with a Gaussian of half-width σ_w. The effect of this broadening in k space is to impose an exponential localization in r space, substantially lessening the interference of neighboring peaks and the accompanying distortion in apparent peak position and magnitude.[31] This choice minimizes the loss of structural resolution, but necessitates the use of analysis techniques which are unaffected by residual oscillations in the peak function. A similar effect can be achieved by using the following window function: a square window at each end of which has been appended one-half of a Hanning function.[228]

The data shown in Fig. 22 have been transformed using a square window, $k = (3.9\ \text{Å}^{-1}, 14.2\ \text{Å}^{-1})$, with a Gaussian broadening of half-width $\sigma_w = 0.7$ Å^{-1}. The absence of data for negative k has resulted in a φ which is complex. The real part and the magnitude of $\varphi(r)$ for the K edge of crystalline Ge are shown in Fig. 22a. The signals from the first three shells of atoms are clearly evident to the right, while the small peaks near the origin result from the incomplete removal of σ^0. The peaks in φ extend for some distance beyond 5 Å, as expected on the basis of the sharp structure shown in $k\chi$ in Fig. 21a. Note that the Fourier transform has nearly separated from one another the signals from the principal peaks, as intended. The $\varphi(r)$ shown for a-Ge in Fig. 22b is strikingly different. The single frequency of oscillation evident in Fig. 21b has resulted in a single structural peak in φ, as anticipated, corresponding to the nearest neighbors. Note that each of the peaks

[227] For a discussion, see J. Waser and V. Schomaker, *Rev. Mod. Phys.* **25**, 671 (1953).
[228] G. D. Bergland, *IEEE Spectrum* **6**, 41 (1969).

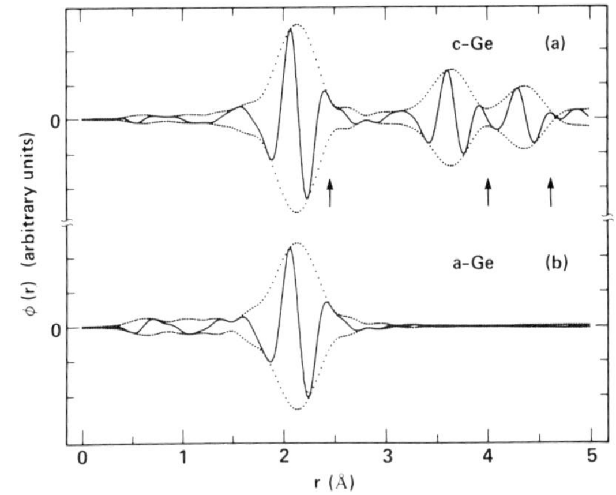

FIG. 22. The real part (solid line) and the magnitude (dotted line) of the Fourier transform $\varphi(r)$ of the EXAFS on the Ge K-shell absorption in (a) c-Ge and (b) a-Ge at 77°K. The units of the vertical scales are arbitrary but identical. The data were transformed using a square window with k between 3.9 and 14.2 Å^{-1}, broadened by convolution with a Gaussian of half-width $\sigma_w = 0.7$ Å^{-1}. The corresponding near-neighbor distances in c-Ge are shown by vertical arrows in the lower part of (a).

in φ is shifted to lower r from the position of the corresponding peak in $p_{\alpha\beta}$, which are indicated for c-Ge by the vertical arrows in the lower portion of Fig. 22a. The shift arises from the k dependence of the phase shift $\eta_{l'}$ of the excited atom α and from that of the phase of t_β in the expression for Λ [Eq. (3.28)]. The shift is characteristic of the atom species involved (α and β) and depends somewhat on the transform range.

A substantial amount of information can be gleaned from even the most qualitative comparison between these two φ's. It is convenient, however, to introduce into this discussion those quantitative comparison techniques which have proven essential to the interpretation of the EXAFS in more complicated situations.

d. Extraction of Structural Information in r Space

Most of the early EXAFS studies relied upon somewhat qualitative comparisons among data sets. The evolution of EXAFS data reduction into well-defined procedures such as those presented here has resulted in increasingly reproducible end products, the $k\chi$ or φ. This has stimulated the development of equally well-defined procedures for the structural analysis of those quantities. The analysis which is described here is accomplished in r space and is based upon Eq. (1.5). The extraction of structural information after a portion of φ has been Fourier transformed back into k space will be discussed in the next subsection. The variations which are observed among the φ's obtained from different systems are assumed to arise solely from variations in the $p_{\alpha\beta}(r)$ (presuming, of course, that the identities of the excited and backscattering atoms are unchanged).[229] Thus the shift or broadening of a peak in $p(r)$ appears as a shift of the peak in φ along the r axis or a convolution of it with a broadening function (such as a Gaussian), respectively. The unique peak function to be associated with scattering from atom species β about excited species α, $\xi_{\alpha\beta}$, is extracted from the φ measured using a structurally known sample, or standard. The φ obtained from a structurally unknown system can then be analyzed to yield its $p_{\alpha\beta}(r)$, to the extent that those unknown $p_{\alpha\beta}$ can be represented as a linear combination of shifted and broadened peaks from the standards. More generally, the $p_{\alpha\beta}$ in the unknown system and that in a standard need only be such that one can be obtained from the other through convolution with a suitable function. It is essential to this approach that the variations in $\xi_{\alpha\beta}$ from system to system be insignificant. This establishes the minimum requirement that the identical window function be used in the $k \rightarrow r$ Fourier transform of data from the unknown system and from the standards. There are also several possible fundamental sources of variations in $\Lambda_{\alpha\beta}$ (and hence $\xi_{\alpha\beta}$) from system to system, including changes in electronic configuration with composition or crystal structure, as well as the many-body effects discussed in Section 4,a. Addressing this issue experimentally, Citrin *et al.*[230] have demonstrated the "transferability" of the phases found in Λ, and hence in ξ. The invariance of the ξ is further supported by the success of comparisons between data and calculated spectra.[231] Furthermore, Stern *et al.*[53,54] have studied extensively the dependence of the amplitude of Λ on chemical environment and concluded that it is insensitive, except as regards the electron mean-free-path length (which can vary by a factor of 2). These and other studies are discussed at length by Lee *et al.*[9] It is still not obvious, however, that the variations in $\xi_{\alpha\beta}$ from system to system are so small as to be neglected as more and more detail is demanded from the analysis of EXAFS data.

The application of this procedure for structural analysis is quite straightforward. The relevant ξ's are extracted from appropriately chosen standards as numerical functions (i.e., a sequence of values for the real and imaginary parts corresponding to selected values of r). This process is greatly facilitated by a properly selected window function. The choice described earlier lessens

[229] The *ansatz* is supported qualitatively by the great similarity of the four peaks in the data shown in Fig. 22. Specifically, the shape of the magnitude of φ and that of its real part are nearly identical in all four instances.

[230] P. H. Citrin, P. Eisenberger, and B. M. Kincaid, *Phys. Rev. Lett.* **36,** 1346 (1976).

[231] For example, see Refs. 40, 46, and 67.

the range of the peak function in r space, allowing the necessary separation of otherwise overlapping contributions from near neighbors at more than one distance. A simulated $\varphi_m(r)$ for the unknown system can be calculated from these $\xi_{\alpha\beta}$ and a model for the $p_{\alpha\beta}(r)$. The shape of the peaks in the model $p_{\alpha\beta}$ need not be constrained, except out of knowledge of the system in question. In particular, the peaks can be allowed to exhibit the sort of asymmetry shown in Fig. 12 and discussed at length in Section 6,d. The model parameters in $p_{\alpha\beta}(r)$ are varied until the "best" fit to the measured φ is found, corresponding to a minimum in

$$R^2 = (2N)^{-1} \sum^N \left(\frac{[\mathrm{Re}(\varphi - \varphi_m)]^2}{(\mathrm{Re}\,\varphi)^2 + (\mathrm{Re}\,d\varphi/dr)^2} + \frac{[\mathrm{Im}(\varphi - \varphi_m)]^2}{(\mathrm{Im}\,\varphi)^2 + (\mathrm{Im}\,d\varphi/dr)^2} \right), \quad (10.1)$$

where the sum extends over all N points in the range of the structural feature being fit. This measure of the quality of fit is a standard fractional difference least squares for the real and imaginary parts individually, except for the $d\varphi/dr$ terms in the denominators. These terms have been introduced to enhance the sensitivity of R to variations in the shape of the peaks in $p(r)$, at the expense of some of its extraordinary sensitivity to position.

It is also possible to carry out the analysis procedure above by substituting a calculated EXAFS spectrum for the experimental spectrum from a standard. The analysis is somewhat better defined in that no assumption need be made about the p corresponding to the calculated spectrum. It can be given any convenient shape (e.g., a δ function at a reasonable distance). The requirement that ξ vary little between the standard and the unknown is replaced with the requirement that the calculated ξ accurately represent that for the unknown. As discussed in Section 5, it is necessary to adjust the zero of conduction electron energy E_0 to compensate for inaccuracies in the calculation. This is somewhat more cumbersome in r space than in k space, since the calculated $k\chi$ corresponding to each new E_0 must be Fourier transformed before comparison with the unknown φ.

The procedure described above can now be applied to the structural analysis of φ's similar to those shown in Fig. 22. A φ for c-Ge is shown in Fig. 23, together with a $\xi_{\mathrm{Ge-Ge}}$ extracted from the contribution of nearest-neighbor Ge atoms. Using this $\xi_{\mathrm{Ge-Ge}}$, the entire φ for the crystal may be simulated with great accuracy, as shown in Fig. 23c.[31,232] For the first three near-neighbor distances, this process yields the known separations in the radial positions and the relative widths of the peaks in $p_{\mathrm{Ge-Ge}}$. In addition, the averaged electron mean-free-path length is determined to be 8 Å. Turning now to the φ for a-Ge, a comparison between that structural peak and

[232] T. M. Hayes and S. H. Hunter, *in* "The Structure of Non-Crystalline Materials" (P. H. Gaskell, ed.), p. 69. Taylor & Francis, London, 1977.

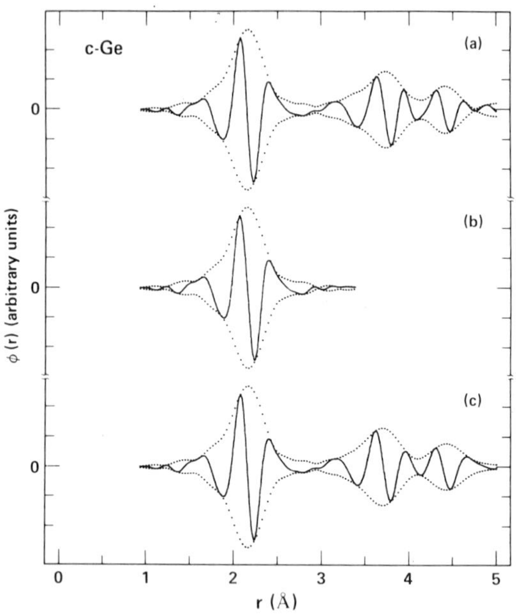

FIG. 23. (a) The real part (solid line) and the magnitude (dotted line) of the Fourier transform $\varphi(r)$ of the EXAFS on the Ge K-shell absorption in c-Ge at 77°K. (b) The $\xi_{\text{Ge-Ge}}(r)$ extracted from the nearest-neighbor contribution to (a). (c) The model $\varphi(r)$ constructed from $\xi_{\text{Ge-Ge}}$ and a model $p_{\text{Ge-Ge}}$ consisting of three Gaussian peaks of adjustable position, amplitude, and width. The peak positions agree with crystallographic data, while the amplitudes can be used with the known number of nearest neighbors to estimate the electron mean-free-path length. The units of the vertical scales are arbitrary but identical. The data were transformed using a square window with k between 4.89 and 16.09 Å^{-1}, broadened by convolution with a Gaussian of half-width $\sigma_w = 1.0\ \text{Å}^{-1}$. [Taken from Fig. 2 of T. M. Hayes, P. N. Sen, and S. H. Hunter, *J. Phys. C* **9**, 4357 (1976).]

the $\xi_{\text{Ge-Ge}}$ extracted from c-Ge shows that the nearest-neighbor distances are identical, but that the a-Ge peak has been broadened by an additional $\Delta\sigma = (\sigma_a^2 - \sigma_c^2)^{1/2} = 0.04$ Å.[232] The number of nearest neighbors in a-Ge appears to be greater than four. The source of this unphysical result is an experimental difficulty arising from the presence in the incident X rays of higher-order diffracted beams from the monochromator, an effect discussed in Section 8. Finally, the absence of structure beyond the nearest-neighbor contribution in a-Ge implies that more distant peaks in $p_{\text{Ge-Ge}}$ are at least 0.2 Å broader than the first. The high sensitivity of EXAFS peaks to the broadening of radial structure is discussed in detail in Section 6. These results for c- and a-Ge strongly support the underlying thrust of this analysis—the structure of an unknown system may be determined in many cases through a detailed comparison between the EXAFS data from the unknown and that from a suitably chosen standard.

The foregoing EXAFS data analysis procedures have been applied to a wide variety of systems. As long as the data reduction follows a well-defined, detailed prescription, there has been no evidence for irreproducibility or for systematic errors affecting the structural interpretation. It should be obvious, moreover, that most of the principles set forth and conclusions drawn in this section apply equally well to comparisons in k space between data and calculated spectra, or between two data sets, to which we now turn.

e. Extraction of Structural Information in k Space

While it is possible to analyze $k\chi$ directly without any Fourier transform, this is seldom done. Even when only one peak in p contributes to $k\chi$, as in the case of a-Ge, a Fourier transform into r space can be used to remove high- or low-frequency "noise" from $k\chi$, as well as low-frequency contributions due to an improper extraction of σ^0. This is often referred to as "Fourier filtering." The transform of a portion of φ back into k space requires, of course, the use of another window function. This presents no problem in those cases where the peak or peaks being back-transformed are well separated from other features in φ. On the other hand, if the $r \to k$ window function must be adjusted to eliminate contributions from other features in φ which overlap with the peak of interest, the result in k space will unavoidably vary with the extent and shape of the window function. Moreover, the analysis will be complicated by the following effect. Consider two EXAFS spectra which differ only in that one peak in p in one of them is significantly broader than in the other, such that those contributions to $k\chi$ differ by a factor of $\exp(-2k^2\Delta\sigma_r^2)$. This leads, of course, to peaks in φ which have different widths. If isolating these peaks from others in the same spectrum preparatory to a $r \to k$ transform requires use of a window function which does not include all of each peak, then the two Fourier-filtered k-space spectra will no longer be related by the factor of $\exp(-2k^2\Delta\sigma_r^2)$. This effect is illustrated in Fig. 24. It arises because the broadening of the peak in p redistributes its signal in r space, resulting in some components being moved into and some out of the $r \to k$ window function. Thus the k-space analysis must proceed by broadening the peak *before* the $r \to k$ transformation, so as to preserve the r-space content of a peak, *unless* that peak is effectively isolated in r space.

Most of the considerations mentioned in the subsection on r-space analysis apply equally well to a complete k-space analysis and will not be mentioned again. As noted earlier, the adjustment of E_0 required in comparisons with calculated spectra is less cumbersome in k space. The least-squares fit proceeds as before, except that a shift in the position of a peak in p results

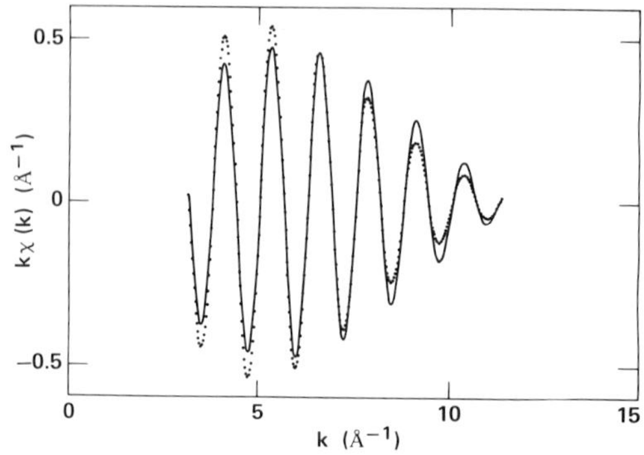

FIG. 24. The EXAFS spectra $k\chi(k)$ resulting from the application of two distinct procedures to a single model EXAFS spectrum given by $k\chi_m(k) = \sin(5k)$. The spectrum represented by a solid line results from the following procedure: Multiply $k\chi_m$ by $\exp(-0.02k^2)$ to simulate thermal broadening with peak half-width 0.1 Å; transform $k \to r$ with a square window function, $k = (3.1\ \text{Å}^{-1}, 11.3\ \text{Å}^{-1})$; transform $r \to k$ with a square window, $r = (2.2\ \text{Å}, 2.8\ \text{Å})$. The spectrum represented by a dotted line is processed in precisely the same fashion, *except* that it is multiplied by $\exp(-0.02k^2)$ *only* after both transforms. This second procedure is analogous to that which might be used to analyze a thermally broadened EXAFS spectrum in k space, using the same sample at low temperatures as a "standard." The broadening operation does not commute with the transforms because the $r \to k$ transform window function deletes a portion of the r-space peak. The spectra differ as a consequence. A fitting procedure would incorrectly interpret the differences between the spectra shown here in terms of differences in the structural parameters.

in a change in the phase of $k\chi$ by $2k\Delta r$, while broadening contributes a factor similar to $\exp(-2k^2\Delta\sigma_r^2)$. One must, of course, choose a measure of the quality of fit other than that given by Eq. (10.1).

Some special techniques have been developed for k-space analysis which have no obvious analog in r space. For example, Stern *et al.* proposed in an early paper[93] that the number of near neighbors and the width of the peak in p can be determined relative to a known standard by a suitable ratio of the amplitude of the back-transformed $k\chi$ to that of the standard. Martens *et al.*[233] have described a method by which "beats" in the EXAFS can be used to determine the splitting of two close near-neighbor distances if the atom species are identical. In a later paper, Martens *et al.*[234] suggested a method by which the difference in the phases of two $k\chi$'s can be used to determine independently the difference in E_0 *and* the difference in near-

[233] G. Martens, P. Rabe, N. Schwentner, and A. Werner, *Phys. Rev. Lett.* **39,** 1411 (1977).
[234] G. Martens, P. Rabe, N. Schwentner, and A. Werner, *Phys. Rev. B: Solid State* [3] **17,** 1481 (1978).

neighbor distance from one spectrum to the other (assuming that the two samples have identical excited and backscattering atoms). This works as well in comparisons among experimental spectra as in comparisons with calculated spectra. The reader is referred to the literature for a discussion of other approaches.[9,40,46,67,235]

Finally, Lengeler and Eisenberger[104,105] have published two exhaustive studies in which they examined the complications which can arise when the EXAFS contributions are compared between neighbors at different separations or in different chemical environments. In particular, they focused on the many-electron amplitude reduction, the final-state electron mean free path, peak shape variations, and other aspects of the transferability of phase and amplitude. They concluded that it is possible, with proper care, to determine radial separations to within 1%, the number of neighbors to within 10%, and the width of a peak to within 20%.

V. Experimental Studies

A complete survey of the extensive literature of EXAFS experimental studies would be well beyond the scope of this article. In the following sections, the range of applications of EXAFS spectroscopy is indicated by representative studies. Most of those studies whose primary purpose was to demonstrate a particular measurement technique or to establish a procedure for interpreting EXAFS spectra have already been discussed in the theory and experiment sections. In the following sections, we summarize the results of studies of specific materials, with emphasis on new insight which has been gained into the physical properties of the materials at issue.

11. Crystalline Solids

In general, crystalline materials are best studied using diffraction techniques, which can reveal the translational symmetry and, with greater effort, the locations of the atoms within the unit cell. There are nonetheless many special situations in crystals for which EXAFS has substantial advantages over diffraction as a probe of local structure. These studies exploit two characteristic features of EXAFS which distinguish it from diffraction: It is both a local and a spectroscopic probe—that is, it probes the identity and position of the near-neighbor atoms of only the specific element to which one tunes the incident X-ray energy. We discuss several of these special situations in this section.

[235] D. R. Sandstrom, *J. Chem. Phys.* **71**, 2381 (1979).

a. Elemental Materials and Simple Compounds

A variety of crystals that are structurally well known have been studied using EXAFS. The general purpose of these studies was to demonstrate a particular aspect of the EXAFS technique, rather than to obtain structural information. These range from the early observation of the extended fine structure on Mg, Cr, and Fe compounds in 1920,[11] through the discussion of a Fourier-transform technique to analyze the data on Ge and Cu,[93] to the recent examinations of many-body effects on the EXAFS amplitude[54] and amplitude transferability[105] using a host of compounds. The results of most of these studies have already been discussed in the theory and experiment sections. One class of these studies does, however, provide new insight into the structure of these compounds, as well as yield information on the EXAFS technique itself. These are investigations of the width of the peaks in the atom–atom pair correlation functions.

As derived in Section 3 and discussed in some depth in Section 6, the structural information in the EXAFS are the pairwise distributions of near-neighbor atoms of species β about the excited atom α, $\rho_{\alpha\beta}(\mathbf{r})$. These are called the interatomic pair correlation functions or, after spherical averaging, the radial distribution functions, $p_{\alpha\beta}(r)$ [see Eqs. (3.24), (3.31), and (6.2)]. The width of a near-neighbor peak in these distributions is determined by near-neighbor correlations. In contrast, the Debye–Waller factor derived from diffraction measures the correlation of an atom with its lattice site. As a consequence, the width of the peak in $\rho_{\alpha\beta}(\mathbf{r})$ or $p_{\alpha\beta}(r)$ can be significantly smaller than that suggested by the Debye–Waller factor. Denote by σ_j the root-mean-square deviation of the distance between the excited atom and a neighboring atom j, the half-width of a peak in ρ or p. Beni and Platzman[94] have shown that $\sigma_j^2 = 2(\text{MSD} - \text{DCF})$, where MSD is the mean-square displacement that enters the Debye–Waller factor and DCF is a displacement correlation function. This latter term ensures that only out-of-phase thermal motions along the vector connecting the atoms contribute to σ_j. Beni and Platzman derived a specific expression for σ_j in the Debye approximation for a monatomic cubic crystal. A general result of this calculation is that the DCF reduces σ_j significantly, since DCF/MSD approaches 0.4 for temperatures at and above the Debye temperature. They also compared the calculated width of the first-neighbor peak in Zn with the data of Brown *et al.*[86] Since Zn is hexagonal close-packed (hcp) and thus anisotropic, their expression for cubic crystals was generalized by introducing a directionally dependent Debye temperature. The calculated widths show a large anisotropy and are in semiquantitative agreement with the experimental results of $\sigma_\perp \approx 0.20$ Å and $\sigma_\parallel \approx 0.12$ Å, respectively perpendicular and parallel to the basal plane at room temperature.

Böhmer and Rabe[97] have obtained EXAFS data on the K edges in metallic

Cu and Co and of Rb and Sr in RbCl, SrS, SrF_2, and $SrCl_2$ from 80 to 420°K. They compared their results for the temperature variation of σ_j with those they calculated from a Debye model. Good agreement was obtained for Cu, Co, and RbCl; less satisfactory agreement was obtained for SrS, SrF_2, and $SrCl_2$, owing to a large uncertainty in the measured σ_j, an uncertain Debye temperature, or possibly an incorrect estimate of the correlation. They also pointed out that the DCF/MSD ratio depends strongly on temperature and on the particular shell of neighbors considered. This ratio is large for the first shell and is smaller for distant shells, consistent with the motions of nearest neighbors being more highly correlated than those of more distant neighbors. The ratio also increases with increasing temperature, approaching 0.38 for the first-neighbor shell in a face-centered cubic (fcc) lattice, in agreement with Beni and Platzman.[94] This correlation reduces the EXAFS peak width relative to the Debye–Waller width. In Cu, for example, the width of the first shell σ_1 is about 17% smaller than the Debye–Waller width deduced from diffraction at 80°K. Their values for the widths in Cu are in good agreement with the widths obtained earlier by Stern *et al.*[93] and Gurman and Pendry.[67]

Greegor and Lytle[98] have examined the temperature variation of the EXAFS in Cu from 10 to 683°K. They extracted the peak width relative to that at 10°K and compared these values with a Debye model with no correlations, the Debye model with correlations of Beni and Platzman,[94] and the lattice dynamical models of Sevillano *et al.*[96] Good agreement with the latter two was obtained. The first model, however, overestimated the peak width for the first shell by $\approx$25% at 77°K, and by 35% at 295°K. This indicates the importance of correlations in reducing the EXAFS peak widths below the Debye–Waller width.

Rabe *et al.*[100] have studied the temperature dependence of the first- and second-neighbor peaks in the pair correlation function for c-Ge. They concluded that the Debye model of Beni and Platzman[94] overestimates the temperature dependence of the peak widths. This disagreement is particularly striking in the case of the first-neighbor peak, for which the predicted σ^2 increases more than four times faster than is observed. This is not particularly surprising, however, given the existence of fairly rigid covalent nearest-neighbor bonds in these materials, the effects of which on the near-neighbor correlations are not expected to be reproduced by a Debye model.

Stern *et al.*[54] have made EXAFS measurements on the K edges of the transition metals in $MnCl_2$, $FeCl_2$, and $CoCl_2$, on the K edges of both constituents of CuBr, ZnSe and GaAs, on the K edges of pure Cu and Ge, and on the L_{III} edge of Pt. These measurements were made as a function of temperature in order to extract the peak widths in a study of many-body effects on the EXAFS amplitudes. For Cu and Pt, excellent agreement between experiment and the theory of Sevillano *et al.*[96] is obtained for the widths of the first three shells of neighbors. For the dichlorides, an Einstein model gives good agreement with the temperature variation of the widths. For the tetrahedral materials, CuBr, ZnSe, GaAs, and Ge, the first- and second-neighbor peak widths were examined. An Einstein oscillator model gave a good fit to the first-neighbor peaks, where the high-frequency optical modes dominate the phonon spectra. A Debye model gave a better fit for the second neighbor. Anharmonic effects are important for CuBr for temperatures above 80°K, so a Gaussian peak shape could not be used, consistent with the discussion in Section 11,b. Of all these dichlorides and tetrahedral materials, *a priori* calculations are available only for Ge. Good agreement between theory and experiment is obtained for this material.

b. *Superionic Conductors*

Superionic conductors are crystalline solids which display ionic conductivities typical of those found for molten salts [i.e., $\approx 1\ (\Omega\ \mathrm{cm})^{-1}$].[236] Whereas most ordinary solids exhibit ionic conductivities $\approx 10^{-8}\ (\Omega\ \mathrm{cm})^{-1}$ at moderate temperatures, the superionic conductors achieve these high conductivities well below their melting points. These systems consist of two or more elemental components, only one of which is mobile. They are naturally characterized structurally in terms of mobile and immobile sublattices. The immobile ions form a complex network of channels through which the mobile ions move. This network is not rigid, since these ions execute large vibrations about their lattice sites. The positions of the immobile ions define characteristic voids which are populated to varying degrees by the mobile ions and through which these ions move. In superionic AgI for example, the iodine forms a body-centered cubic (bcc) lattice with the mobile Ag cations occupying and moving through tetrahedral voids.

Structural studies of these systems are invaluable, since they can yield not only the location of the mobile ions in the superionic phase, but also information on the conduction path from one site to another. Diffraction studies yield a great deal of information about the immobile-ion sublattice, but cannot provide as much information about the mobile ions due to the much greater disorder in their distribution. As a consequence, EXAFS studies can fill an important role by providing the location of the mobile ions relative to the immobile ones.[237] These systems also provide an excellent example of non-Gaussian peak shapes, as discussed in detail in Section 6,d.

[236] See, for example, the review by J. B. Boyce and B. A. Huberman, *Phys. Rep.* **51,** 189 (1979), and references contained therein.

[237] For a comparison of the information content of EXAFS with that of diffraction for these materials, see J. B. Boyce and T. M. Hayes, *in* "Physics of Superionic Conductors" (M. B. Salamon, ed.), Chapter 2. Springer-Verlag, Berlin and New York, 1979.

In the initial EXAFS study of AgI,[238] it was found that structural models previously proposed to explain the diffraction results were inconsistent with the EXAFS data; rather, it was determined that the Ag ions occupied the tetrahedral regions of the iodine bcc lattice and that the conduction path was from tetrahedron to tetrahedron through their shared faces. Later neutron diffraction data on single crystals of superionic AgI have supported these conclusions and also yielded detailed Ag ion density maps.[239] Mobile-ion density contours have also been obtained from the EXAFS data,[240] and shown to be in good agreement with those from diffraction. Those contours were deduced using an excluded volume model, which we now describe as an example of the physical modeling of a pair distribution function.[241]

Since the superionic conductors are dynamically disordered to a high degree, the usual Gaussian peak shape would not be appropriate for the anion–cation pair correlation function, as discussed in Section 6,d. The actual distribution function is expected to be highly asymmetric, due to the large anharmonicity of these materials, motivating the proposal of an excluded volume model.[242] In this model, the dominant anion–cation interaction is core–core repulsion. The Coulomb and polarization interactions give rise to the long-range order of the immobile-ion sublattice and prevent the mobile ions from approaching one another closely. These forces are expected to cancel out, however, along the highly symmetric pathways of the mobile ions,[243] so that variations in the potential will be quite small, as observed. The resulting anion–cation pair potential V_{a-c} is the core–core repulsion, modeled by a hard-sphere interaction,

$$V_{a-c}(r) = \begin{cases} \infty, & r < r_{ex}\,, \\ 0, & r > r_{ex}\,. \end{cases} \tag{11.1}$$

In this expression, the excluded radius $r_{ex} = r_a + r_c$, where r_a and r_c are the effective hard-sphere radii of the anion and cation. The resulting mobile cation distribution is uniform throughout the solid, except for exclusion from a sphere around each anion site. This distribution is then convoluted with a Gaussian to account for the slight softness of the actual core–core repulsion, as well as the thermal vibrations of the anions about their lattice sites. Given the average location of the immobile ions from diffraction

[238] J. B. Boyce, T. M. Hayes, W. Stutius, and J. C. Mikkelsen, Jr., *Phys. Rev. Lett.* **38,** 1362 (1977).

[239] R. Cava, F. Reidinger, and B. J. Wuensch, *Solid State Commun.* **24,** 411 (1977).

[240] J. B. Boyce, T. M. Hayes, and J. C. Mikkelsen, Jr., *Phys. Rev. B: Condens. Matter* [3] **23,** 2876 (1981).

[241] See Section 6,d for a discussion of this model in the context of non-Gaussian peak shapes.

[242] T. M. Hayes, J. B. Boyce, and J. L. Beeby, *J. Phys. C* **11,** 2931 (1978).

[243] W. H. Flygare and R. A. Huggins, *J. Phy. Chem. Solids* **34,** 1199 (1973).

results, this model completely determines the mobile-ion locations using two parameters only: r_{ex} and the Gaussian width.

This model provides an excellent fit to the EXAFS data on AgI in both the low-temperature wurtzite phase and the high-temperature superionic phase. In fact, the fit using this excluded volume model is significantly better than that using competing models, including the Strock model, the various displaced site models, and the anharmonic oscillator models.[240] The Ag-I pair distribution functions which result from such a fit to the EXAFS data at three temperatures are shown in Fig. 25. It is seen that the nearest-neighbor peak in the pair distribution function is already asymmetric at 22°C, and becomes more so with increasing temperature. A long tail develops at 198°C in the superionic phase, consistent with the Ag ion distribution spreading out from the tetrahedral locations throughout tetrahedral faces into neighboring tetrahedra. It is possible to generate a map of cation density from these data, shown at 302°C in Fig. 26.

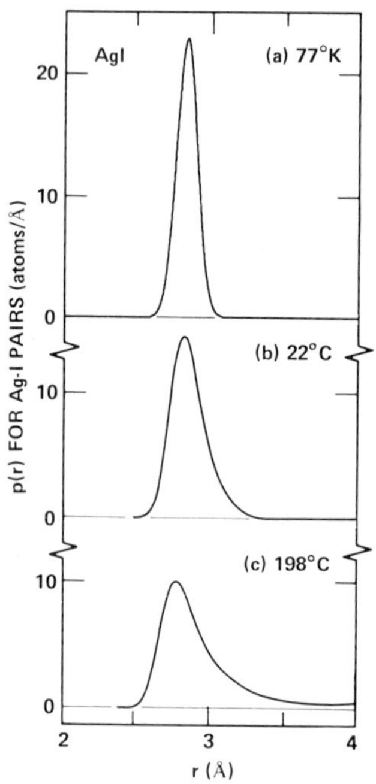

FIG. 25. First-neighbor pair correlation function for AgI at (a) 77°K and (b) 22°C in the normal β phase and at (c) 198°C in the superionic α phase. The narrow Gaussian of half-width 0.06 Å at 77°K becomes broad and asymmetric at elevated temperatures. This trend is described accurately by the excluded volume model. [Taken from Fig. 1 in T. M. Hayes, J. B. Boyce, and J. L. Beeby, *J. Phys. C* **11**, 2931 (1978).]

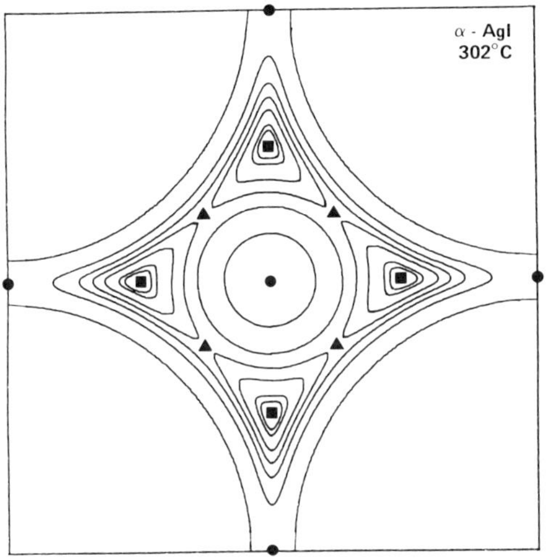

FIG. 26. A contour plot of the cation density in the superionic α phase of AgI at 302°C. The plot is for a (100) plane which includes the conduction path in the $\langle 110 \rangle$ directions from one tetrahedron through a trigonal face into another tetrahedron. The various Strock sites are labeled as follows: □, tetrahedral; ○, octahedral; and △, bridging face site. Moving away from the tetrahedral centers, the contours correspond to ρ/ρ_0 = 0.95, 0.9, 0.7, 0.5, 0.3, 0.1, and ≈ 0. Note that the cation density peaks at the tetrahedral sites but also spreads substantially through the tetrahedral faces. [Taken from Fig. 10 in J. B. Boyce, T. M. Hayes, and J. C. Mikkelsen, Jr., *Phys. Rev. B: Condens. Matter* [3] **23,** 2876 (1981).]

This model also provides a good description of the structure of superionic CuI.[244] This material differs from AgI in that the iodine forms an fcc immobile-ion lattice rather than a bcc lattice. In the fcc structure the tetrahedra share faces with octahedra rather than with other tetrahedra. In CuI, it was determined that the preferred Cu conduction path is from a tetrahedral location through a face into an octahedral location, and again to a tetrahedron. The softened hard-sphere interaction provides a better description of the structure than the competing models in this case, as well as in the other two Cu conducting halides, CuBr and CuCl.[245]

Other superionic conductors studied using EXAFS include the Ag conductor $RbAg_4I_5$[246] and the Cu conductor Cu_2Se.[247] In both of these mate-

[244] J. B. Boyce, T. M. Hayes, J. C. Mikkelsen, Jr., and W. Stutius, *Solid State Commun.* **33,** 183 (1980).

[245] J. B. Boyce, T. M. Hayes, and J. C. Mikkelsen, Jr., *Solid State Commun.* **35,** 237 (1980).

[246] W. Stutius, J. B. Boyce, and J. C. Mikkelsen, Jr., *Solid State Commun.* **31,** 539 (1979).

[247] J. B. Boyce, T. M. Hayes, and J. C. Mikkelsen, Jr., *in* "Fast Ionic Transport in Solids" (J. B. Bates and G. C. Farrington, eds.), p. 497. North-Holland, Amsterdam, 1981.

rials, it was found that mobile-ion correlations are important and affect both the structure and conduction.

The EXAFS data and the excluded volume model also provided a starting point for a calculation of the temperature-dependent dc conductivity of these materials. The mobile cations are modeled as a Boltzmann gas moving in an effective one-ion potential as prescribed by the softened hard-sphere model. The calculated conductivities of AgI[248] and the cuprous halides[108] compare well with the measured conductivities.

EXAFS data on superionic AgI and CuI have been compared with molecular dynamics (MD) calculations on these materials.[249,250] Since the EXAFS depends sensitively on the details of the near-neighbor peak in the interatomic pair correlation function, as discussed in Section 6, it can provide valuable information on the short-range portion of the pair potentials. These pair potentials serve as the starting point for the MD calculations. It was found[101] that the cation–anion pair distribution function derived using MD begins at too short a cation–anion separation (by about 0.1 Å) and is too broad (by an additional Gaussian half-width of about 0.15 Å) to be consistent with the EXAFS data on both AgI and CuI. The two cation–anion pair distribution functions are shown in Fig. 27. Later MD calculations have reduced this discrepancy by a factor of 2.[251]

c. *Mixed-Valence Materials*

Mixed-valence systems typically involve rare-earth atoms with nonintegral occupation of the 4f electron levels, implying the coexistence of two different valence states of the rare earth, $4f^n$ and $4f^{n+1}$.[252] It is anticipated that there are structural manifestations of this coexistence, since the ionic radii of the rare-earth atoms change substantially when the 4f occupation changes by one electron. In addition, experiments on mixed-valence rare-earth systems indicate that the width of the f band is approximately a phonon frequency (i.e., $\approx 10^{13}$ s^{-1}). As a result, there is the possibility of a strong electron–lattice coupling. This coupling could take the form of a large local dynamic disorder in the near-neighbor spacings due to the valence fluctuations. The near-neighbor peak in the pair distribution function would then be quite broad, corresponding to the large difference in the near-neigh-

[248] T. M. Hayes and J. B. Boyce, *J. Phys. C* **13,** L225 (1980).

[249] P. Vashishta and A. Rahman, *Phys. Rev. Lett.* **40,** 1337 (1978).

[250] P. Vashishta and A. Rahman, *in* "Fast Ion Transport in Solids" (P. Vashishta, J. N. Mundy, and G. K. Shenoy, eds.), p. 527. North-Holland, New York, 1979.

[251] P. Vashishta, private communication.

[252] See, for example, the review by J. M. Lawrence, P. S. Riseborough, and R. D. Parks, *Rep. Prog. Phys.* **44,** 1 (1981), and references contained therein.

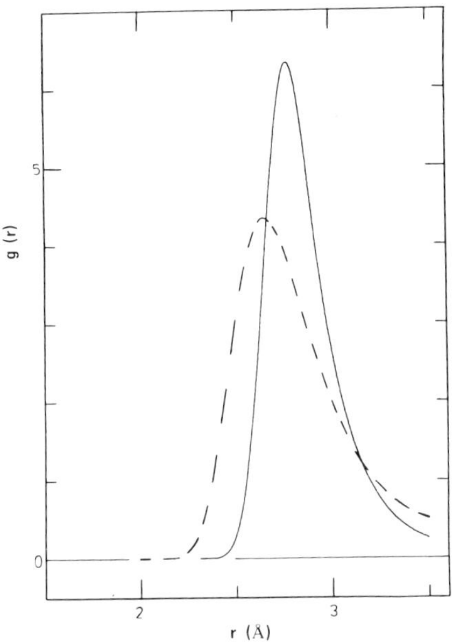

FIG. 27. The Ag–I pair correlation function $g_{Ag\text{-}I}(r)$ deduced from EXAFS data on AgI at 471°K in the superionic phase (full curve) [T. M. Hayes, J. B. Boyce, and J. L. Beeby, *J. Phys. C* **11,** 2931 (1978)]. Also shown is the comparable function determined in a molecular dynamics (MD) calculation at 430°K (broken curve) [P. Vashishta and A. Rahman, *Phys. Rev. Lett.* **40,** 1337 (1978); *in* "Fast Ion Transport in Solids" (P. Vashishta, J. N. Mundy, and G. K. Shenoy, eds.), p. 527, North-Holland, New York, 1979]. Note that the MD result becomes nonzero for a smaller value of r and increases less sharply than does the result deduced from EXAFS data. [Taken from Fig. 1 in T. M. Hayes and J. B. Boyce, *J. Phys. C* **13,** L731 (1980).]

bor spacings for the two valence states (i.e., $\approx$0.15 Å). The possibility also exists that the lattice will follow the f occupation and adopt a double-peaked distribution with the two characteristic near-neighbor spacings ($\Delta r_1 \approx 0.15$ Å), one for each of the two valence states. If, on the other hand, the lattice does not follow the electronic fluctuations, then the near-neighbor environment may adopt a well-defined single distance, intermediate between that of the two different valence states.

EXAFS is ideally suited to address this problem, owing to its local nature. The charge fluctuations on the ions are presumed to be uncorrelated in space and time, and so any possible lattice distortions will be uncorrelated. As a result, Bragg scattering studies would yield an average near-neighbor spacing, even if local lattice distortions did exist.

X-Ray absorption studies can also yield information on the valence state of the rare-earth ions by probing the near edge structure. In fact, this is the way that the mixed-valence state was first observed in SmB_6.[253] There is a chemical shift of the X-ray absorption edge to ≈8 eV higher binding energy when the Sm is trivalent rather than divalent. The L_{II} and L_{III} edges often exhibit a large peak at the edge, the "white" line, which can serve as an edge marker. The position of this white line depends on valence, so that the typical chemical shifts of ≈8 eV are easily resolved in an X-ray absorption experiment. The nonintegral valence can be determined from the ratio of the intensities of the two peaks.

Two such absorption studies, covering both the near edge and the EXAFS regions, have been performed on mixed-valence materials: one on TmSe[116] and the other on $Sm_{1-c}Y_cS$.[117,118] We consider first the edge results. Figure 28 shows the near edge structure for the Sm L_{III} edge of SmS and $Sm_{0.8}Y_{0.2}S$. SmS is single valent, with Sm in the 2+ state (i.e., $4f^6$ electronic configuration), and so shows only a single peak at the edge. Mixed-valent $Sm_{0.8}Y_{0.2}S$ exhibits a double-edge structure which varies dramatically with temperature. This temperature variation is consistent with that of the valence deduced from lattice constant measurements.[254] Its large change is due to the "explosive" first-order transition this material undergoes between 20°C and 77°K. The actual magnitude of the valence determined from the edge, however, differs somewhat from that derived from the lattice constant values, being 2.5 at 300°K and 2.2 at 77°K, compared with 2.7 and 2.4°K, respectively. It should be noted that this large change in the edge structure with temperature argues against the double-peaked structure being due to shake-off processes and for it being due to the mixed-valence state.

In TmSe, the Tm L_{III} edge yields a temperature-independent valence of 2.6, to be compared with 2.75 from the lattice constant and 2.55 from the Curie constant. Other edge (but not EXAFS) studies of mixed-valence materials have been reported. These include investigations of the rare-earth L_{III} edges in $EuCu_2Si_2$, $YbCu_2Si_2$, and Sm_4Bi_3[255] and the Cu K edge in copper complexes of penicillamine and cysteamine.[256]

The EXAFS studies of both TmSe and $Sm_{1-c}Y_cS$ led to the same conclusion concerning the electron–photon coupling. In both materials, the chalcogenide near neighbors to the mixed-valence rare-earth ions adopt a single average distance rather than a statically or dynamically distorted environment. The near-neighbor distributions are very sharp, indicating that the breathing shell charge fluctuations of the rare-earth ion are only

[253] E. E. Vainshtein, S. M. Blokhin, and Yu. B. Paderno, *Sov. Phys.—Solid State (Engl. Transl.)* **6**, 2318 (1965).

[254] L. J. Tao and F. Holtzberg, *Phys. Rev. B: Solid State* [3] **11**, 3842 (1975).

[255] T. K. Hatwar, R. M. Nayak, B. D. Padalia, M. N. Ghatikar, E. V. Sampathkumaran, L. C. Gupta, and R. Vijayaraghaven, *Solid State Commun.* **34**, 617 (1980).

[256] H. L. Nigam, B. D. Srivastava, and J. Prasad, *Solid State Commun.* **28**, 1001 (1978).

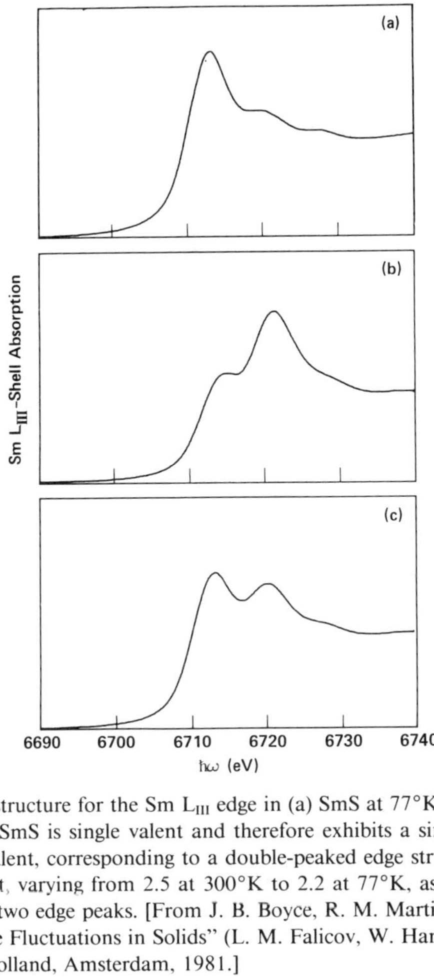

FIG. 28. Near edge structure for the Sm L_{III} edge in (a) SmS at 77°K and $Sm_{0.8}Y_{0.2}S$ at (b) 300°K and (c) 77°K. SmS is single valent and therefore exhibits a single peak at the edge. $Sm_{0.8}Y_{0.2}S$ is mixed valent, corresponding to a double-peaked edge structure. The valence is temperature dependent, varying from 2.5 at 300°K to 2.2 at 77°K, as determined from the relative heights of the two edge peaks. [From J. B. Boyce, R. M. Martin, J. W. Allen, and F. Holtzberg, *in* "Valence Fluctuations in Solids" (L. M. Falicov, W. Hanke, and M. B. Maple, eds.), p. 427. North-Holland, Amsterdam, 1981.]

weakly coupled to the phonons. Also, the single-peak distribution shows that a strongly coupled polaron system with two near-neighbor spacings does not form. In both materials, large changes would be expected for a strong electron–lattice coupling, corresponding to the differences in the 2+ and 3+ near-neighbor spacings. $\Delta r_1 \approx 0.16$ Å for TmSe and ≈ 0.18 Å for $Sm_{1-c}Y_cS$. Such large differences could easily have been detected, but were not. An example of a fit to the data of $Sm_{0.75}Y_{0.25}S$ at 300°K (valence ≈ 2.44) is shown in Fig. 29. The solid line is the Fourier-filtered EXAFS data on the Sm L_{III} edge for the first shell of six S neighbors to the Sm ions. The dots are a least-squares fit to the data using SmS as the single-

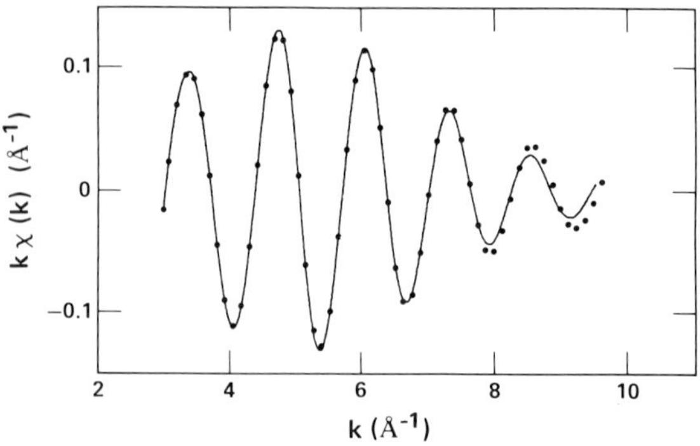

FIG. 29. The Fourier-filtered EXAFS data on the Sm L_{III} edge (solid line) for the first shell of S neighbors to the Sm atoms in $Sm_{0.75}Y_{0.25}S$ at 300°K. The dots are a least-squares fit to the data, yielding a single Sm–S near-neighbor spacing of 2.84 Å, intermediate between the Sm^{3+}–S spacing of 2.80 Å and the Sm^{2+}–S spacing of 2.98 Å. [Taken from Fig. 3 in J. B. Boyce, R. M. Martin, J. W. Allen, and F. Holtzberg, *in* "Valence Fluctuations in Solids," (L. M. Falicov, W. Hanke, and M. B. Maple, eds.), p. 427. North-Holland, Amsterdam, 1981.]

valent standard. The fit yielded a single near-neighbor spacing of 2.84 Å, intermediate between the Sm^{3+}–S spacing of 2.80 Å and the Sm^{2+}–S spacing of 2.98 Å. Also, the width of the near-neighbor peak is comparable to that in the single-valent SmS standard. No large static or dynamic distortions occur. The fit essentially reproduces the data, with small deviations only at high k, where $k\chi(k)$ is small. The conclusion of these studies is that the lattice adopts a single near-neighbor distance corresponding to the hybrid mixed-valence wave function, rather than a distorted near-neighbor environment corresponding to a polaronic state.

d. Spin Glasses

Spin glasses are characterized by a prominent cusp in the magnetic susceptibility occurring at low temperatures ($\approx$15°K). The microscopic origin of this behavior is generally attributed to randomly coupled atomic spins.[257] The archetypal material is a dilute solution of transition metal atoms in a noble metal host. These atomic spins are conceived as being distributed randomly on the crystalline lattice. A regular but spatially oscillating pairwise interaction such as that of Ruderman, Kittel, Kasuya, and Yosida (the RKKY interaction) will then lead to spin–spin interactions of varying sign and magnitude, random in a sense. It is this random spin–spin interaction which leads to the term "spin glass," not a lack of atomic crystallinity, although amorphous materials can also be spin glasses (see below). Typical

[257] See, for example, a review by J. A. Mydosh, *J. Magn. Magn. Mater.* **7**, 237 (1978).

systems include Au and Cu with 0.5–10% of the magnetic moment-bearing atom (Fe or Mn). A critical question for this type of magnetism is whether the magnetic ions form local metallurgical clusters. If these systems were ideal solid solutions, the Fe or Mn atoms would substitute randomly for Au or Cu in the fcc lattice of the host. Canonical spin glasses like CuMn and AuFe change their magnetic properties dramatically after cold working, thermal treatment, and so on, however, and these changes are correlated with precipitation and the formation of magnetic clusters.[258] Often the magnetic ions are not very soluble in the nonmagnetic host. So the question of possible magnetic clusters in these materials has to be addressed before the spin glass properties can be explained.

The local and spectroscopic nature of EXAFS makes it well suited to investigate this question. A few such studies have been performed, the first of which was on AuMn.[259] The cusp in its magnetic susceptibility is enhanced by annealing at high temperatures. To address the question of possible chemical clustering or phase separation during this process, the EXAFS on the Mn K edge of annealed and unannealed $Au_{0.95}Mn_{0.05}$ was investigated. The experiments were performed in transmission, despite the strong background absorption from the Au atoms evident in Fig. 30. Evaluation of Eq. (9.5) shows that fluorescence and transmission are equally suitable, given the 5 at. % concentration of Mn, provided that background fluorescence is small and a large solid angle of fluorescence detection is used (i.e., $\Omega/4\pi \approx 0.1$). An analysis of the transmission data using Mn metal and Au_2Mn as known standards showed no evidence for changes in the Mn nearest neighbors with annealing, despite the large enhancement of the cusp. Using this result, a model was presented which explains the annealing dependence of the magnetic properties of AuMn spin glasses.

Another crystalline spin glass system that has been studied is $La_{1-c}Gd_cOs_2$.[260] This is a particularly interesting one, since, in addition to being a spin glass, it is also a superconductor up to $c \approx 0.16$ ($T_c \approx 9$°K). It is a difficult metallurgical system to prepare, and the possibility of precipitation of the magnetic Gd ions exists. An EXAFS study of this spin glass is made difficult not only by the poor signal-to-background ratio, as for AuMn, but also by the complex structure of the nonmagnetic $LaOs_2$ host. $LaOs_2$ crystallizes in both the cubic (C15) and hexagonal (C14) Laves phases, but special heat treatment can produce the cubic phase which was used for the EXAFS study. This study involved investigating the L-edge EXAFS on each of the three components of $La_{1-c}Gd_cOs_2$ for $c = 7$ and 14% and com-

[258] See the review by P. A. Beck, *Prog. Mater. Sci.* **23,** 1 (1978).

[259] T. M. Hayes, J. W. Allen, J. B. Boyce, and J. J. Hauser, *Phys. Rev. B: Condens. Matter* [3] **22,** 4503 (1980).

[260] J. B. Boyce and K. Baberschke, *Solid State Commun.* **39,** 781 (1981).

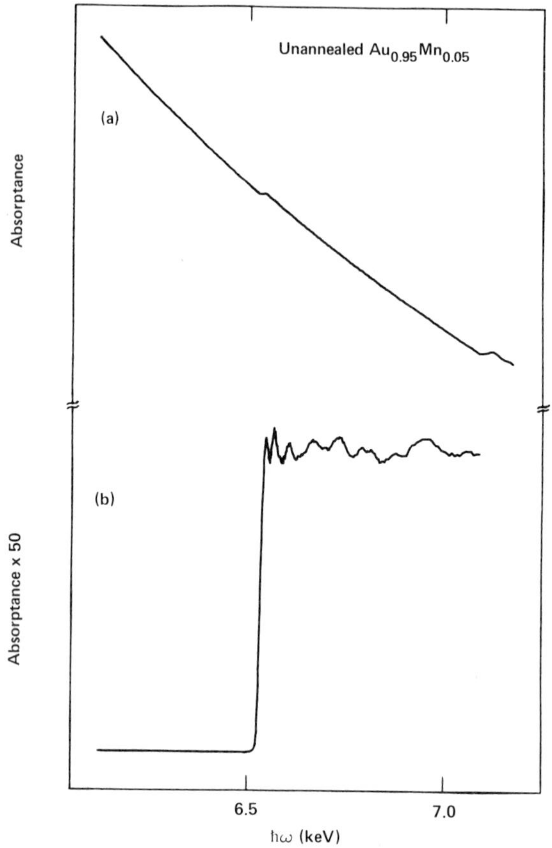

FIG. 30. (a) X-Ray absorptance as a function of photon energy $\hbar\omega$ for an unannealed sample of $Au_{0.95}Mn_{0.05}$ in the vicinity of the Mn K edge at 6.54 keV. The feature at 7.11 keV is due to iron impurities in the aluminum substrate. (b) The absorptance due to the Mn K edge alone, after subtracting the prethreshold absorption. [Taken from Fig. 2 in T. M. Hayes, J. W. Allen, J. B. Boyce, and J. J. Hauser, *Phys. Rev. B: Condens. Matter* [3] **22,** 4503 (1980).]

paring these results with those on known structural standards, $LaOs_2$, $GdOs_2$, and Gd metal. From this comparison it was determined that Gd substitutes predominately for La in the C15 cubic Laves phase of $LaOs_2$. The magnetic properties of this superconductor cannot be attributed to small Gd or $GdOs_2$ clusters, therefore, but, rather, must be explained by a majority of the magnetic Gd ions being randomly distributed on the La sites.

An EXAFS study has also been performed on a concentrated spin glass, amorphous (a-)MnSi.[261] This material provides an interesting contrast with

[261] T. M. Hayes, J. W. Allen, J. B. Boyce, and J. J. Hauser, *Phys. Rev. B: Condens. Matter* [3] **23,** 4691 (1981).

the crystalline spin glasses discussed above, in which the random magnetic interaction is provided by a random distribution of dilute spins on a lattice. In this system, the magnetic disorder is provided by the amorphous structure of the material. Crystalline (c-) and a-MnSi have quite different magnetic properties: c-MnSi is a helical antiferromagnet below 29°K, with a very long period ($\approx$180 Å), while a-MnSi is a spin glass with a cusp temperature $\approx$ 22°K. The EXAFS study involved a comparison of the Mn K-edge EXAFS on a-MnSi with that on known structural standards, c-MnSi and Mn metal. The analysis yielded that the nearest neighbors of Mn in a-MnSi consist of five to six Si atoms, with no evidence for a single close Si neighbor or a well-defined shell of Mn neighbors such as are found in c-MnSi. A model based on the structure was proposed to explain the different magnetic behavior of the two materials.

e. Solid Solutions

The local, spectroscopic nature of the EXAFS has been exploited to study a variety of dilute alloys and pseudobinary compounds. The EXAFS on the absorption edge of the solute atom is typically used to determine the local distortions that exist about this atom and whether or not precipitation has occurred. If precipitates have formed, the number and type of near neighbors can be determined and used to specify the nature of the precipitate phase. These studies are similar to those discussed in Section 11,d on spin glasses, where the local structure so determined was related to the magnetic properties of the alloys. Here we discuss some investigations of nonmagnetic, dilute, binary alloy systems (AlCu, CuFe, AlZn, TiCu, and AlMg), followed by a brief discussion of some studies of pseudobinary compounds and a study of dilute samples of Fe monomers and dimers isolated in argon.

(i) AlCu. This alloy system is especially interesting due to the various phases and precipitates that form depending on the heat treatment and the Cu concentration.[262] Using a well-known heat treatment, a 2% solid solution of Cu in Al, for example, undergoes the following precipitation sequence, leading to the formation of the Al_2Cu θ phase: supersaturated solid solution (α phase) $\rightarrow$ Guinier–Preston (GP) zones $\rightarrow$ θ'' phase $\rightarrow$ θ' phase $\rightarrow$ θ phase. We consider the results of four separate EXAFS studies of these phases.[104,263–265]

Fontaine *et al.*[263] investigated the α and θ' phases and the GP zones in

[262] For a discussion see, for example, P. G. Shewmon, "Transformations in Metals," Chapter 7. McGraw-Hill, New York, 1969.

[263] A. Fontaine, P. Lagarde, A. Naudon, D. Raoux, and D. Spanjaard, *Philos. Mag.* [*Part*] *B* **40,** 17 (1979).

[264] S. K. Prasad, S. P. Singhal, H. Herman, J. A. del Cueto, and N. J. Shevchik, *Scr. Metall.* **13,** 549 (1979).

[265] H. Maeda, T. Tanimoto, H. Terauchi, and M. Hida, *Phys. Status Solidi A* **58,** 629 (1980).

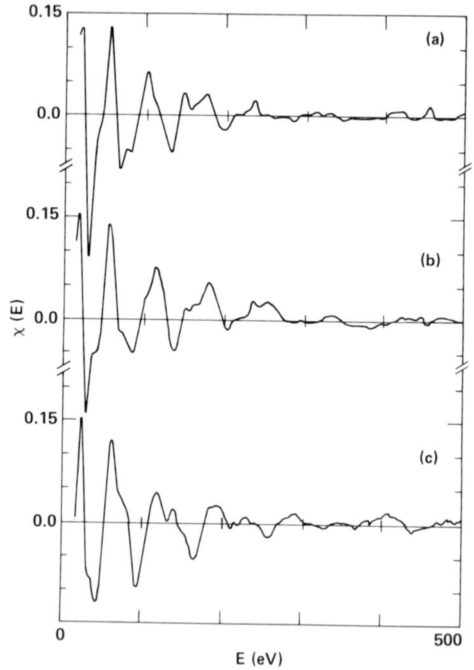

FIG. 31. EXAFS spectra $\chi(E)$ as a function of final-state electron energy E from a sample of 2.5 at. % Cu in Al in three states: (a) the solid solution as prepared (α phase), (b) after annealing for 50 hr at 90° C ($\approx$60% of the Cu atoms in GP zones), and (c) after annealing for 50 hr at 240° C ($\approx$60% of the Cu atoms in a θ' phase). These spectra indicate that the near-neighbor environments of the Cu atoms are significantly different in these three states. [Taken from Fig. 2 in A. Fontaine, P. Lagarde, A. Naudon, D. Raoux, and D. Spanjaard, *Philos. Mag.* [*Part*] *B* **40,** 17 (1979).]

samples having 2 and 2.5 at. % Cu. The EXAFS spectra from these three forms differ significantly, as shown in Fig. 31 for the alloy with 2.5 at. % Cu. To analyze these data, Fontaine and co-workers used calculated phase shifts and backscattering amplitudes, and shifted the energy threshold for a best fit. The θ' phase was used as a structural standard to test the analysis procedure, since its structure is well known from X-ray diffraction. Reasonable agreement was found. One difficulty in the analysis is that, for both the θ' phase and GP zone samples, half the Cu atoms remain dissolved in solid solution, while the other half forms precipitates. It was necessary to determine this fraction from magnetic susceptibility measurements and remove its contribution from the EXAFS data before the structure of the α phase and GP zones could be determined. Firstly, it was found for the α phase solid solution that a large relaxation of the Al matrix occurs around the Cu solute atoms. A contraction of 0.13 ± 0.02 Å from the Al spacing of 2.86 Å is obtained, with the uncertainty due largely to the energy threshold adjustment. Secondly, the GP zones were determined to have an average composition of $Al_{50}Cu_{50}$, with an *assumed* in-plane spacing equal to the Al–Al spacing in the Al matrix and a contracted Cu–Al distance perpendicular to the plane. The spacing between the zone and the neighboring (100) planes was determined to be contracted by $\approx$17% from the 2.02 Å spacing of the Al host, to $\approx$1.68 Å. The original model of Preston[266] and Guinier[267] had zones consisting of a single (100) Cu-rich plane surrounded by Al planes which move closer to the zone than the normal plane spacing of the Al host. The EXAFS result is consistent with this single-plane model, except for the 50–50 composition of the zone.

Lengeler and Eisenberger[104] studied the α phase in a 0.5% Cu sample, the θ' and θ phases in a 2% Cu sample, and the θ phase of $Al_{2.05}Cu$. To analyze the data, they used the backscattering amplitudes and scattering phase shifts calculated by Teo and Lee,[40] but introduced an adjustable amplitude factor for each shell of neighbors to account for inelastic electron scattering losses. The procedure was tested on Cu metal and gave good fits to the data beyond $k = 7$ Å^{-1}. The accuracy of this procedure was estimated to be 1% in interatomic distances, 15% in coordination numbers, and 20% in peak widths. This analysis procedure was then applied to the <u>Al</u>Cu alloys. The results on the θ and θ' phases are in agreement with the structures derived from Bragg scattering within the above uncertainties. For the α phase, the contraction of the near-neighbor spacing around a Cu solute atom was determined to be 0.07 ± 0.03 Å (i.e., 2.79 versus 2.86 Å for the host). The difference between this result and the larger 0.13 ± 0.02 Å contraction determined by Fontaine *et al.*[263] may be due to different heat treatments of the samples. In addition, Lengeler and Eisenberger found evidence for Cu–Cu clustering in the α phase of their 0.5% alloy. Specifically, their fits to the data improved when 0.4 Cu atoms were added to the first neighbor shell at a Cu–Cu distance of 2.48 Å.

The results of Maeda *et al.*[265] on a supersaturated solid solution of 1.9% Cu in Al also indicate clustering of the Cu atoms. They used the functional forms suggested by Teo and Lee and co-workers[77,78] for the phase-shift and amplitude functions, but determined the required 14 parameters through fits to data from Cu and θ Al_2Cu standards. The local structure of the alloy was then determined in an eight-parameter least-squares fit to the data. They found the near-neighbor shell of each Cu atom to have three Cu atoms and nine Al atoms, with a Cu–Cu spacing of 2.67 Å and a Cu–Al spacing of 2.93 Å. The large number of parameters used and the approximate nature

[266] D. G. Preston, *Proc. R. Soc. London, Ser. A* **167,** 526 (1938).
[267] A. Guinier, *Ann. Phys. (Paris)* [11] **12,** 161 (1939).

of the parameterized functions for the phase shifts and amplitudes make it difficult to assess the uncertainties in these results.

(*ii*) *AlZn and AlMg*. The studies of the EXAFS on the K edge of the solute atoms in AlZn[268] and AlMg[269] are similar to those of AlCu discussed above. The analysis procedure is that of Fontaine *et al.*[263] It is found that random solid solutions are formed in AlZn up to 2.6 at. % Zn. The Zn–Al spacing is only 0.02 Å smaller than the Al–Al nearest-neighbor spacing in Al metal, to be contrasted with the −0.13 Å change in AlCu. This contraction and that in AlCu are nonetheless three times larger than predicted using the continuum elastic medium theory. This indicates that the continuous medium assumption is not applicable in the vicinity of the solute atoms. Evidence of clustering is found for Zn concentrations above 2.6%. For example, 3.2 Zn and 8.8 Al neighbors are found in the first shell of Zn neighbors in the 6.8% sample.

For the AlMg alloys, no evidence of Mg clustering or Guinier–Preston zone formation was found in 3 and 7% samples.[269] Large expansions of the lattice about the Mg atoms were determined, ≈0.08 Å. As in the other Al alloys, this change is much larger than predicted using Vegard's law and indicates that the continuous elastic theory is not able to account for displacements of the host near neighbors to a solute atom.

(*iii*) *CuFe*. The local structure around Fe solute atoms in Cu in both reduced and oxidized dilute alloys has been determined by Hastings *et al.*[166] In addition to determining the structure of a 75-ppm alloy, this study also demonstrated the use of a crystal filter assembly to measure the fluorescence signal in the presence of a large background signal ($I_b/I_f \approx 10^4$), as discussed in Section 9,a. The Fe fluorescence signal was analyzed in k space using the amplitudes and phase shifts calculated by Teo and Lee,[40] with a scale factor determined from model compounds. For the reduced alloy, the results give 12 ± 2 Cu neighbors to Fe at 2.54 ± 0.1 Å. No evidence for Fe clustering was observed. The fact that the Fe–Cu distance and coordination are near those of pure Cu (i.e., 12 neighbors at 2.56 Å) indicates that Fe atoms substitute for the Cu atoms and do not cluster. For the oxidized alloy, Fe atoms were determined to cluster, and to have Cu and oxygen near neighbors as well. Interatomic distances different from those in characteristic Fe compounds were determined in the oxidized alloy, indicating the presence of large local strains.

(*iv*) *TiCu*. A dilute alloy of 0.5 at. % Cu in Ti has been studied by

[268] J. Mimault, A. Fontaine, P. Lagarde, D. Raoux, A. Sadoc, and D. Spanjaard, *J. Phys. F* **11,** 1311 (1981).

[269] D. Raoux, A. Fontaine, P. Lagarde, and A. Sadoc, *Phys. Rev. B: Condens. Matter* [3] **24,** 5547 (1981).

[270] M. Marcus, *Solid State Commun.* **38,** 251 (1981).

Marcus[270] to investigate the fact that the diffusion of Cu in Ti is characterized by a low activation energy and a low preexponential factor. Due to the lack of a reference compound with Cu surrounded by Ti in the first coordination shell, the Cu K-edge fluorescence data was analyzed using three different methods which involved various combinations of calculated and measured phase shifts and amplitudes. The methods were generally consistent and indicated that the Cu is mostly substitutional in the Ti lattice, with little Cu clustering, if any. An indication of a small local contraction of the Cu–Ti spacing compared with the average Ti–Ti near-neighbor spacing was also found. An analysis of the temperature variation of the EXAFS peak widths showed that the Cu–Ti peak width is similar to that of the Ti–Ti peak in Ti metal. Both of these widths are larger by a factor of 2.5 than that in Cu metal, indicating softer short-range forces. These results suggest that the diffusion of Cu atoms in Ti metal is similar to that of Ti in Ti metal, and that the anomalous nature of their diffusion may well be due to the softness of the short-range forces in these materials.

(v) Pseudobinary Compounds. EXAFS studies of three pseudobinary compounds have been reported. The results on $(La_{1-c}Gd_c)Os_2$ have been discussed in Section 11,d on spin glasses.[260] The magnetic Gd ions were determined to substitute for the La in this cubic Laves phase material for $c = 7$ and 14%, with little or no clustering.

One aspect of the $(Sm_{1-c}Y_c)S$ system has already been discussed in Section 11,c on mixed-valence compounds. The local distortion about the Sm and Y ions has also been addressed.[118] It is found that large distortions in the near-neighbor spacings do occur for this rock salt material when Y substitutes for Sm. For example, in the $c = 25\%$ sample at 300°K, the alloy lattice constant is 5.66 Å, implying an average Sm–S and Y–S near-neighbor spacing of 2.83 Å. The Y–S spacing from the EXAFS is 2.77 Å, however, 0.06 Å smaller than implied by the lattice constant, but 0.04 Å greater than in YS. The Sm–S distance is determined to be 2.82 Å, in good agreement with that from the alloy lattice constant, but 0.14 Å less than in SmS. Accordingly, there is evidence for large local distortions, even though the weighted mean of the Y–S and Sm–S distances is in good agreement with the average spacing from the lattice constant.

A study of GaAs–Ga_2Se_3 solid solutions has been made.[271] An analysis of the data on the K edge of each of the constituents of this zinc blende structure solid solution indicates that clustering does occur, unlike the studies discussed above. A descriptive chemical formula for this system is $Ga_{(3-c)/3}(As_{1-c}Se_c)$, corresponding to the introduction of Ga vacancies as Ga_2Se_3 is added to GaAs. It is found that the Ga vacancies are not randomly distributed on the zinc blende lattice but, rather, are chemically ordered

[271] J. C. Mikkelsen, Jr. and J. B. Boyce, *Phys. Rev. B: Condens. Matter* [3] **24,** 5999 (1981).

about the Se atoms. This nonrandom distribution of vacancies has been related to the clustering models proposed to explain the electrical behavior of GaAs heavily doped with Se.

(vi) Fe in Argon. Rare-gas matrix isolation is used to study atoms and molecules which are unstable if not isolated from one another.[272] In these studies, the local structure about the isolated species is of particular interest. Montano and Shenoy[273] have used EXAFS to study dilute samples of Fe isolated in an Ar matrix. EXAFS data on the Fe K edge on samples of 0.1 at. % Fe were acquired at 4.2° K. At this low concentration of Fe, only monomers and dimers are detectable using Mössbauer spectroscopy. Three peaks are observed in the EXAFS data in real space. They were analyzed using the parameterized phase shifts and amplitudes of Teo and Lee and co-workers.[77,78] An extrapolation from Cl was made, since those for Ar were not available. Since the peaks in the pair correlation function overlap substantially, a method utilizing the interference beats in k space[233] was used to determine the peak separations. From this analysis and a comparison with Mössbauer results on similar samples, the three peaks were identified as follows: one Fe atom at 1.87 ± 0.13 Å, eight Ar atoms at 2.82 ± 0.08 Å, and an undetermined number of Ar atoms at 3.72 ± 0.09 Å. The large uncertainty in these results is due to the overlap of these peaks, the weak signal in this dilute sample, the large number of parameters needed to specify the near-neighbor environment, and the approximate nature of the parameterized phase shifts and amplitudes. The first peak would indicate a substantial contraction of the Fe–Fe distance in the dimers from the 2.48-Å spacing in bulk iron. The second peak is attributed to Ar around the dimers and/or iron in an interstitial site. The third peak is identified as due to substitutional Fe with a distance close to the Ar–Ar spacing of 3.75 Å. A high degree of disorder of the matrix is also evident from the EXAFS results.

f. A15 Compounds

The A15 compounds A_3B are of great interest owing to their high superconducting transition temperatures $T_c \approx 20°K$.[274] It is found that these materials are generally unstable, however, so that special preparation methods are required to obtain the metastable material. The value of T_c depends sensitively on preparation conditions and is typically highest in well-ordered stoichiometric A15 compounds, which are difficult to obtain. Most samples

[272] See, for example, A. M. Bass and H. P. Broida, eds., "Formation and Trapping of Free Radicals." Academic Press, New York, 1960.

[273] P. A. Montano and G. K. Shenoy, *Solid State Commun.* **35,** 53 (1980).

[274] See, for example, the reviews by M. Weger and I. B. Goldberg, *Solid State Phys.* **28,** 1(1973); L. R. Testardi, *Rev. Mod. Phys.* **47,** 637 (1975).

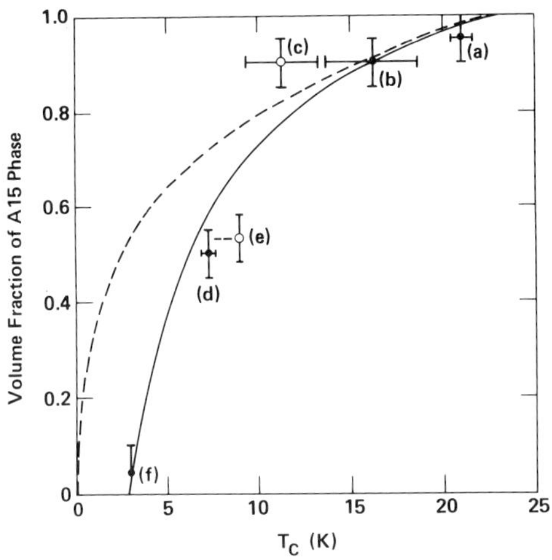

FIG. 32. The correlation between the superconducting transition temperature T_c and the volume fraction of crystalline A15 phase (as deduced from EXAFS spectra) for several samples of Nb_3Ge. The data points are labeled after the sample notation of Brown *et al.* [G. S. Brown, L. R. Testardi, J. H. Wernick, A. B. Hallak, and T. H. Geballe, *Solid State Commun.* **23,** 875 (1977)]. Solid circles are given by Brown *et al.*, while open circles are given by G. S. Brown (unpublished). Substrate temperatures T_s are 1050 (sample *a*), 1148 (samples *b* and *c*), 973 (samples *d* and *e*), and 800°K (sample *f*). The solid and broken lines result from the theory of C. S. Pande and R. Viswanathan [*Solid State Commun.* **26,** 893 (1978)] applied to as-grown and neutron-damaged Nb_3Ge, respectively. Note that the nonlinear relationship between T_c and the volume fraction of A15 phase (as T_s is varied) is well accommodated in the theory. [Taken from Fig. 1 in C. S. Pande and R. Viswanathan, *Solid State Commun.* **26,** 893 (1978). Copyright 1978, Pergamon Press, Ltd.]

have some imperfections as prepared, such as nonstoichiometry, crystalline defects, or two-phase regions. As a result, EXAFS is a useful probe to characterize these high-T_c superconductors. We discuss three such studies.

Brown *et al.*[275] have used the fluorescence EXAFS above the Ge K edge to study the local environment about the Ge atoms in Nb_3Ge thin films prepared at various deposition temperatures by two different techniques, sputtering and electron beam coevaporation. It is found that sputtering onto a substrate held at $T_s = 1050°K$ yields samples which are 90–100% crystalline A15 phase with $T_c \approx 20°K$, but that a lower substrate temperature of 800°K leads to samples which are "quasi-amorphous" with T_c as low as 3°K. The correlation between T_c and the volume fraction of crystalline A15 phase in these samples (as deduced from the EXAFS measurements) is shown in Fig. 32. Note that most of the reduction in T_c occurs as T_s is

[275] G. S. Brown, L. R. Testardi, J. H. Wernick, A. B. Hallak, and T. H. Geballe, *Solid State Commun.* **23,** 875 (1977).

lowered from 1050 to 1000°K. Brown *et al.*[275] found that the Nb nearest neighbors of Ge in the amorphous phase are at 2.66 Å, roughly 0.21 Å less than in the A15 crystalline phase. Their analysis of the EXAFS data suggested further that samples intermediate in T_s consist of mixtures of the A15 and quasi-amorphous phases. Both T_c and the volume fraction of the A15 phase increase steadily with T_s. They concluded, however, that the emergence of the A15 phase with increasing T_s occurs for values of T_s which are too low for that emergence to be correlated with the rapid change in T_c with substrate temperature. The coexistence of these two phases is supported by a later electron diffraction study by Pande and Viswanathan.[276] Specifically, these authors concluded that the mixed-phase samples consist of small regions of highly disordered quasi-amorphous material in a matrix of ordered A15 phase. They argued further that the volume fraction of the amorphous phase which was deduced by Brown *et al.*[275] can indeed account for the observed suppression of T_c if the effect of these small regions of low-T_c material on the superconductivity of the high-T_c matrix is treated properly. The results of their theory are shown in Fig. 32.

Knapp *et al.*[277] have studied the effects of radiation damage on the A15 materials Nb_3Ge and V_3Ga. For the V_3Ga of their study, neutron irradiation lowers T_c from 14 to 4.5°K, yet only small changes are observed in the EXAFS on the Ga K edge. They estimated that the anti-site disorder expected in the irradiated material consists of 20% or less of the Ga atoms on V sites. For Nb_3Ge, α-particle irradiation was used to lower T_c from 21 to 7°K. It is proposed that the radiation causes the Ge–Nb near-neighbor peak at 2.87 Å in the unirradiated A15 sample to split symmetrically into two peaks, one at 2.77 Å and the other at 2.97 Å. This large splitting is proposed to account for the large Debye–Waller factors obtained from X-ray diffraction on similarly irradiated Nb_3Sn.

An investigation of the temperature variation of the EXAFS on the Ge K edge in Nb_3Ge prepared by chemical vapor deposition has also been performed.[278] Evidence for superlattice formation at 486°K exists in this material from electron diffraction measurements.[279] Such a transition could be similar to the low-temperature martensitic transformation observed in other A15 materials such as Nb_3Sn and V_3Si. No anomalous variation of

[276] C. S. Pande and R. Viswanathan, *Solid State Commun.* **26,** 893 (1978).

[277] G. S. Knapp, R. T. Kampwirth, P. Georgopoulos, and B. S. Brown, *in* "Superconductivity in d- and f- Band Metals" (H. Suhl and M. B. Maple, eds.). Academic Press, New York, 1980.

[278] T. Claeson, J. B. Boyce, and T. M. Geballe, *Phys. Rev. B: Condens. Matter* **25,** 6666 (1982).

[279] P. H. Schmidt, E. G. Spencer, D. C. Joy, and J. M. Rowell, *in* "Superconductivity in d- and f-Band Metals" (D. H. Douglass, ed.), p. 431. Plenum, New York, 1976; J. M. Rowell, P. H. Schmidt, E. G. Spencer, P. D. Dernier, and D. C. Joy, *IEEE Trans. Magn.* **MAG-13,** 644 (1977).

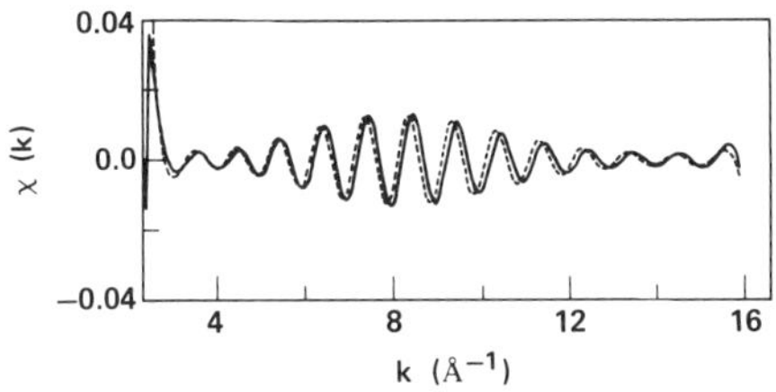

FIG. 33. The contribution to the Nb K-edge EXAFS spectra $\chi(k)$ from Nb second nearest neighbors in $NbSe_2$ (solid line) and in an intercalated sample, $Rb_{0.28}NbSe_2$ (broken line). The phase difference corresponds to an increase of 0.03 Å in the Nb–Nb separation in the Rb-intercalated sample. The anomalies at the extremes of these curves result from the window function used in the $k \rightarrow r$ Fourier transform. [Taken from Fig. 4c in A. J. Bourdillon, R. F. Pettifer, and E. A. Marseglia, *J. Phys. C* **12,** 3889 (1979).]

the EXAFS near-neighbor peak width was observed, however. The square of that width varies linearly with temperature, as expected for a simple Einstein oscillator. A small distortion of the Nb chains within the resolution of the EXAFS ($\approx$0.01 Å) could not, however, be ruled out.

g. Intercalation Compounds

There is a class of layered compounds which can be intercalated with a variety of materials, including rare gases, alkali metals, transition metals, and organic amines. The intercalate is weakly bound in the "van der Waals gap" between the two-dimensional layers of the host. These hosts can be either graphite[280] or layered transition metal dichalcogenides[281] (i.e., MX_2, with M representing a transition metal and X representing S or Se). Examples of both types of intercalation compounds have been studied using EXAFS. We first consider the MX_2 intercalates.

Bourdillon *et al.*[282] have studied the system $NbSe_2$ intercalated with Rb. They measured the EXAFS on the Rb and Nb K edges at 77°K in $Rb_{0.28}$:$NbSe_2$. The Rb EXAFS was quite weak, indicating large static or dynamic disorder of the Rb in the $NbSe_2$ planes. Large Rb vibrations are to be expected from the weak Rb–$NbSe_2$ binding, consistent with the weak EXAFS signal. The Nb K-edge EXAFS in $Rb_{0.28}$:$NbSe_2$ was compared with that in pure $NbSe_2$. Essentially no change in the Nb–Se near-neighbor spacing was found on intercalation. The Nb–Nb intralayer near-neighbor dis-

[280] For a review of graphite intercalation compounds, see J. E. Fisher and T. E. Thompson, *Phys. Today* **31**(7), 36–45 (1978).

[281] For a review of transition metal dichalcogenide intercalation compounds, see F. R. Gamble and T. H. Geballe, *in* "Solid State Chemistry" (B. B. Hannay, ed.), Vol. 3. Plenum, New York, 1976.

[282] A. J. Bourdillon, R. F. Pettifer, and E. A. Marseglia, *J. Phys. C* **12,** 3889 (1979).

tance was observed to change, however, as illustrated in Fig. 33. A detailed analysis led Bourdillon *et al.*[282] to conclude that the Nb–Nb separation increases by $\approx$0.03 Å upon intercalation. This is identical to the change observed in the lattice constant along the *a* axis, which lies in the $NbSe_2$ planes. The *c* axis, which is perpendicular to the planes, has a large increase due to the intercalate pushing these planes apart, as one would expect. The results are that the strong Nb–Se bonds are not altered by the charge transfer from the Rb, but that the metal atoms, which receive this charge, move further apart. Furthermore, the observed structural distortions associated with the superlattice formation in $NbSe_2$[283] are of the same magnitude as the distortions upon intercalation.

Two EXAFS investigations of intercalated graphite have been performed. The first,[284] on Br_2 in graphite, studied the Br K edge in samples containing 0.27 and 0.75 mol % Br_2. The bromine in both cases intercalates as a Br_2 molecule. The lower concentration corresponds to a coverage of 1.3 monolayers of Br_2 on the Grafoil surface, just above the threshold for intercalation. The carbon contribution to the EXAFS on the Br K edge is sharply damped beyond $k = 6$ Å^{-1}, owing to the rapid decline of its backscattering amplitude with increasing *k*. As a result, this study was able to concentrate on the Br backscattering contribution to the EXAFS. The backscattering amplitude and phase shift were obtained from Br_2 gas, and the structural parameters of the intercalated Br_2 were determined in a least-squares fit to the data. Two types of Br_2 were found for the lower-concentration sample: One was identified as an adsorbed species[285] with its Br–Br spacing increased by 0.03 Å from the 2.28-Å spacing in Br_2 gas, as discussed in Section 14 on surfaces; the other species, which constitutes 63% of the total bromine, has a highly stretched Br–Br distance, 0.25 Å larger than in bromine gas. From the polarization dependence of the EXAFS, it is found that this species is randomly oriented relative to the graphite planes. On increasing the bromine concentration to 0.75 mol %, it is found that the amount of this stretched Br_2 remains constant and that all the additional Br_2 goes into a slightly stretched version of the free molecule, with its bond lengthened by 0.06 Å. From the polarization dependence of the EXAFS for this latter component, it was determined that it lies flat on the graphite planes and is therefore identified as the intercalated species. The nature of the other greatly stretched component is not known, but it is speculated to be associated with a defect or edge site, and may play a role in initiating the intercalation process by forcing the graphite planes apart. The increase in the bond length

[283] D. E. Moncton, J. D. Axe, and F. J. Di Salvo, *Phys. Rev. B: Solid State* [3] **16,** 801 (1977).
[284] S. M. Heald and E. A. Stern, *Synth. Met.* **1,** 249 (1979–1980).
[285] S. M. Heald and E. A. Stern, *Phys. Rev. B: Solid State* [3] **17,** 4069 (1978).

for each of these three Br_2 species is attributed to a weakening of the Br–Br bond caused by a charge transfer of electrons from the graphite to the bromine. It is estimated that this transfer is ≈ 0.16 electron for the intercalated species and 0.6 electron for the expanded species.

The other graphite intercalate study was performed on a series of potassium–graphite intercalation compounds.[286] The EXAFS on the potassium K edge was obtained at 20°C and at 77°K on stage 1–4 compounds. The structure of the stage 1 compound, KC_8, is known from X-ray diffraction[287] to consist of potassium occupying a site over the center of a carbon hexagon, with 12 carbon neighbors at 3.06 Å and 12 at 3.94 Å. The potassium ions are ordered in the graphite planes. For stages 2–4, KC_{12n}, where n is the stage number, the potassium ions are disordered in the planes. The stage 1 data served as a structural standard to analyze the data on the other stage compounds. It was found that the distance to the first-nearest-neighbor carbon atoms is similar in all four stages. As a result, it was concluded that the potassium atoms in stages 2–4 are well localized over the centers of the carbon hexagons, as they are in stage 1. Thus the potassium intraplane disorder for these higher-stage compounds is best characterized by a lattice-gas model. From the variation of the K–C near-neighbor spacing with stage, it was determined that the carbon–intercalate–carbon layer spacing is not constant but rather depends on the stage number, in contradiction with a long-standing classical model. Combining this distance information with the peak widths, it was concluded that the carbon planes do not remain flat, but undergo puckering distortions. These distortions are periodic for stage 1, but random for the higher stages.

h. Anisotropic Materials

The polarization dependence of the EXAFS can be used to study anisotropic materials. This requires a highly polarized source, presumably a synchrotron source, for the reasons given in Section 8. The structure in the direction $\hat{\mathbf{r}}_j$ from the excited atom is probed according to $(\hat{\epsilon} \cdot \hat{\mathbf{r}}_j)^2$ for K-edge EXAFS, where $\hat{\epsilon}$ is the polarization vector of the X rays. Many studies of anisotropic materials have been performed, some of which are discussed in the other sections. These include a large number of investigations of surfaces and intercalated compounds, which obviously fall into the anisotropic category. Polarization-dependent measurements are vital in characterizing their structures. Surfaces are discussed in Section 14, and intercalates in Section 11,g. In this subsection, we discuss three studies of anisotropic

[286] N. Caswell, S. A. Solin, T. M. Hayes, and S. J. Hunter, *Physica B (Amsterdam)* **99**, 463 (1980).

[287] W. Rudorff and E. Schultze, *Z. Anorg. Allg. Chem.* **227**, 156 (1954).

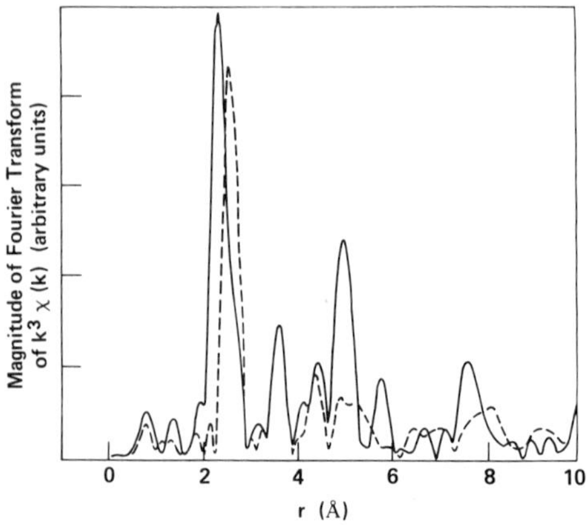

FIG. 34. Magnitude of the Fourier transform of $k^3\chi(k)$ from a single-crystal sample of hexagonal close-packed Zn metal in two orientations: the photon electric field vector $\hat{\epsilon}$ parallel to the *a* axis (solid line) and parallel to the *c* axis (broken line). The data were transformed over $k = 2–15$ Å^{-1}. The nearest-neighbor atoms at 2.67 Å do not contribute to the latter spectrum, so that the first structural peak in *r* space arises solely from neighbors at 2.89 Å. This ability to probe the near-neighbor environment in selected directions by using polarized X rays is a significant attribute of EXAFS spectroscopy. [Taken from Fig. 2 in G. S. Brown, P. Eisenberger, and P. Schmidt, *Solid State Commun.* **24,** 201 (1977). Copyright 1977, Pergamon Press, Ltd.]

materials that demonstrate the use of the directionality of the EXAFS and present experimental and data analysis methods.

Single crystals of Zn have been studied[86] using synchrotron radiation which is 93% polarized (after double Bragg reflection from the monochromator). The structure of Zn is hexagonal close-packed (hcp) with a c/a ratio of 1.83, far from the ideal close-packed $c/a = 1.633$. As a result, the nearest-neighbor shell is split by 0.22 Å, with six atoms *in* the basal plane at 2.67 Å and three atoms *above* and three *below* the basal plane at 2.89 Å. By measuring the EXAFS with the X-ray polarization parallel to the *c* axis (and therefore perpendicular to the basal plane), only the more distant shell is observed. This is illustrated in Fig. 34. For polarization along the *a* axis, the signal from the atoms at 2.67 Å is about five times larger than that from the atoms at 2.89 Å. These signals can be deconvolved easily, and shell separations obtained which are in agreement with the known structure. This study thereby experimentally demonstrated the orientation dependence of the EXAFS.

Another study of the polarization dependence of the EXAFS was performed by Heald and Stern.[288] They examined the EXAFS on the Se K edge in single crystals of the layered material 2H-WSe_2, using synchrotron

[288] S. M. Heald and E. A. Stern, *Phys. Rev. B: Solid State* [3] **16,** 5549 (1977).

radiation. Their results correlated well with the known anisotropy of this material, and their conclusions were similar to those of the previous study. In addition, they obtained some important results concerning the EXAFS on the L_{II} and L_{III} edges. Absorption at these edges involves excitation of the 2p core electrons, and therefore involves both s and d final states. As a result, the expressions for the EXAFS are more complicated for the L edge than for the K edge, as discussed in Section 3. The results of this study confirm the theoretical prediction that the p $\rightarrow$ d transition dominates and the p $\rightarrow$ s transition is small. Investigating the EXAFS on the L_{III} edge of W in 2H-WSe_2, it was determined that the ratio of the final s state contribution to the total absorption to that of the d state is about 0.02. This result implies that the L-edge EXAFS for polycrystalline materials, where the cross term between the s and d final states averages to zero, can be treated in a manner similar to that of K-edge EXAFS. This is discussed in Section 3. In addition, Heald and Stern verified the expressions for the polarization dependence of the EXAFS on the L_{II} and L_{III} edges. This dependence is $(1 + 3(\hat{\epsilon} \cdot \hat{\mathbf{r}}_j)^2)/2$, to be contrasted with the $3(\hat{\epsilon} \cdot \hat{\mathbf{r}}_j)^2$ variation for K edges, as discussed in Section 6,b. This study also examined the near edge structure and its anisotropy for both 2H-WSe_2 and 1T-TaS_2. The importance of the thickness effect[289] in such edge studies was pointed out.

A third study using the polarization dependence of the EXAFS has been performed on single crystals of GeS,[87] which has a layered structure with an orthorhombic unit cell. The X-ray source was again a synchrotron, with the radiation 94% polarized after two Bragg diffractions from a Si(220) monochromator. The *b* axis, which is perpendicular to the planes of the layers, was aligned parallel to the incident X-ray beam so that rotation about the *b* axis could align the polarization vector at any angle from parallel to perpendicular to the *a* axis. In this configuration, the angular dependence could be measured without a variation of the effective sample thickness. This avoided spurious changes in the EXAFS due to the thickness effect. The EXAFS on the Ge K edge was studied as a function of this angle and analyzed to yield the radial distribution functions about the Ge atoms. They were used in turn to obtain the three-dimensional arrangement of the atoms in this layered crystal. The results are in good agreement with X-ray diffraction data.

i. Materials at High Pressure

A few EXAFS studies of crystalline materials at high pressures have been performed.[149-151] Such studies are useful, since the angular range available to measure scattered radiation in a diffraction experiment is often severely

[289] L. G. Parratt, C. F. Hempstead, and E. L. Jossem, *Phys. Rev.* [2] **105,** 1228 (1957).

limited by the high-pressure anvils. In addition, high-pressure phase transitions can convert a single crystal into a polycrystal, thereby limiting the information accessible to diffraction techniques. The first high-pressure EXAFS study, by Ingalls *et al.*,[149] examined the EXAFS on the Fe K edge in FeS_2 up to 64 kbar and in FeF_2 up to 21 kbar, using a modified Bridgman anvil. The Fe–Fe distance in both materials was observed to decrease with pressure by an amount which agrees well with the change in lattice constant deduced from X-ray diffraction. The Fe–S and Fe–F near-neighbor distances, on the other hand, were originally observed to have little or no change in this pressure range, indicating that the Fe–S and Fe–F bonds are significantly less compressible than the Fe–Fe bonds. More recent analysis of this data indicates, however, that the Fe–S and Fe–F distances do indeed change with pressure and scale with the Fe–Fe separation.[290] In addition, the widths of the near-neighbor S and F peaks are observed to decrease with increasing pressure. This implies a decrease in the thermal disorder with pressure.

Ingalls *et al.*[151] have studied the EXAFS on the K edge in Ge up to 52 kbar of pressure, and the Br K edge in NaBr up to 21 kbar, using a boron nitride and diamond anvil cell. A difficulty with these experiments is that the diamond crystal causes Bragg scattering out of the beam, introducing large spikes in the absorption data.[150] These spurious peaks are larger than the EXAFS, as illustrated in Fig. 35, and must be removed. A detailed analysis of the data yielded nonetheless the first-neighbor distance in both materials and the second-neighbor distance in NaBr. In all cases, the distances are in good agreement with existing P–V data. This work also shows that the phase shifts and backscattering amplitudes are independent of pressure, consistent with the notion of chemical transferability. These results as well as some recent measurements on the alkali halides, CuBr, SmSe and Ga, have been reviewed by Ingalls *et al.*[290]

12. Disordered Solids

The ability to quantify long-range order is a principal advantage of diffraction probes. It is irrelevant, however, in studies of disordered solids and liquids, both of which are characterized by the complete absence of any long-range order. Structural studies of these materials usually focus on the spatial relationship between an atom and its nearest or next nearest neighbors, which is often quite well defined in an EXAFS spectrum (see Section

[290] R. Ingalls, J. M. Tranquada, J. E. Whitmore, E. D. Crozier, and A. J. Seary, *in* "EXAFS Spectroscopy: Techniques and Applications" (B. K. Teo and D. C. Joy, eds.), p. 127. Plenum, New York, 1981.

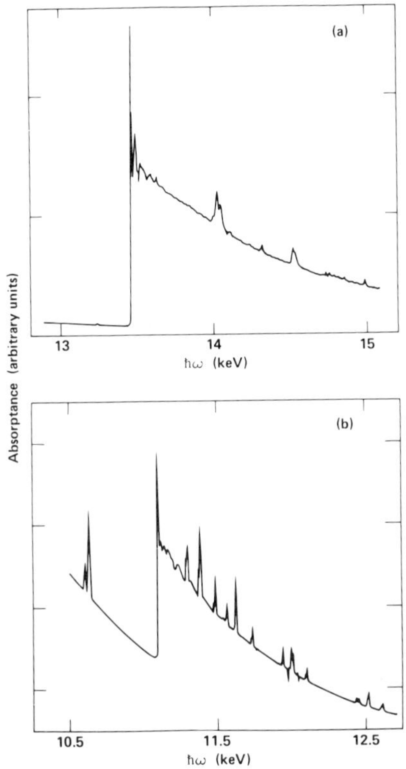

FIG. 35. X-Ray absorptance as a function of photon energy $\hbar\omega$ obtained at high pressures: (a) Br K edge in NaBr at 21 kbar and (b) K edge in Ge at 19 kbar. Both spectra exhibit substantial peaks arising from Bragg diffraction by the diamond anvil. [Taken from Fig. 2 in R. Ingalls, E. D. Crozier, J. E. Whitmore, A. J. Seary, and J. M. Tranquada, *J. Appl. Phys.* **51**, 3158 (1980).]

6 for a detailed discussion). As a consequence, EXAFS spectroscopy has found wide application to disordered materials. A few of these studies are discussed in this and the following sections.

a. Elemental Amorphous Materials

The structural information content of an EXAFS spectrum of an elemental amorphous solid is similar in kind to that of a diffraction spectrum, owing to the absence of Bragg scattering from such a material. The salient difference is that the ranges of k (or q) which can be accessed tend to be somewhat complementary. As was discussed in detail in Section 6, low-k information is very difficult to obtain from an EXAFS spectrum, but it is relatively easy to measure the scattering factor to values of k higher than those in a diffraction experiment. It would be especially interesting, therefore, to combine these two measurements, obtaining in effect the coordination number from diffraction and the detailed shape of the nearest-neighbor peak in the pair correlation function from EXAFS.[291] This has not really been accomplished, however, as most EXAFS studies of elemental amorphous materials have been directed toward establishing the theoretical basis for the interpretation of EXAFS spectra. This includes studies of amorphous (a-)Ge,[26,31,232,292] a-As,[293] and a-Se.[232]

The high sensitivity of an EXAFS spectrum to disorder can be used to good advantage in many circumstances. A good example of this is an interesting EXAFS study of the dependence of the short-range structure of evaporated films of Ge on the substrate temperature during deposition T_s by Evangelisti *et al.*[294] For $T_s = 130°C$, at which substrate temperature the films are believed to be truly amorphous, these authors found that the peaks in the radial distribution function (RDF) corresponding to neighbors beyond the first are not significantly above the noise level, as is often observed in an amorphous material. For $T_s = 390°C$, they found that the spectrum is indistinguishable from that of crystalline (c-)Ge, and were able to distinguish seven peaks in the first 8 Å of the RDF. The peaks beyond the first increase steadily with increasing T_s between these limits, with no evidence for a sharp transition at any well-defined temperature. The authors argued that these changes with increasing T_s can best be explained on the basis of the growth of microcrystallites at the expense of an amorphous matrix, and were able to estimate the microcrystallite size from the second- and third-neighbor peaks.

b. Binary Amorphous Materials

EXAFS studies of binary systems can take advantage of the ability of this technique to probe the environment of each atom species individually. There have been several studies of amorphous III–V compounds, with emphasis on GaAs and GaP. Del Cueto and Shevchik[295] measured the EXAFS spectra associated with the Ga and As K-edges in c- and a-GaAs and concluded that the nearest neighbors of As move away by $\approx$0.04 Å in the glass relative to the crystal, while those of Ga are unchanged. They attributed

[291] As discussed in Section 6, valuable insight into the short-range portion of the interatomic interactions can be obtained from the near-neighbor peak shape.

[292] D. E. Sayers, F. W. Lytle, and E. A. Stern. *J. Non-Cryst. Solids* **8–10,** 401 (1972).

[293] J. C. Knights, T. M. Hayes, and J. C. Mikkelsen, Jr., *Phys. Rev. Lett.* **39,** 712 (1977).

[294] F. Evangelisti, M. G. Proietti, A. Balzarotti, F. Comin, L. Incoccia, and S. Mobilio, *Solid State Commun.* **37,** 413 (1981).

[295] J. A. Del Cueto and N. J. Shevchik, *J. Phys. C* **11,** L829 (1978).

this to the contributions of bonds between like atoms, or so-called "wrong" bonds (i.e., As–As nearest-neighbor bonds, which are expected to be longer than As–Ga bonds), even though this contradicts the deductions of an earlier photoemission experiment.[296] Theye, Gheorghiu, and co-workers[297,298] performed a series of EXAFS measurements on the Ga and As K edges of c- and flash-evaporated a-GaAs and GaP. The resulting spectra are shown in r space in Fig. 36. In contradiction with the prior study, these authors found that the neighbors of both Ga and As in a-GaAs move away relative to c-GaAs, but by quite small amounts ($\approx$0.01 Å). They also argued that the nearest-neighbor peaks in both cases in a-GaAs are too narrow to contain two different atom species, which they expected would exhibit significantly different bond lengths. From these results, they concluded that there is no evidence to support the existence of "wrong" bonds. They did note, however, that both Ga and As appear to have reduced coordination in a-GaAs ($\approx$3.2 neighbors, instead of 4 as in the crystal). This could be an indication of experimental difficulties resulting from inhomogeneous samples (see Section 8,e), or perhaps an indicator of an asymmetric nearest-neighbor peak. As discussed at length in Section 6, asymmetric peaks can lead to the deduction of coordination numbers and distances which are smaller than the actual ones. Theye, Gheorghiu, and co-workers[297,298] observed similar effects in the EXAFS spectra on the Ga K edges of c- and a-GaP. In addition, however, they noted that the amplitude of the EXAFS is anomalously large in a-GaP at large k. They interpreted this as weak evidence that Ga might have some Ga nearest neighbors in a-GaP, since the Ga backscattering amplitude is much larger than that of P at large k.

There have been a series of EXAFS studies of binary chalcogenide glasses. Sayers *et al.*[292,299] have reported measurements on amorphous GeO_2. Vitreous SiO_2 and GeO_2 had already been studied extensively using X-ray and neutron diffraction techniques,[300] so that there was no question about the identity or position of the nearest neighbors of Ge in a-GeO_2. The new result of the EXAFS study of Sayers *et al.* was the conclusion that the distribution of Ge second neighbors is too broad to be consistent with a microcrystallite model for the amorphous phase. It has been argued,[301] however, that the second-neighbor structural information extracted from the EXAFS data in this study was known already from previous diffraction studies. Furthermore, the reasoning which leads from that structural infor-

[296] N. J. Shevchik, J. Tejeda, and M. Cardona, *Phys. Rev. B: Solid State* [3] **9,** 2627 (1974).

[297] M.-L. Theye, A. Gheorghiu, and H. Launois, *J. Phys. C* **13,** 6569 (1980).

[298] A. Gheorghiu and M.-L. Theye, *J. Non-Cryst. Solids* **35–36,** 397 (1980).

[299] D. E. Sayers, E. A. Stern, and F. W. Lytle, *Phys. Rev. Lett.* **35,** 584 (1975).

[300] See discussion in A. C. Wright and A. J. Leadbetter, *Phys. Chem. Glasses* **17,** 122 (1976).

[301] J. Karle and J. H. Konnert, *Phys. Rev. Lett.* **36,** 823 (1976).

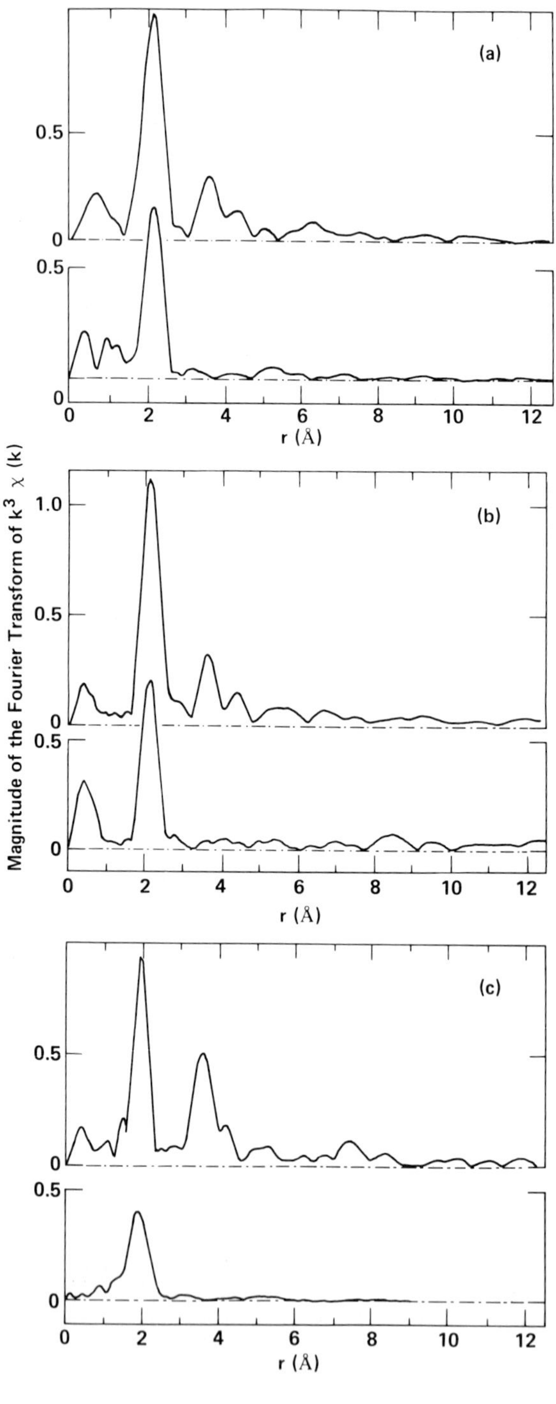
(a)
0.5
0
0.5
0
0 2 4 6 8 10 12
r (Å)
Magnitude of the Fourier Transform of k³ χ (k)
(b)
1.0
0.5
0
0.5
0
0 2 4 6 8 10 12
r (Å)
(c)
0.5
0
0.5
0
0 2 4 6 8 10 12
r (Å)

mation to the conclusions of Sayers *et al.* has been challenged[300] on the grounds that a broad distribution is not inconsistent with quasi-crystallinity.

Sayers *et al.* have also measured the EXAFS in amorphous GeSe,[292,302] $GeSe_2$, As_2S_3, As_2Se_3, and As_2Te_3.[302] The emphasis in these studies is on the detection of bonds between like atoms, or "wrong" bonds. These were, in fact, observed in evaporated a-As_2Se_3. This observation has been confirmed and quantified in a recent study[303] of a-As_2S_3, As_2Se_3, and $GeSe_2$ in three forms: as-evaporated thin films, annealed films, and bulk glasses. Both a-As_2S_3 and a-As_2Se_3 show evidence for like-atom bonds in the as-deposited state. The number of these wrong bonds decreases substantially on annealing, resulting in EXAFS spectra much closer to those from the bulk glass. These results and those of infrared and Raman spectroscopy have been combined in a model for the annealing dependence of the optical properties of these films.[304]

Pettifer and McMillan[305] have examined glassy As_2O_3 and found no evidence for the presence of As_4O_6 molecules, in contrast with diffraction evidence for the existence of the analogous molecules in a-As_2S_3 and As_2Se_3.[300] In a later and more detailed analysis of the same data, Gurman and Pettifer[74] concluded that the (first-neighbor) As–O and the (second-neighbor) As–As distances are unchanged between glassy and crystalline As_2O_3, implying that the As–O–As bond angle is also unchanged. They found no evidence for nearest-neighbor As–As bonds in either spectrum. Finally, they noted that the widths of the various peaks in the radial distribution of neighbors about As are monotonically increasing with distance in the glass, but not in the crystal. They associated deviations from monotonic behavior with the known molecular structure of the crystalline form, and concluded that this is further evidence for the absence of "well-defined" As_4O_6 molecules in the glass.

In another study, Pettifer *et al.*[66] measured the EXAFS for glassy As_2Te_3.

[302] D. E. Sayers, F. W. Lytle, and E. A. Stern, *in* "Amorphous and Liquid Semiconductors" (J. Stuke and W. Brenig, eds.), p. 403. Taylor & Francis, London, 1974.

[303] R. J. Nemanich, G. A. N. Connell, T. M. Hayes, and R. A. Street, *Phys. Rev. B: Condens. Matter* [3] **18,** 6900 (1978).

[304] R. A. Street, R. J. Nemanich, and G. A. N. Connell, *Phys. Rev. B: Condens. Matter* [3] **18,** 6915 (1978).

[305] R. F. Pettifer and P. W. McMillan, *Philos. Mag.* [8] **35,** 871 (1977).

FIG. 36. Magnitude of the Fourier transform of $k^3\chi(k)$ for crystalline and amorphous samples as follows: (a) GaAs at the Ga K edge and (b) at the As K edge and (c) GaP at the Ga K edge. The upper curve in each part is for the crystalline sample, while the lower one is for the amorphous sample. Note how quickly the correlations vanish with increasing r beyond the nearest-neighbor peak in the amorphous samples, even though analysis shows that the nearest-neighbor peak itself is changed only slightly. [Taken from Fig. 4 in M.-L. Theye, A. Gheorghiu, and H. Launois, *J. Phys. C* **13,** 6569 (1980).]

The data were originally interpreted in terms of a split shell of three Te about As in the glass. This splitting and the sharpness of the peaks in $p_{\mathrm{As-Te}}(r)$ relative to the crystalline forms were interpreted as evidence for the glass being much more rigid and covalent than the crystals. In a later study, Pettifer and Gurman[306] concluded that those data are better explained on the basis of "wrong bonds" (i.e., As nearest neighbors of As). This is one of only two examples identified by Pettifer[55] in which the nearest-neighbor distribution in a glass has significant deviations from that in an analogous crystal. The other example is a-As, in which the nearest-neighbor distance is 0.06 Å shorter than in the crystal.[307]

In a study of a-As_2S_3 and As_2Se_3, Pettifer[55] found that the Gaussian half-width σ of the nearest-neighbor peak is between 0.05 and 0.06 Å. A simple estimate of the thermal contribution to this width, based on Raman spectra, leads to an estimate of less than 0.03 Å for the "static" contribution to σ. This is significantly smaller than the static width of 0.05–0.08 Å deduced similarly from diffraction measurements on the crystalline phases.[308] Possible explanations for this include (*a*) that the glass is actually more "ordered" on a nearest-neighbor basis than is the crystal and (*b*) that the conclusions of the diffraction studies are in error. Pettifer believed that the second explanation is the more likely, although he did not rule out the first.

A study of the nearest neighbors of arsenic dopants in a-Si–H alloys[293] took particular advantage of the atom species selectivity of an EXAFS measurement. This study was motivated by observations of changes in electrical conductivity characteristic of the substitutional doping of crystalline semiconductors when phosphorus, boron, or arsenic is incorporated in thin films of amorphous "silicon" prepared by the plasma decomposition of silane.[309] These conductivity changes were surprising, because it had long been argued that the amorphous phase would allow for local structural relaxation around impurities so as to satisfy their bonding requirements and thereby prevent their serving as donors or acceptors. A knowledge of the nearest-neighbor environment of the impurities would help in understanding this phenomenon. The low concentration of impurities in these films, on the order of 1–10 at. %, argued against the use of a standard diffraction technique. Arsenic-doped amorphous silicon films plasma deposited from silane–arsine mixtures were chosen for the EXAFS study because the As

[306] R. F. Pettifer and S. J. Gurman, unpublished, discussed in Ref. 55.

[307] J. Bordas, S. J. Gurman, and R. F. Pettifer, unpublished, discussed in Ref. 55.

[308] If the temperature-dependent portion of the nearest-neighbor peak width has been removed properly, any residual width in the crystalline phase must be due to "zero-point" motion and distortions in the lattice.

[309] W. E. Spear and P. G. LeComber, *Solid State Commun.* **17**, 9 (1975); J. C. Knights, *Philos. Mag.* [8] **34,** 663 (1976).

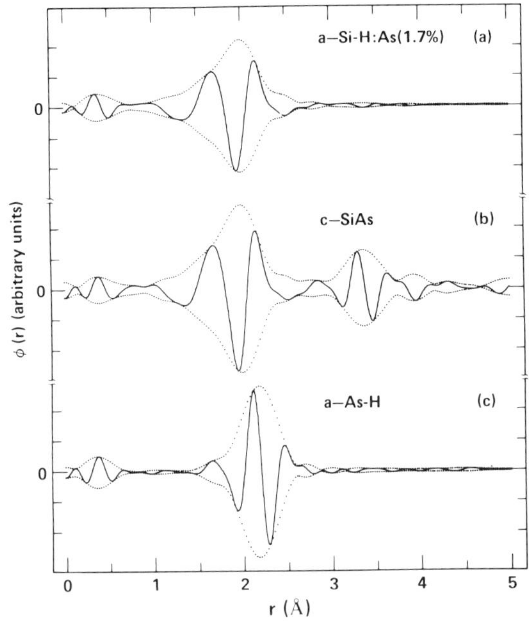

FIG. 37. The real part (solid line) and the magnitude (dotted line) of the Fourier transform of $k\chi(k)$ from the As K-shell absorption in (a) 1.7 at. % As in amorphous Si–H, (b) crystalline SiAs, and (c) amorphous As–H. The units of all three vertical scales are arbitrary but identical. The data were all transformed using a square window between $k = 3.75$ and 13.85 Å^{-1}, broadened by convolution with a Gaussian of half-width $\sigma_w = 0.7$ Å^{-1}. Notice that the contributions from Si backscattering atoms [the features at ≈ 2 Å in parts (a) and (b)] are significantly unlike those from As backscattering atoms [at ≈ 3.3 Å in part (b) and at ≈ 2.15 Å in part (c)]. [Taken from Fig. 1 in J. C. Knights, T. M. Hayes, and J. C. Mikkelsen, Jr., *Phys. Rev. Lett.* **39,** 712 (1977).]

K edge is at a convenient energy. The Fourier transforms of the EXAFS spectra on the As K-shell absorption are shown in Fig. 37 for an amorphous arsenic-doped Si–H sample and for two standards, crystalline SiAs and amorphous As. Analysis of these and other spectra revealed the degree and nature of H incorporation in the Si–H samples. They contain hydrogen in excess of 13 at. % and are properly described as doped silicon–hydrogen alloys. The environment about As was shown to be radially disordered at the second-nearest-neighbor distance, indicated by the absence of structure beyond 3 Å in Fig. 36a. Furthermore, the nearest-neighbor peak (at ≈ 2 Å) in Fig. 36a is nearly identical to the nearest-neighbor peak in Fig. 36b, which is known to arise from Si atoms and is quite dissimilar to either the second peak (at ≈ 3.3 Å) in Fig. 36b or the peak at ≈ 2.15 Å in Fig. 36c, both of which are due to As atoms. Detailed analysis backs up the obvious conclusion that these spectra reflect a preference for unlike atom bonds. In other words, there was no evidence of significant As clustering. Finally, the striking enhancement of electronic conductivity at low As concentrations was shown to correlate with the onset of fourfold coordination of the As, providing the first direct evidence for substitutional doping in an amorphous semiconductor.

Oyanagi *et al.*[310] studied the Ge and Ni K edges in co-sputtered amorphous $Ge_{100-x}Ni_x$ alloys for several values of x between 7 and 55. They concluded that Ge has three to four Ge nearest neighbors at all compositions. As x increases from 7, Ni atoms initially substitute for Ge. Beginning at $x \approx 23$, the coordination of Ge increases with increasing Ni concentration due to Ni nearest neighbors *in addition to* the four Ge atoms. This coincides with the transition from semiconducting to metallic behavior. At $x \approx 55$, each Ge has approximately four Ge and four Ni nearest neighbors. The authors observed that the Ge–Ni separation in the metallic phase is at least 0.13 Å less than expected from the sum of the metallic radius of Ni and either the metallic or covalent radius of Ge. A similar foreshortening of the Ge–metal separation had been observed in an earlier study of Pd–Ge metallic glasses, as discussed below. Finally, they noted that the intensity of the "white line" at the Ge K edge increases with increasing Ni content, which they interpreted as evidence for charge transfer from Ge to Ni atoms.

c. *Multicomponent Glasses*

Looking at a rather more complicated system, Hunter *et al.*[311] have done an extensive EXAFS study of glassy alloys of Cu in As_2Se_3. They observed three consequences of increasing the Cu content in these alloys from 5 to 25%: The Se coordination increases from 2 to 4, the number of As–Se bonds decreases, and the number of metal–metal bonds increases. This information goes substantially beyond the results of diffraction studies on this system.[312]

In a study of glassy $As_2S_3Se_3$, Pettifer[55] concluded that each As atom has two S and one Se nearest neighbors. While this totals three nearest neighbors, as one might expect for As, the presence of As–Se bonds is somewhat surprising, since that bond is believed to be less stable than an As–S bond.

[310] H. Oyanagi, K. Tsuji, S. Hosoya, S. Minomura, and T. Fukamachi, *J. Non-Cryst. Solids* **35–36,** 555 (1980).

[311] S. H. Hunter, A. B. Bienenstock, and T. M. Hayes, *in* "The Structure of Non-Crystalline Materials" (P. H. Gaskell, ed.), p. 73. Taylor & Francis, London 1977; S. H. Hunter, A. B. Bienenstock, and T. M. Hayes, *in* "Amorphous and Liquid Semiconductors" (W. E. Spear, ed.), p. 78. University of Edinburgh, Edinburgh, 1977.

[312] K. S. Liang, A. Bienenstock, and C. W. Bates, *Phys. Rev. B: Solid State* [3] **10,** 1528 (1974).

Note that the stoichiometry is such that all As–Se bonds could, in principle, have been excluded.

Pettifer *et al.*[313] have studied the Zn environment in a series of Na_2O–ZnO–SiO_2 glasses. They deduced that Zn is always surrounded by four nearest-neighbor O atoms, evidence that Zn is a network former in these glasses.

The addition of small amounts of Ti to silica glasses can be used to reduce thermal expansion to nearly zero. Sandstrom *et al.*[314] studied the Ti environment in silica glasses to which had been added 3.4, 7.5, and 9.5 wt. % TiO_2. They deduced that the Ti^{4+} ions are primarily fourfold coordinated with O. They also found that a small fraction of the Ti atoms are sixfold coordinated with O, and that this fraction increases with Ti content (to $\approx$18% sixfold coordinated at 9.5 wt. % TiO_2). The sixfold coordinated Ti is marked by a longer bond length ($\approx$2.1 Å rather than 1.8 Å, as in the fourfold coordinated site). This interpretation is consistent with changes in the near edge structure and with other, non-X-ray measurements.

Calas *et al.*[315] have studied the chemical state of Fe and Fe–O bonding in four different silicate glasses, examining the position and shape of the absorption edge as well as the EXAFS.

d. Thermal Effects

Interesting information can also come from a study of the temperature dependence of the EXAFS spectrum. Of particular interest in this connection is the effect of temperature on the distribution of nearest-neighbor bond lengths. Crozier *et al.*[147] have monitored the EXAFS spectra of amorphous As_2Se_3 between 100 and 773°K, including the melting point at 627°K. They found no discontinuity in the width of the distribution of nearest-neighbor bond lengths on melting and no startling change in the derivative of that quantity with temperature.

Rabe *et al.*[100] have studied the temperature dependence of the first-neighbor peak in the pair correlation function for both c- and a-Ge. They concluded that the first-neighbor distance is unchanged but that the temperature-independent portion of the peak width is greater in a-Ge than in c-Ge, presumably due to significant static disorder.

Wong and Lytle[316] measured the EXAFS spectrum on the Zn K edge in glassy $ZnCl_2$ at several temperatures, from 100°K, through the glass tran-

[313] R. F. Pettifer, P. W. McMillan, and J. Bordas, unpublished, discussed in Ref. 55.

[314] D. R. Sandstrom, F. W. Lytle, P. S. P. Wei, R. B. Greegor, J. Wong, and P. Schultz, *J. Non-Cryst. Solids* **41,** 201 (1980).

[315] G. Calas, P. Levitz, J. Petiau, P. Bondot, and G. Loupias, *Rev. Phys. Appl.* **15,** 1161 (1980).

[316] J. Wong and F. W. Lytle, *J. Non-Cryst. Solids* **37,** 273 (1980).

sition and melting temperatures, to 673°K. They concluded that the Zn coordination is ≈ 5 at all temperatures measured (to be compared with 4 in the crystalline α phase), and that the half-width σ of the nearest-neighbor distribution increases with temperature rather sharply: $\Delta\sigma(700°K - 100°K) \equiv [\sigma^2(700°K) - \sigma^2(100°K)]^{1/2} \approx 0.10$ Å. In sharp contrast, the coordination of Ge in a-GeO_2 is always ≈ 4, as in the crystalline α phase, and $\Delta\sigma(700°K - 100°K) \approx 0.03$ Å. The nearest-neighbor distance is unchanged in both materials. The authors viewed these differences as manifestations of the greater ionic character of the Zn–Cl bond (i.e., less covalent and therefore less rigid), supporting their idea that $ZnCl_2$ is a weakened structural analog of SiO_2 and GeO_2.

e. *Amorphous Metals*

Examining the relationship between structural disorder and superconductivity, Brown *et al.*[275] undertook an EXAFS study of several Nb_3Ge thin films, characterized by various degrees of crystallinity. Their results are discussed in Section 11,f.

A study of Pd–Ge metallic glasses[317] relied heavily on the ability of an EXAFS measurement to identify the near-neighbor atom species. One of the more thoroughly studied types of glass-forming metallic alloys consists of a transition or noble metal host to which has been added approximately 20 at. % of a metalloid, such as B, P, Si, or Ge. The melting point is very low at this composition, often as much as 700° K below that for the pure metal. For a narrow range of compositions about this eutectic, the alloys can be quenched rapidly from the melt to form glasses. Sputtered amorphous films can also be prepared, usually over a wider composition range. These transition metal–metalloid glasses have been discussed extensively in reviews.[318] The sharpness of the eutectic and the narrowness of the glass-forming region testify to the crucial role of the metalloids in these effects. It is not likely that the binary alloys are merely random arrangements of the two components but, rather, that there is an unusual chemical ordering in the glass and probably also in the melt. To study this ordering, Hayes *et al.*[317] measured the EXAFS spectra on the Ge K-shell absorption in two amorphous Pd–Ge alloys and in a crystalline compound of known structure, reproduced in Fig. 38. It is obvious from Fig. 38 that the nearest-neighbor environments of Ge in arc-quenched glassy $Pd_{78}Ge_{22}$ and in sputtered amorphous $Pd_{80}Ge_{20}$ are essentially identical. Detailed analysis of these spec-

[317] T. M. Hayes, J. W. Allen, J. Tauc, B. C. Giessen, and J. J. Hauser, *Phys. Rev. Lett.* **40,** 1282 (1978).

[318] For example, see G. S. Cargill III, *Solid State Phys.* **30,** 227 (1975); P. Chaudhari and D. Turnbull, *Science* **199,** 11 (1978).

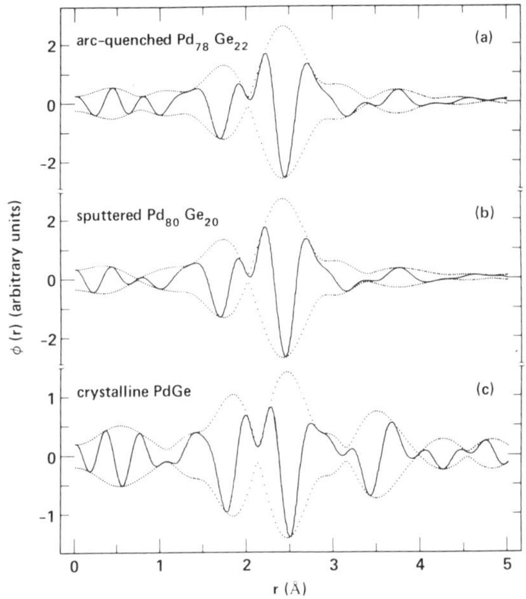

FIG. 38. The real part (solid line) and the magnitude (dotted line) of the Fourier transform of $k\chi(k)$ from the Ge K-shell absorption in (a) arc-quenched glassy $Pd_{78}Ge_{22}$, (b) sputtered amorphous $Pd_{80}Ge_{20}$, and (c) polycrystalline PdGe. The units of all three vertical scales are arbitrary but identical. The data were all transformed using a square window between $k = 3.5$ and 9.9 Å^{-1}, broadened by convolution with a Gaussian of half-width $\sigma_w = 0.7$ Å^{-1}. Notice that the EXAFS from the two amorphous samples are essentially identical, despite significantly different preparation methods. [Taken from Fig. 1 in T. M. Hayes, J. W. Allen, J. Tauc, B. C. Giessen, and J. J. Hauser, *Phys. Rev. Lett.* **40,** 1282 (1978).]

tra showed that each Ge atom in the amorphous samples has eight to nine nearest-neighbor Pd atoms at a mean separation of 2.49 Å, distributed with a full width at half maximum (FWHM) less than 0.24 Å. This finding of a surprisingly narrow spread in r_{Pd-Ge} has since been supported by computer modeling of the structure of this glass.[319] In addition, it is consistent with the conclusions reached by Cargill in his analysis[320] of X-ray diffraction data on a-Nb_3Ge. Specifically, Cargill found that the FWHM of the first peak in p_{Nb-Ge} must be <0.15 Å, or less than one-third of the FWHM of the first peak in p_{Nb-Nb}. Such an unusually narrow peak implies that the coupling between the metalloid and transition metal atoms is more rigid than that which would result from the weak central forces normally associated with

[319] D. S. Boudreaux, *Phys. Rev. B: Condens. Matter* [3] **18,** 4039 (1978).

[320] G. S. Cargill III, *in* "Liquid and Amorphous Metals" (E. Luscher and H. Coufal, eds.), p. 161. Sijthoff & Noordhoff, Alphen aan den Rijn, Netherlands, 1980.

simple metals. Hayes *et al.*[317] also concluded that each Ge atom in the amorphous Pd–Ge alloys has considerably fewer that one Ge nearest neighbor, well under the two Ge atoms expected from a random arrangement of atoms. This chemical ordering is further strong evidence for the operation of a special interaction between the transition metal and the metalloid atoms. Finally, it was argued that the regularity of Ge–Pd separations is consistent only with a model in which the metalloids have a substantial influence on the structure of the metallic glass. This brought into question the applicability of those models based on a structure determined by the metal atoms alone.

Stern *et al.*[321] performed an interesting EXAFS study of several amorphous rare-earth–transition-metal alloys: a-$DyFe_2$, $TbFe_2$, and $HoFe_2$. They found that the Fe–Fe separation in a-$DyFe_2$ is similar to that found in the crystal, but that the Dy–Fe distance is shortened by $\approx$0.4 Å. A similar result was obtained for a-$TbFe_2$, where the Tb–Fe distance is shortened by $\approx$0.3 Å. This suggests the influence of chemical effects similar to those observed in the Pd–Ge glasses discussed above. Note, however, that they also concluded that the coordination number of the rare earths has dropped to approximately two-thirds of its crystalline value. A reduced coordination number could be a consequence of asymmetry in the nearest-neighbor peak shape. As we noted in Section 6, such asymmetry can lead to significant errors in a Gaussian peak analysis such as this one, with the deduced coordination number and distance both being too small. Thus the rare-earth–transition-metal distance may not be so dramatically foreshortened as the analysis by these authors suggests.

Pettifer *et al.*[322] have studied the Nb environment in a glassy alloy of Nb and Ni. They observed that the nearest-neighbor distance is shorter than in Nb metal, which they took for evidence that Nb is principally coordinated with Ni. They noted further that the nearest-neighbor peak is quite broad, which they attributed tentatively to the disordering effects of the rapid quenching necessary to produce these glasses.

Wong and co-workers[323] have studied the EXAFS on the Fe and Ni K edges in a-$Fe_{40}Ni_{40}B_{20-x}P_x$, with x ranging from 0 to 20. They used a somewhat primitive analysis method in which changes in the height of the nearest-neighbor peak in the Fourier transform φ of the measured spectrum are attributed to changes in the half-width σ of the peak in the radial distribution

[321] E. A. Stern, S. Rinaldi, E. Callen, S. Heald, and B. Bunker, *J. Magn. Magn. Mater.* **7**, 188 (1978).

[322] R. F. Pettifer, J. Bordas, I. Donald, B. G. Lewis, and H. A. Davis, unpublished, discussed in Ref. 55.

[323] J. Wong, F. W. Lytle, R. B. Greegor, H. H. Liebermann, J. L. Walter, and F. E. Luborsky, *Rapidly Quenched Met., Proc. Int. Conf., 3rd, 1978* p. 345 (1978).

function (RDF), and assumed the peak height to be proportional to σ^{-1}. This is not correct in general for φ, although it might be a reasonable *ansatz* in analyzing the RDF itself. Using this method, these authors have examined changes in $\Delta\sigma(300°K - 77°K) \equiv [\sigma^2(300°K) - \sigma^2(77°K)]^{1/2}$ as P is added. They observed that the addition of P increases the $\Delta\sigma$ associated with both Fe and Ni, but has a greater effect on the Fe environment. Specifically, $\Delta\sigma_{Ni} < \Delta\sigma_{Fe}$ when $x = 0$, but $\Delta\sigma_{Fe} < \Delta\sigma_{Ni}$ when $x = 20$. Assuming that the smaller $\Delta\sigma$ corresponds to the more rigid "bond" between metal and metalloid (rather than the less rigid coupling between two metal atoms), they concluded that B coordinates strongly with Ni and P with Fe. Furthermore, they noted that the nearest-neighbor peaks sharpen with annealing. A rapid drop in the Fe nearest-neighbor peak width with annealing at $\approx$200° C is tentatively associated with the known embrittlement of these glasses.

In a separate study of a-$Fe_{80}B_{10}Ge_{10}$, Wong *et al.*[324] concluded that there are no Ge–Ge nearest-neighbor pairs in this amorphous metal, in agreement with the previously discussed study of amorphous Pd–Ge alloys.

In presenting the results of EXAFS studies on several glassy metallic alloys, Haensel *et al.*[107] have stressed the potential importance of asymmetric nearest-neighbor peak shapes, discussed at length in Section 6. They proposed a specific peak shape with a sharp onset and an exponential falloff, and used this model in the analysis of EXAFS spectra from several glassy metals. For the Fe environment in a-$Fe_{80}B_{20}$, they found that the assumption of a Gaussian peak shape leads to deduced values of approximately one-half the number of neighbors determined by diffraction and a Fe–Fe distance which is $\approx$0.1 Å too short. Use of their shape leads to agreement with diffraction results. A similar conclusion has been reached by De Crescenzi *et al.*[325] Cargill[326] has observed that a peak shape as asymmetric as that proposed by these authors for p_{Fe-Fe} in a-$Fe_{80}B_{20}$ would shift the diffraction peak position nearly as much as the EXAFS peak position and therefore cannot explain the observed discrepancy between the two measurements. From an analysis of EXAFS data on the transition metal K edges of a-$Co_{76}P_{24}$, c-Fe_3P, c-Fe_2P, and c-Co_2P, Cargill concluded[326] that the most likely explanation for the discrepancy is that the EXAFS peak arises from a very sharp peak in p_{Co-P} (rather than a peak in p_{Co-Co}, in analogy with the other analyses), the position of which *is* consistent with diffraction data. He presumed that the peak in p_{Co-Co} is simply too broad to contribute significantly to the EXAFS data, a possibility discussed in Section 6,d. Cargill's explanation is consistent with the deductions of his diffraction study[320] of

[324] J. Wong, H. H. Liebermann, and R. P. Messmer, unpublished.

[325] M. De Crescenzi, A. Balzarotti, F. Comin, L. Incoccia, S. Mobilio, and N. Motta, *Solid State Commun.* **37**, 921 (1981).

[326] G. S. Cargill III, *Proc. Int. Conf. Rapidly Quenched Met., 4th, 1981* (in press).

a-Nb_3Ge and with the notion of an unusually rigid coupling between the transition metal and the metalloid atoms.

Haensel *et al.*[107] found that data from the Zr and Cu K edges in $Zr_{54}Cu_{46}$ cannot be understood without using asymmetric peak shapes for the Cu–Cu and Cu–Zr correlations (although a Gaussian peak appears to be suitable for the Zr–Zr correlations). Finally, they examined the Ni and Fe environments in a-$Fe_{40}Ni_{40}B_{20}$, and found that these two environments require peak shapes which are quite distinct from one another. This suggests that the Ni and Fe environments are fundamentally different from one another, in agreement with the study of a-$Fe_{40}Ni_{40}B_{20-x}P_x$ by Wong *et al.*[323] mentioned earlier.

13. Liquids

The motivations behind and the advantages of EXAFS studies of the structure of liquids are essentially the same as for disordered solids. The principal difference is quantitative: The greater structural disorder in a liquid will reduce to varying degrees the amount of information in EXAFS spectra, as is discussed in detail in Section 6. Interesting studies of liquids include studies of molten materials, which are similar to disordered solids, and of aqueous solutions. Studies of the *internal* structure of molecules in a solvent do not touch upon the intrinsic structure of liquids and will be discussed in Section 16. As will become apparent in the following discussion, EXAFS spectra of liquids offer an unusual opportunity to gain insight into the short-range portion of the interatomic interactions which underlie many of the properties of condensed matter.

a. Melts

The solid and liquid phases of As_2Se_3 are all viewed as composed of (weakly) interacting molecular units. Crozier *et al.*[147] have measured the As and Se K-edge EXAFS in amorphous and liquid As_2Se_3 from 100°K through the melting temperature at 627°K until 773°K. They observed a strong EXAFS signal out to the limit of their measurement, ≈ 12 Å^{-1}, even in the liquid at 658°K. The nearest neighbors of As (Se) are believed to be Se (As) atoms at all temperatures. The mean nearest-neighbor As–Se distance, the coordination number, and the width of that distribution vary smoothly with temperature, showing no more than a small discontinuity at the melting temperature. Similar results were obtained by Wong and Lytle[316] in a study of amorphous and liquid $ZnCl_2$ discussed in Section 12.

The above studies did not consider the possibility that the nearest-neighbor peak shape in a liquid may be other than a Gaussian. One might anticipate, however, that the absence of even intermediate-range structure in

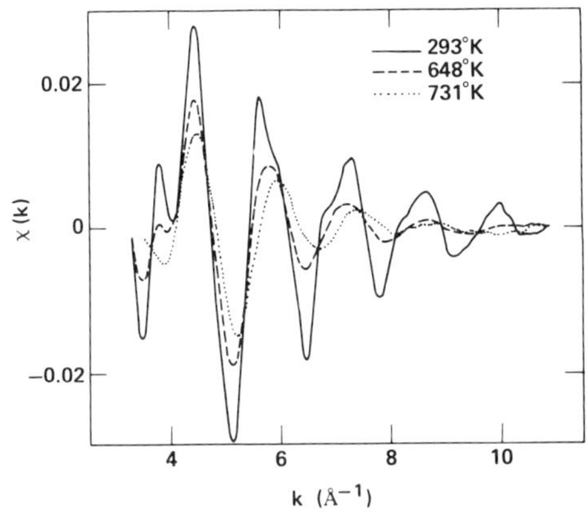

FIG. 39. EXAFS spectra $\chi(k)$ for Zn at three temperatures: polycrystalline at 293 (solid line) and at 648°K (dashed line), liquid at 731°K (dotted line). Note that the EXAFS oscillations extend to the end of the data range even in the liquid sample, and that they are only slightly weaker in the liquid state than in the solid state at similar temperatures. This indicates strong nearest-neighbor correlations in the melt. [Taken from Fig. 3 in E. D. Crozier and A. J. Seary, *Can. J. Phys.* **58,** 1388 (1980).]

a liquid might well lead to peak shapes which are quite far from a symmetric Gaussian. Crozier and Seary[103] have explored this question in depth for solid and liquid Zn between 273 and 782°K. The samples were polycrystalline hexagonal close-packed (hcp) Zn, with a melting point of 692°K. At each temperature, they observed strong EXAFS oscillations out to the end of their data range at 10 Å^{-1}, as shown in Fig. 39. These spectra imply that the nearest-neighbor distance is fairly well defined, even in the liquid at 731°K (i.e., it corresponds to a sharp feature in the radial distribution function). This observation is more surprising for liquid Zn than it is for liquid As_2Se_3 or $ZnCl_2$, since, unlike Zn, both of those materials are viewed as forming "molecular" liquids. It is strong evidence that EXAFS data can indeed provide interesting information on the nearest-neighbor distribution in liquids. The shape of this distribution depends primarily on the short-range portion of the interatomic pair interaction potentials, since the sum of the long-range interactions with distant neighbors is effectively averaged to a constant potential by the absence of intermediate- and long-range order. We conclude that EXAFS studies of liquids hold promise as a sensitive probe of these pair interaction potentials, upon which depend many properties of condensed matter.

Secondly, Crozier and Seary[103] found that the asymmetry of the nearest-neighbor peak is significant in hcp polycrystalline Zn, and essential to understanding the liquid data as well. They found that the anharmonic oscillator model proposed for solid phase Zn by Eisenberger and Brown[91] is barely able to accommodate the large asymmetry found for the nearest-neighbor peak in the solid at high temperatures, and is clearly inadequate for the liquid. Crozier and Seary analyzed the EXAFS spectra from liquid Zn using two alternative models for that very asymmetric nearest-neighbor peak shape. Their results agreed quite well with the pair correlation functions measured for the liquid using X-ray diffraction. In agreement with the earlier studies of more covalent materials, they found that the nearest-neighbor peak widths in the hot solid and in the melt are similar (even though the shape of the peak is somewhat more asymmetric in the latter case). Finally, they observed that an analysis on the basis of Gaussian peak shapes would lead to a dramatically underestimated mean Zn–Zn separation at high temperatures ($\approx$0.25 Å too small at 782°K), consistent with the discussion in Section 6.

b. *Aqueous Solutions*

There are two structural issues which have been of particular interest in aqueous solutions: What is the nature of the hydration of the ions? and is there an intermediate-range cation–anion order in concentrated solutions? Interest in the latter question was substantially stimulated by the neutron diffraction study of a concentrated aqueous solution (5.5 *M*) of $NiCl_2$ by Howe *et al.*[327] (A 1 *M* solution contains 1 g molecular weight of solute per liter of solvent). They observed a peak in the Ni^{2+}–Ni^{2+} correlation function at $\approx$6 Å, and took this to be evidence for relatively long-range order. Subsequent analysis of this data by Quirke and Soper[328] has cast doubt, however, upon the deduction of significant long-range order. They treated the ions *together with* their hydration spheres of water molecules as structural entities, modeling them as hard spheres with diameters $\approx$5 Å. The Ni^{2+}–Ni^{2+} peak then arises naturally from a Percus–Yevick treatment of the three-component fluid of these hard spheres (hydrated cations and anions plus excess water molecules), even though that model fluid is arguably without long-range order. As a consequence, the neutron diffraction data of Howe *et al.*[327] cannot be said to yield strong evidence for the existence of true long-range order. Many diffraction and EXAFS studies have yielded valuable information, however, about the short-range ordering found in aqueous solutions.

Sandstrom *et al.*[329] measured the Ni K-edge EXAFS in a 0.1 *M* aqueous solution of $NiCl_2$ and compared this with the EXAFS spectrum from a solid salt, $Ni(NO_3)_2 \cdot 6H_2O$. (The 0.1 *M* solution corresponds to an atomic con-

[327] R. H. Howe, W. S. Howells, and J. E. Enderby, *J. Phys. C* **7**, L111 (1974).
[328] N. Quirke and A. K. Soper, *J. Phys. C* **10**, 1803 (1977).
[329] D. R. Sandstrom, H. W. Dodgen, and F. W. Lytle, *J. Chem. Phys.* **67**, 473 (1977).

centration $\approx$0.002 for the Ni atoms if the H atoms are neglected.) In agreement with diffraction studies,[330] they found that the nearest-neighbor peak about Ni in solution corresponds to 6 H_2O molecules, with the oxygen atoms at $\approx$2.06 Å, essentially identical to the solid salt. The backscattering strength of the neighboring hydrogen atoms is negligible for $k \geq 2$ Å^{-1}, while that of the oxygen atoms falls off very rapidly for $k \geq 5$ Å^{-1}. Principally for this reason, the EXAFS data from the solution extend out only to 7 or 8 Å^{-1}. Finally, they identified a second feature in the spectrum of the solution as a second hydration sphere. While this feature is not much above the noise level, its identification is consistent with the tentative identification of a second hydration sphere in a diffraction study [331] of a 1 *M* aqueous solution of $[Cr(H_2O)_6]Cl_3$.

In a study of more concentrated aqueous solutions of $NiCl_2$ (i.e., 2.78 and 3.74 *M*), Sandstrom[235] found that the first hydration sphere about Ni^{2+} is essentially identical with the solid salt in these solutions as well, again in agreement with diffraction studies.[330] He also found strong evidence for a coordination sphere about Ni containing $\approx$3 Cl^- ions at $\approx$3.1 Å. Furthermore, he found no evidence for Cl ions in the first hydration sphere, at variance with the hypothesis of Fontana and co-workers.[332] It should be noted that Sandstrom has substantiated the results of his study by including in his paper a detailed description of the data analysis procedure.

Aqueous solutions of $CuBr_2$ have been the subject of numerous EXAFS studies. Eisenberger and Kincaid[333] studied dilute solutions ($\approx$0.1 *M*) and determined that the nearest-neighbor peak about Cu consists of O at 1.97 ± 0.08 Å, and that about Br consists of O at 3.14 ± 0.1 Å.[334] Fontaine *et al.*[335] measured the EXAFS spectra on the Cu and Br K edges of $CuBr_2$ in aqueous solutions between 0.5 and 4.5 *M* (and in the solid salt to calibrate their analysis), as shown in Fig. 40. Apart from a more rapid decrease in signal with increasing *k*, the spectrum from the more concentrated solutions is quite similar to that from the solid salt. Analysis of that spectrum led Fontaine *et al.*[335] to conclude that there are too many Cu neighbors of each Br ion for the solution to consist of simple hydrated Cu ions and $CuBr_4$

[330] For example, see A. K. Soper, G. W. Neilson, J. E. Enderby, and R. A. Howe, *J. Phys. C* **10,** 1793 (1977).

[331] R. Caminiti, G. Licheri, G. Piccaluga, and G. Pinna, *J. Chem. Phys.* **65,** 3134 (1976).

[332] For example, see M. P. Fontana, G. Maisano, P. Migliardo, and F. Wanderlingh, *J. Chem. Phys.* **69,** 676 (1978).

[333] P. Eisenberger and B. M. Kincaid, *Chem. Phys. Lett.* **36,** 134 (1975).

[334] This asymmetry in distance reflects the intrinsic polarizability of the water molecule: Cations attract the O while anions attract the H.

[335] A. Fontaine, P. Lagarde, D. Raoux, M. P. Fontana, G. Maisano, P. Migliardo, and F. Wanderlingh, *Phys. Rev. Lett.* **41,** 504 (1978).

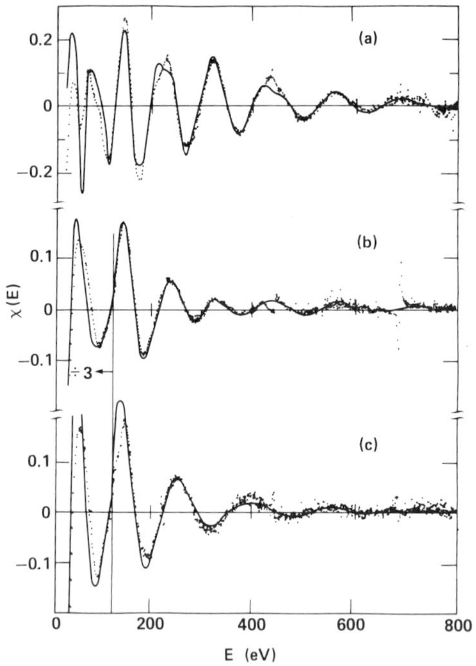

FIG. 40. EXAFS spectra $\chi(E)$ as a function of final-state electron energy E for (a) powdered crystalline $CuBr_2$ and (b) 4.5 and (c) 0.5 M $CuBr_2$ aqueous solutions. The solid line in each case represents a fitted calculated spectrum. The vertical scale of parts (b) and (c) is compressed by a factor of 3 below 120 eV. Note the similarity between the spectra from the solid salt and the more concentrated solution, apart from the more rapid decrease in the latter signal with increasing E. [Taken from Fig. 1 in A. Fontaine, P. Lagarde, D. Raoux, M. P. Fontana, G. Maisano, P. Migliardo, and F. Wanderlingh, *Phys. Rev. Lett.* **41,** 504 (1978).]

anion complexes (as had been suggested by Fontana *et al.*[332]). They hypothesized instead that half of the Cu and Br ions in the 4.5 M solution are in $CuBr_2$ chains composed of Br rectangles, with a Cu ion in the center of each rectangle. The remaining ions are isolated from one another by hydration spheres of H_2O, as in other solutions. Finally, they concluded that no more than half of the Cu and Br ions can be simple hydrated ions, even as the concentration is lowered to 0.5 M, and that the remainder must be in some unspecified complexes.

This study was followed by EXAFS measurements[336] on several concentrations of aqueous solutions of various metal bromides and chlorides: Cu, Zn, Ni, and Sr bromide (both edges), and Ni and Cu chloride (metal

[336] P. Lagarde, A. Fontaine, D. Raoux, A. Sadoc, and P. Migliardo, *J. Chem. Phys.* **72,** 3061 (1980).

edge only). The work on $ZnBr_2$ provides the strongest evidence for the existence of extensive complexes in the solutions. Lagarde *et al.*[336] found that the EXAFS spectra of an 8.08 *M* aqueous solution of $ZnBr_2$ strongly resembled that of the solid salt. Analyzing it, they determined that each Br ion has, on the average, 1.75 ± 0.25 near-neighbor Zn ions. They inferred from this that 85% of the Zn and Br ions are in complexes formed of linked Br tetrahedra with a centered Zn, such that each Br has two Zn near neighbors and each Zn has four Br near neighbors. Since the Zn–Br distance is constant while the coordination number decreases with decreasing concentration, they inferred that the fraction of ions in such complexes (rather than as simple hydrated ions) decreases with concentration, but that the character of neither the complexes nor the hydration changes with concentration. Furthermore, they observed no change in the spectrum with temperature, even though the hydration is believed to change character at $\approx 60°C$. The ideas about the structure of concentrated aqueous solutions that have been suggested by these studies are somewhat controversial and certainly stimulating.

Finally, there is a study by Sano *et al.*[337] of Cu hydration in a 0.32 *M* aqueous solution of $Cu(ClO_4)_2$. They identified a Cu–O distance of 1.94 Å, in rough agreement with earlier studies, and possibly also a second hydration sphere at 2.46 Å.

14. Surfaces and Adsorbates on Surfaces

Those diffraction techniques which have proven so useful in defining the atomic scale structure of bulk solids are not especially helpful when applied to the study of surface structures. Neutrons and X rays are intrinsically bulk probes. Due to the long mean-free-path lengths of neutrons and X rays in condensed matter, only a negligible portion of the diffracted signal is derived from the surface region in a conventional experiment.[338] This problem does not occur in the diffraction of low-energy electrons, which have a mean-free-path length typically ≈ 10 Å. Over the last several years, low-energy electron diffraction (LEED) has become the principal direct probe of surface structures. Unfortunately, the strong scattering of low-energy electrons by atoms leads to the predominance of multiple-scattering events in the diffracted spectrum, greatly complicating the analysis of LEED data. Fur-

[337] M. Sano, K. Taniguchi, and H. Yamatera, *Chem. Lett.* p. 1285 (1980).

[338] In recognition of the virtues of a diffraction study, there have been recent attempts to get around the long mean-free-path length of X rays by performing the diffraction experiment at a glancing angle, so as to take advantage of total external reflection [W. C. Marra, P. Eisenberger, and A. Y. Cho, *J. Appl. Phys.* **50**, 6927 (1979); S.-L. Weng, A. Y. Cho, W. C. Marra, and P. Eisenberger, *Solid State Commun.* **34**, 843 (1980)].

thermore, only ordered surfaces generate a LEED signal which is sufficiently strong to enable the extraction of structural information, given the current state of the art. For these and other reasons, there has been substantial interest in adapting the EXAFS technique to the study of the structure of surfaces.

Our understanding of adsorption on surfaces and of surface reconstruction would be greatly augmented by the sort of structural information which EXAFS probes directly: the position and identity of the nearest neighbors of the excited atom. The directionality of the EXAFS probe of near neighbors can be particularly useful in surface studies, since a macroscopic asymmetry has been imposed on the system (i.e., the surface itself). If θ is the angle between the polarization vector of the X rays $\hat{\epsilon}$ and the vector from the excited atom to a neighboring atom, then the contribution of that atom to the EXAFS spectrum is proportional to $3 \cos^2 \theta$ for the $s \rightarrow p$ transitions characterizing excitation of K and L_I core electrons and $(3 \cos^2 \theta + 1)/2$ for the $p \rightarrow d$ transitions that dominate the excitation spectrum of L_{II} and L_{III} core electrons (see Section 3 for a discussion). These factors average to unity for a randomly oriented bulk polycrystalline sample, but will not in surface studies, since the angle between $\hat{\epsilon}$ and the surface normal has a single value determined by the macroscopic orientation of the sample. It is useful, accordingly, to measure the EXAFS spectrum with two polarizations: $\hat{\epsilon}$ parallel to and nearly perpendicular to the surface normal. In the analysis of such data, it is common practice to define an *apparent* or *effective* number of neighbors N_a for a given site, which is simply the sum of the angular factor over the actual θ's determined by $\hat{\epsilon}$ and the specific geometry of the proposed site. N_a is then that number of neighbors which would yield an EXAFS signal of equal strength if found in a randomly oriented bulk polycrystalline sample, and might be either smaller or larger than the actual number of neighbors of the surface site. The various surface EXAFS detection schemes are discussed in Section 9.

a. *Iodine on Ag and Cu*

In an extensive series of experiments over several years, Citrin, Eisenberger, and Hewitt have studied the environment of adsorbed iodine in ordered monolayer and submonolayer coverages on single-crystal Ag and Cu. Their first experiments concerned the ⅓ monolayer $(\sqrt{3} \times \sqrt{3})R30°$ phase of adsorbed iodine on an Ag $\{111\}$ surface.[182] They measured the EXAFS on the iodine L_{III} absorption cross section by detecting the $L_{III}M_{IV,V}M_{IV,V}$ Auger electron yield. As discussed in Section 9,b, the Auger spectrum is exactly proportional to the absorption spectrum. At 110°K, they were able to detect EXAFS oscillations nearly 300 eV above the L_{III} edge. Through comparison with the room-temperature EXAFS spectrum of bulk γ-AgI, they concluded that the I–Ag distance on the surface $\approx 2.87 \pm 0.03$ Å, 0.04–0.10 Å larger than in the bulk compound. The geometry of the experiment was such that the polarization vector of the X rays $\hat{\epsilon}$ was $\approx 20°$ from the surface normal. They estimated the apparent number of nearest-neighbor Ag atoms expected for an iodine on each of three possible Ag $\{111\}$ surface sites: directly above one Ag atom, in a bridge site with two nearest-neighbor Ag, and in a threefold-coordinated hollow. Neglecting in their analysis the effects of possible differences in the shape of the peak in $p(r)$ and in the electron mean-free-path length λ between the surface phase and the bulk sample, they concluded that the measured amplitude is consistent with only the threefold-coordinated hollow adsorption site. These conclusions are completely consistent with the results of an earlier LEED study,[339] but yielded a presumably more accurate bond length (i.e., 2.87 ± 0.03 Å instead of 2.80 ± 0.14 Å).

The next system studied by Citrin *et al.*[340] was the analogous ⅓ monolayer $(\sqrt{3} \times \sqrt{3})R30°$ phase of iodine adsorbed on a Cu $\{111\}$ surface. They noted that the absence of any contribution from a nearest-neighbor I–I peak in the EXAFS spectra from the Ag and Cu $\{111\}$ surfaces suggests that I_2 dissociates on adsorption in both cases. Through comparison with bulk CuI, they deduced that the I–Cu distance is 2.69 ± 0.02 Å, again 0.05–0.09 Å larger than in the bulk compound. In attempting to identify the adsorption site from the measured amplitude of the EXAFS, they found a relatively serious inconsistency. The *apparent* number of nearest neighbors N_a was much too large to be compatible with any of the sites: 3.4–4.2 Cu atoms in contrast to the maximal $N_a = 2.87$ expected for the threefold-coordinated hollow site. While they tentatively attributed this to the electron mean-free-path length λ being longer on the surface than in the bulk, it could also be caused by a different peak shape on the surface than in the bulk, as discussed below.

Citrin *et al.* also examined the c(2×2) phase of iodine adsorbed on a Cu $\{110\}$ surface.[340] This experiment was performed with $\hat{\epsilon}$ parallel to the surface. If the substrate is annealed at a "high" (but unspecified) temperature prior to adsorption, only Cu nearest neighbors of I contribute to the EXAFS spectrum, as shown in Fig. 41a. When the substrate is "incompletely" annealed at a lower temperature ($\approx 400°$C) following Ar ion sputtering but prior to adsorption, the EXAFS is dominated by nearest-neighbor I–I pairs, as shown in Fig. 41b. This suggests that iodine adsorbs dissociatively on the annealed Cu $\{110\}$ surface to form a ½ monolayer c(2×2) phase, but adsorbs as a molecule on the incompletely annealed Cu $\{110\}$ surface to

[339] F. Forstmann, W. Berndt, and P. Büttner, *Phys. Rev. Lett.* **30,** 17 (1973).

[340] P. H. Citrin, P. Eisenberger, and R. C. Hewitt, *Surf. Sci.* **89,** 28 (1979).

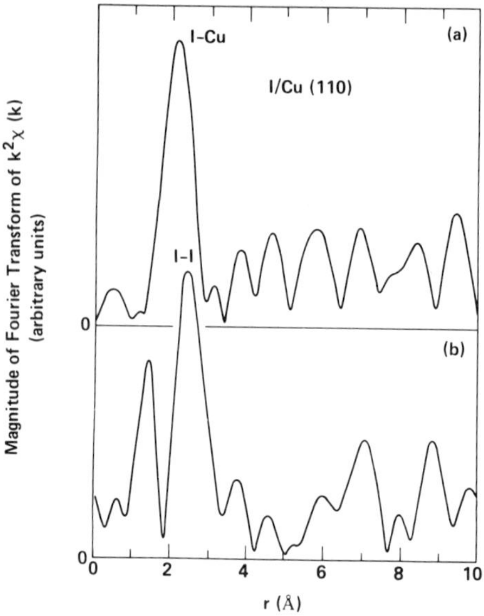

FIG. 41. Magnitude of the Fourier transform of $k^2\chi(k)$ from the L_{III}-shell X-ray absorption of iodine adsorbed in a c(2 × 2) phase on Cu{110} surfaces prepared in two ways: (a) annealed at a "high" (but unspecified) temperature and (b) "incompletely" annealed at a lower temperature (≈400°C). In both cases, the surface was oriented parallel to the direction of polarization of the photon electric field vector. Analysis showed that the feature at ≈2.1 Å in (a) arises from Cu nearest neighbors of the excited iodine atoms, while the features at ≈1.5 and ≈2.5 Å in (b) are due to I neighbors at a single nearest-neighbor distance. This is evidence that iodine adsorbs dissociatively on the well-annealed Cu {110} surface, but as I_2 molecules on the "incompletely" annealed surface. This study relied on the distinctive nature of the EXAFS contributions from different neighboring atom species. [Taken from Fig. 3 in P. H. Citrin, P. Eisenberger, and R. C. Hewitt, *Surf. Sci.* **89**, 28 (1979).]

form a 1 monolayer c(2 × 2) phase. Furthermore, when the EXAFS is measured for the incompletely annealed sample with $\hat{\epsilon}$ nearly perpendicular to the surface, the spectrum is dominated by I–Cu nearest-neighbor pairs, rather than I–I. This implies that the I_2 molecule lies parallel to the surface.

Finally, Citrin *et al.*[341] examined the iodine environment in the ⅓ monolayer $(\sqrt{3} \times \sqrt{3})R30°$ phase of iodine adsorbed on a Cu {111} surface (as before) and the ¼ monolayer p(2 × 2) phase of iodine adsorbed on a Cu {100} surface. They revised the I–Cu bond length determined for the {111} surface to 2.66 ± 0.02 Å and determined it to be 2.69 ± 0.02 Å for the {100} surface, larger than in the bulk by ≈0.04 and 0.07 Å, respectively. In determining the apparent number of neighbors N_a through comparison

[341] P. H. Citrin, P. Eisenberger, and R. C. Hewitt, *Phys. Rev. Lett.* **45**, 1948 (1980).

with the bulk compound, they again neglected any possible differences in the shape of the peak in $p(r)$ and in the electron mean-free-path length λ. They placed principal emphasis in their conclusions, however, on the ratio of N_a determined with $\hat{\epsilon}$ perpendicular to the surface to N_a determined with $\hat{\epsilon}$ parallel to the surface. This, they reasoned, would minimize the effects of unknown variations in peak shape and λ. On the basis of this analysis, they concluded that the threefold-coordinated hollow adsorption site is most consistent with the data on the Cu $\{111\}$ surface, and the fourfold-coordinated hollow site with that on the Cu $\{100\}$. It is also noteworthy that they were led to similar conclusions from analysis of all three variants of electron yield: Auger, partial, and total (see Section 9,b for a discussion). Both the slightly longer I–Cu distance on the $\{100\}$ surface and the preferred hollow site are consistent with LEED studies of O, S, Se, and Te on Ni, Cu, and Ag $\{111\}$ and $\{100\}$ surfaces.[342]

In connection with these studies, it should be noted that bulk AgI and CuI have been studied extensively using EXAFS spectroscopy, as discussed in detail in Sections 6,d and 11,b. The nearest-neighbor peaks in the Ag–I and Cu–I pair correlation functions have been shown to be very asymmetric, a property which is related to their unusual superionic conductivity. The asymmetry is not negligible at room temperature and, accordingly, will affect the use of these materials at room temperature in the above study as standards to determine the bond length and coordination number. It is obvious from the discussion in Section 6,d that an overestimate of the nearest-neighbor distance and coordination number in the unknown will result from comparing the EXAFS from one of these materials at room temperature with an unknown spectrum which is properly described by a symmetric peak shape. It is noteworthy that the analysis of Citrin *et al.* did lead to values for both of these quantities which *are* in fact larger than in the bulk compounds. The degree of overestimation which actually resulted from the neglect of peak shape in the above analysis depends in detail on the shape of the first peak in the pair correlation function for the adsorbed iodine on the surface relative to that for the bulk sample, and can be determined only through further analysis of the data.

b. Oxygen on Al

Another system which has been the object of extensive EXAFS studies is that of O adsorbed on an Al $\{111\}$ surface. Johansson and Stöhr[343] used partial electron yield to measure the EXAFS on the O K-shell absorption cross section in the 1 monolayer (1×1) phase of O on an Al $\{111\}$ surface.

[342] For a review, see F. Jona, *J. Phys. C* **11,** 4271 (1978).

[343] L. I. Johansson and J. Stöhr, *Phys. Rev. Lett.* **43,** 1882 (1979).

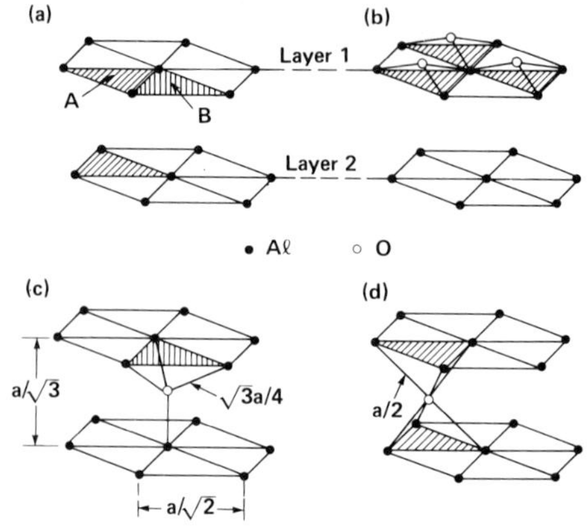

FIG. 42. (a) Model of the first two Al layers for a clean Al(111) surface. Two threefold hollow sites *A* and *B* in the first layer can be distinguished with respect to the second layer. Site *A* has another threefold hollow site underneath, while site *B* is on top of a second-layer Al atom. (b) Proposed geometry of O atoms chemisorbed in a (1 × 1) configuration on Al(111). Adsorption occurs on site *A* in positions which are a continuation of the Al lattice (fcc stacking). (c) Ideal tetrahedral O coordination underneath site *B*. The O–Al bond length is 1.75 Å. (d) Ideal octahedral O coordination underneath site *A*. The O–Al bond length is 2.02 Å. [Taken from Fig. 8 in J. Stöhr, L. I. Johansson, S. Brennan, M. Hecht, and J. N. Miller, *Phys. Rev. B: Condens. Matter* [3] **22,** 4052 (1980).]

Two different films were examined, resulting from 100 and from 150 Langmuir (L) exposures of clean Al {111} surfaces to oxygen (1 L = 1 μTorr sec). They determined the nearest-neighbor O–Al distance to be 1.79 ± 0.05 Å for both exposures. In a subsequent examination of the implications of these data, Stöhr *et al.*[344] cited photoemission studies which concluded that comparable numbers of oxygen atoms must occupy each of two sites after either 100 or 150 L exposures: a "chemisorption site" outside the surface Al layer and an "oxide site" beneath that Al layer. The geometry of an Al {111} surface is shown in Fig. 42, together with possible "chemisorption" and "oxide" sites for oxygen atoms. Stöhr and co-workers found no evidence in their EXAFS spectra for a second O–Al separation (in addition to 1.79 Å), however, at either exposure level or in either of two polarization geometries. They concluded accordingly that the O–Al distance must be the same in these two sites. Furthermore, they noted that the value of 1.79 Å suggests that the tetrahedral interstice in the face-centered cubic Al lattice (Fig. 42c, corresponding to $r_{\text{O-Al}}$ = 1.75 Å) is a more likely candidate for

[344] J. Stöhr, L. I. Johansson, S. Brennan, M. Hecht, and J. N. Miller, *Phys. Rev. B: Condens. Matter* [3] **22,** 4052 (1980).

the "oxide site" than is the octahedral interstice (Fig. 42d, corresponding to $r_{O-Al} = 2.02$ Å).

The O–Al separation of 1.79 Å deduced by Stöhr and co-workers is significantly at variance with the conclusions of LEED studies, which have suggested either 2.12 ± 0.05 Å[345] or 2.20 ± 0.04 Å[346] for the 1 monolayer (1 × 1) phase. Jona and Marcus[347] have reexamined the LEED analysis, however, and concluded that $r_{O-Al} = 1.79$ Å fits the somewhat limited set of LEED data as well as does 2.16 Å. They called for a more extensive set of LEED data to resolve this question. Jona and Marcus also emphasized the importance of characterizing as completely as possible the system being studied. In the absence of proper sample characterization, one could be in the position of laboring to force agreement between the differing results of two separate experiments, which in fact reflect real differences in the samples being measured.

Johansson and Stöhr also identified a second and higher frequency in the EXAFS as coming from a peak in the O–O pair correlation function at ≈2.90 ± 0.05 Å,[343] observing that this distance is consistent with the separation of the threefold-coordinated hollow adsorption sites for the oxygen (≈2.86 Å). This peak occurs principally for $\hat{\epsilon}$ at 45° rather than 11° from the surface normal, suggesting that the vector to the backscattering atom is parallel to the surface. It should be noted, however, that the signal corresponding to $r = 2.90$ Å (i.e., the higher frequency) dominates the EXAFS spectrum at large k, while that corresponding to $r = 1.79$ Å (i.e., the lower frequency) dominates at small k. Since the backscattering amplitude of O falls off much faster with increasing k than does that of Al, why would the O signal dominate at larger k? The identification of the second-neighbor peak by Johansson and Stöhr as O can be correct only if some effect enhances the O signal at large k *but not* at small k. The explanation offered by Stöhr *et al.* for this effect, an anomalously large mean-free-path length for electrons traveling parallel to the surface,[344] is not particularly convincing in the absence of supporting evidence. An alternative explanation is that the peak in the O–O distribution is much narrower than that in the O–Al distribution. In the absence of some strong (possibly covalent) bond, this would be most unexpected for a second neighbor. Thus this identification of a second-neighbor O–O peak by Johansson and Stöhr needs further corroboration.

Stöhr and co-workers[344] have also examined the oxygen on Al {111} system which results from an oxygen exposure of 1000 L and heat treatment

[345] C. W. B. Martin, S. A. Flodström, J. Rundgren, and P. Westrin, *Surf. Sci.* **89,** 102 (1979).
[346] H. L. Yu, M. C. Muñoz, and F. Soria, *Surf. Sci.* **94,** L184 (1980).
[347] F. Jona and P. M. Marcus, *J. Phys. C* **13,** L477 (1980).

for 5 min at 200°C. For this system, they deduced an O–Al separation of 1.88 ± 0.03 Å, just as measured for γ-Al_2O_3.[348] By combining this result with their conclusions about the samples produced by exposure to 100 and 150 L of oxygen, Stöhr *et al.* generated the following model for the oxidation process.[344] At low oxygen exposures, the "oxide site" is the tetrahedral interstice in the face-centered cubic Al lattice (shown in Fig. 42c), which is quite similar to the oxygen site in γ-Al_2O_3. The O–Al separation would be 1.75 Å in an undistorted Al lattice, but is 1.88 Å in the bulk γ-phase oxide. The observed separation of 1.79 Å is a result of these opposing tendencies. When the heavily exposed sample is subjected to heat treatment, the surface is able to reconstruct so as to allow the energetically preferred separation of 1.88 Å.

In another study of oxygen on Al {111}, Bachrach *et al.*[349] found that the surface phases are sensitive to the oxygen pressure during exposure, and not simply to the total exposure. With an oxygen pressure of 2×10^{-7} Torr, they observed an O–Al peak at 2.22 ± 0.1 Å consistent with the result of the LEED analysis, and a second peak at ≈3.2 Å which might be O–O.[350] With 10^{-6} Torr of oxygen, they observed an O–Al peak at 1.92 ± 0.05 Å. This distance is similar to that found by Stöhr *et al.*[344] for the annealed oxide phase (i.e., 1.88 Å). This study serves to underline the caution of Jona and Marcus,[347] that care should be taken to characterize as completely as possible the systems being studied.

Den Boer *et al.*[204] have used extended appearance potential fine structure (EAPFS) spectroscopy to examine the adsorption of oxygen on the {100} surface of Al. This technique is discussed in Section 9,c. They measured the oxygen K-shell EAPFS spectrum from an Al {100} surface after exposure to ≈120 L of oxygen, which extinguishes the LEED spectrum and is believed to correspond to ≈1.5 monolayers of oxygen. They found an O–Al separation of 1.98 ± 0.05 Å. This distance is similar to the longer of the two distances occurring in bulk Al_2O_3, which corresponds to an oxygen atom lying between two Al atoms. They concluded that their data support a model in which the oxygen adatoms occupy sites beneath the top layer of Al atoms.

[348] They also pointed out that the original analysis of an earlier measurement of the O–Al distance in a "natural" oxide of Al [J. Stöhr, D. Denley, and P. Perfetti, *Phys. Rev. B: Condens. Matter* [3] **18,** 4132 (1978)] was in error owing to reliance on calculated phase shifts. A subsequent analysis of that data has led to the deduction of an O–Al distance similar to the 1.88 Å found in γ-Al_2O_3.

[349] R. Z. Bachrach, G. V. Hansson, and R. S. Bauer, *Surf. Sci.* **109,** L560 (1981).

[350] Adsorption of oxygen as the molecule O_2 would also be consistent with their photoemission results.[349]

c. Oxygen on Other Metals

The Cu environment in an oxide formed on a 30-Å film of Cu on exposure to air has been studied by del Cueto and Shevchik.[169] They found that very similar EXAFS spectra result from brief room temperature exposure to air, from heating to 300°C in air, and from CuO, concluding that the thin films oxidize "fully" very rapidly. This study was extended by Fischer *et al.*,[351] who used electron yield to examine the Cu K-shell EXAFS spectra from two samples of Cu metal which had been exposed to oxygen for 1 hr at temperatures of 200 and 300°C, respectively. They found that the spectrum from the first sample strongly resembles that from bulk Cu_2O, while that from the second resembles that from bulk CuO. This result is consistent with the conclusions of other studies of this system.

Stöhr *et al.*[178] reported that the O–Ni distance measured on Ni $\{100\}$ after 40 L of exposure to O_2 is $\approx$2.04 Å, approximately 0.04 Å less than in bulk NiO. Earlier analysis of these same data using calculated phase shifts led to an incorrect O–Ni distance of 1.92 Å.[177,186]

d. Oxygen on Semiconductors

The oxygen environment on a Si $\{111\}$ surface has been probed through partial electron yield EXAFS by Stöhr *et al.*[352] They found an O–Si distance of 1.65 ± 0.03 Å on the Si $\{111\}$ surface after 10^6 L of exposure to excited O_2. This compares with a separation of 1.61 Å found in a SiO_2 thermal oxide. They also concluded that the apparent coordination number is low enough that they are able to rule out certain possible adsorption sites, even after taking into account reservations about the transferability of amplitude from bulk to surface atoms. This uncertainty prevented them, however, from identifying a single most-favored site.

The EXAFS on the O K-shell absorption cross section was studied for 6×10^9 L of O_2 on GaAs $\{110\}$ by Stöhr *et al.*[353] through a partial electron yield measurement. They saw no signal from O–O pairs, from which they inferred that O_2 must dissociate on chemisorption. Analysis to obtain a nearest-neighbor distance is made difficult by lack of a comparable bulk compound. Using an O–Ni phase shift extracted from bulk NiO to approximate the needed O–Ga and O–As phase shifts, they deduced a distance

[351] D. A. Fischer, G. G. Cohen, and N. J. Shevchik, *J. Phys. F* **10,** L139 (1980).

[352] J. Stöhr, L. I. Johansson, I. Lindau, and P. Pianetta, *J. Vac. Sci. Technol.* **16,** 1221 (1979); J. Stöhr, L. I. Johansson, I. Lindau, and P. Pianetta, *Phys. Rev. B: Condens. Matter* [3] **20,** 664 (1979).

[353] J. Stöhr, R. S. Bauer, J. C. McMenamin, L. I. Johansson, and S. Brennan, *J. Vac. Sci. Technol.* **16,** 1195 (1979).

1.70 ± 0.05 Å between the adsorbed O and the surface Ga or As atoms. Calculated phase shifts yield 1.51 ± 0.05 Å, which they regarded as a much less trustworthy value.

The work of Stöhr and collaborators has provided strong evidence that the use of calculated phase shifts can lead to substantial errors in determining the nearest-neighbor positions from a surface EXAFS measurement. This is quite consistent with the conclusions of Pettifer[56] and of Stern *et al.*,[54] that the calculations are inadequate below 200 and 400 eV, respectively. It is not unusual for the EXAFS signal from a surface atom to disappear completely into the noise well before 300 eV above the absorption edge in these difficult experiments.

e. Metals

Bianconi and Bachrach[181] have studied the relaxation of clean Al {111} and {100} surfaces by monitoring the *LVV* Auger electrons at 45 eV. They relied upon the short mean-free-path length of the 45-eV electrons to obtain surface sensitivity. Analyzing a rather short range in k space (one oscillation between 2.6 and 3.6 $Å^{-1}$), they concluded that the Al {100} surface is unchanged from the bulk, while the outermost layer on the Al {111} surface moves *in* by 0.19 ± 0.06 Å (corresponding to a 0.15-Å contraction in the nearest-neighbor Al–Al distance). Earlier LEED studies agreed with the unchanged Al {100} surface, but found that the outermost layer on the Al {111} surface moves *out* by ≈0.06 Å.[354] Jona *et al.*[355] responded to this EXAFS study with a detailed reanalysis of the earlier LEED data, but reaffirmed the original interpretation. The disagreement between the conclusions of the EXAFS and LEED studies has not yet been explained.

f. Bromine on Grafoil

The Br environment in a 0.2 monolayer coverage of Br_2 on Grafoil was studied by Stern.[356] Grafoil is a stacking of many thin sheets of graphite. It has a very large effective surface area per unit volume, but is not very absorptive of X rays at the Br K edge. Thus, it was possible to illuminate a relatively large number of adsorbed atoms with the X-ray beam while measuring the EXAFS using standard transmission techniques. The EXAFS

[354] D. W. Jepsen, P. M. Marcus, and F. Jona, *Phys. Rev. B: Solid State* [3] **5,** 3933 (1972); D. W. Jepsen, P. M. Marcus, and F. Jona, *ibid.* **6,** 3684 (1972); P. M. Marcus, D. W. Jepsen, and F. Jona, *Surf. Sci.* **31,** 180 (1972); M. R. Martin and G. A. Somorjai, *Phys. Rev. B: Solid State* [3] **7,** 3607 (1973).

[355] F. Jona, D. Sondericker, and P. M. Marcus, *J. Phys. C* **13,** L155 (1980).

[356] E. A. Stern, *J. Vac. Sci. Technol.* **14,** 461 (1977).

was measured with the X-ray polarization vector $\hat{\epsilon}$ both perpendicular and parallel to the surface. The EXAFS spectrum is dominated by C backscattering below $k = 4$ Å^{-1}, and by Br backscattering above $k = 6$ Å^{-1}, greatly facilitating the separation of their respective contributions to the EXAFS. Stern deduced a Br–C separation of 2.5 ± 0.3 Å, and Br–Br separations of 2.25 ± 0.03 Å for $\hat{\epsilon}$ perpendicular to the surface, and 2.22 ± 0.03 Å for $\hat{\epsilon}$ parallel to the surface. The Br–Br separation in gaseous Br_2 is 2.28 Å. Furthermore, he noted that the peaks in the Br–Br correlation functions are wider than in the gas ($[\sigma^2_{\text{Br-Br}}\,(\text{surface}) - \sigma^2_{\text{Br-Br}}\,(\text{gas})]^{1/2} = 0.05$ and 0.07 Å, respectively). From the fact that the Br–Br peaks in the gas and on the surface in both polarizations have equal amplitudes, he deduced that the molecules do not dissociate on adsorption and, further, have no preferred orientation relative to the surface normal. The presence of a Br–C peak implies that at least one Br in each molecule is highly correlated with C atoms. The relative intensities in the two polarizations suggest further that the Br–C bond is 42 ± 3° from the surface normal. In summary, Stern concluded that the adsorbed species at 0.2 monolayer coverage consists of Br_2 molecules attached at one end to C atoms, as specified above, but free to "flop around" at the other.

In a later report on this same system, Stern *et al.*[357] reached the same conclusions, except that the Br–C separation was revised to 2.37 ± 0.1 Å. They noted that this distance together with a 42° angle from the normal for the Br–C bond imply that the C–C distance in the plane is 1.58 ± 0.12 Å, at variance with the 1.42-Å spacing known for graphite. Several tentative explanations for this were offered.

The Br environments in 0.6 and 0.9 monolayer coverages of Br_2 on Grafoil have been studied by Heald and Stern.[285] The Br–Br signal measured perpendicular to the surface is very weak, as shown in Fig. 43, suggesting that the Br_2 molecules lie flat on the surface. Specifically, they do not "flop around," as they were deduced to do in the 0.2 monolayer system. The relative strengths of the contributions from the first- and second-nearest-neighbor Br–C peaks points to the Br occupying the center of a C hexagon on the graphite surface. The Br–C separation is ≈2.9 Å, so that the bond direction must be roughly 20° from the surface normal. The Br–Br distance of ≈2.31 Å is slightly stretched from the isolated molecule's bond length of 2.28 Å to accommodate the 2.46 Å separation of the hexagon centers. It was not possible to determine whether the molecules formed a liquid or a solid phase on the surface, because that would have required detecting distant Br–Br correlations, which is unlikely with EXAFS.

[357] E. A. Stern, D. E. Sayers, J. G. Dash, H. Shechter, and B. Bunker, *Phys. Rev. Lett.* **38,** 767 (1977).

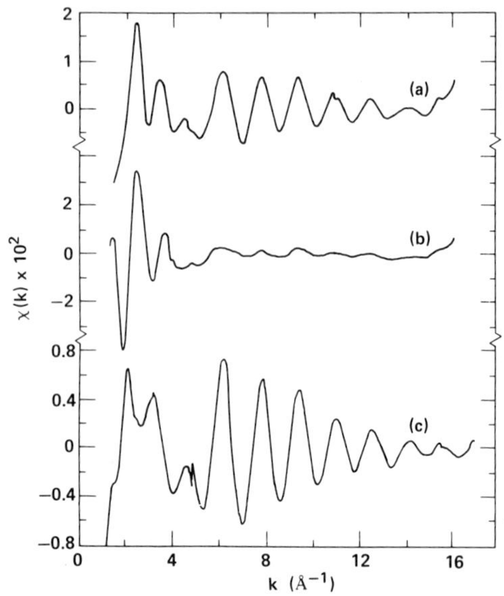

FIG. 43. EXAFS spectra $\chi(k)$ from the Br K-shell absorption for 0.9 monolayer of Br_2 adsorbed on Grafoil at 100°K, with the Grafoil sheets oriented (a) parallel to the photon electric field vector and (b) perpendicular to it, as well as for (c) Br_2 vapor at 293°K. The C backscattering strength declines rapidly with increasing k, so that the oscillations beyond ≈ 6 Å^{-1} are due to Br atoms. The extremely weak signal from Br atoms measured perpendicularly to the Grafoil sheets is strong evidence that the Br_2 molecules lie parallel to the graphite surface. This study relied on the directionality with which the local structure can be probed in EXAFS spectroscopy using polarized X radiation. [Taken from Fig. 3 in S. M. Heald and E. A. Stern, *Phys. Rev. B: Solid State* [3] **17**, 4069 (1978).]

15. Catalytic Systems

Catalysts are of great importance in the chemical industry, since they enable, or at least facilitate, many commercially important chemical reactions. The catalytic agent is highly diluted in the reactants in a typical catalytic reaction, and is often dispersed in a supporting matrix as well. Thus the atom species selectivity of EXAFS spectroscopy makes it an especially attractive structural probe of these systems. Its desirability is enhanced further by the fact that the catalytic agent is often a heavy atom in a (weakly absorbing) light-atom matrix. Moreover, the high intensity of X rays available from a synchrotron makes an EXAFS study feasible on a time scale during which it is possible to stabilize the catalytic reaction in various stages. This allows an examination of the local environment of the active agent in each stage, greatly increasing our knowledge of the various chemical processes which occur. For these reasons, EXAFS spectroscopy has had and will continue to have wide application in the study of catalysts and of catalytic processes. We will confine our discussion to studies of a representative example, however, since the issues raised in catalytic studies tend to be chemical in nature, and are not closely related to the issues central to condensed matter physics.

In homogeneous catalysis, the catalyst and reactants are present in a single phase, typically a liquid solution. In heterogeneous catalysis, on the other hand, they are in separate phases, such as gaseous or liquid reactants in the presence of a solid catalyst. Such solid catalysts are widely used and may be either metals or nonmetals (typically oxides). Since the catalytic action requires intimate contact between the reactants and the active agent, the ideal solid catalyst would have every potentially active atom in a surface site. This goal is actually achieved in large part either by placing the catalyst in a very porous form or by distributing it in very small clusters throughout the internal "surface" of a porous matrix such as alumina or silica. One measure of the success of this process is the fraction of all catalyst atoms which do occupy surface sites, called the dispersion. It often approaches unity in practical systems. The properties of supported metal catalysts have been studied extensively owing to their particular importance[358]; they have also been the object of many EXAFS studies. We will outline the results of a few of these studies below so as to indicate the sort of information which can be determined about these important systems using EXAFS spectroscopy.

Lytle *et al.*[359] studied the Ru K-edge EXAFS spectra for a Ru–silica catalyst containing 1 wt. % Ru in three different stages of catalytic activity: reduced, exposed to oxygen at 25°C, and exposed to oxygen at 400°C. Through comparison with Ru metal and RuO_2, they concluded that the phase at 25°C exposure corresponds to oxygen chemisorbed on the surface of Ru clusters. In the 400°C sample, the Ru–Ru peak disappears and the spectrum resembles that of RuO_2, corresponding to bulk oxidation.

Sinfelt *et al.*[360] studied the Pt L_{III}-edge EXAFS spectra from two Pt–silica catalyts: 1 wt. % Pt with 0.9 dispersion (90% of the Pt atoms on the surface of the Pt clusters) and 10 wt. % Pt with 0.3 dispersion. They observed that the height of the first peak in the Fourier transform, corresponding to Pt–Pt nearest neighbors, decreases steadily through the series from bulk metallic Pt to the 10 wt. % Pt catalyst to the 1 wt. % sample. Noting that the fraction of Pt atoms occupying surface sites increases steadily (with dispersion) in this series, the authors attributed the reduction in peak height to two con-

[358] For a review, see J. H. Sinfelt, *Rev. Mod. Phys.* **51,** 569 (1979).

[359] F. W. Lytle, G. H. Via, and J. H. Sinfelt, *J. Chem. Phys.* **67,** 3831 (1977).

[360] J. H. Sinfelt, G. H. Via, and F. W. Lytle, *J. Chem. Phys.* **68,** 2009 (1978).

sequences of that increase: The average number of neighbors is smaller and the thermal disorder is larger for surface atoms. In particular, they concluded that the average number of Pt–Pt nearest neighbors decreases from 12 in bulk metallic Pt to 7 ± 2 in the 1 wt. % sample.

The same general trends were obtained by Via *et al.*[361] in a series of measurements of the metal L_{III}-edge EXAFS from several 1 wt. % catalysts: Os on silica and Ir and Pt on both silica and alumina. Comparing the catalytic systems with bulk metals, they found that the metal–metal distance is generally within 0.01 Å of the bulk metal value, with the possible exception of Pt–alumina. As before, however, the coordination number is lower than in the bulk, typically 7–10 atoms instead of 12. Similarly, the thermal broadening of the nearest-neighbor distance is larger [i.e., $(\sigma^2_{\text{cluster}} - \sigma^2_{\text{bulk}})^{1/2} \approx 0.05$–$0.06$ Å]. Both differences were attributed, as before, to the large fraction of Pt that occupies the surface sites in these highly dispersed catalysts (a high dispersion implying a small cluster size).

In a later study, Greegor and Lytle[362] reexamined these data as well as data on 0.63 wt. % Cu on silica catalyst, with the aim of obtaining information about the size and shape of the metal clusters. In every case but Pt–alumina, the presence of a strong EXAFS signal from a second coordination shell eliminated the possibility that the metal atoms could be predominantly in flat disklike clusters. In fact, the data on Cu and Ru seem to indicate nearly spherical clusters. This sort of analysis is quite difficult, of course, since it depends heavily upon knowledge of the positions of atoms beyond the first coordination shell. As discussed in Section 6, this information is limited in many EXAFS spectra. It is encouraging, therefore, that these results are consistent with the conclusions of electron microscopy studies.

It has been observed that a mixture of Ru and Cu on silica produces a catalyst with properties very different from either of these elements individually. Such coupling between these metals is especially interesting, because they are essentially immiscible in one another in the bulk. Chemisorption, catalysis, and electron spectroscopy studies have led to the view that the metal cluster consists of a Cu-clad Ru core. Sinfelt *et al.*[363] have provided strong support for this model in an interesting study of a catalyst composed of 1 wt. % Ru and 0.63 wt. % Cu on silica (i.e., a one-to-one atomic ratio of Ru and Cu). They obtained EXAFS spectra from the K-shell absorption of both Ru and Cu in this catalyst. The nearest-neighbor contributions to both are shown as solid lines in Fig. 44. As can be seen from fits to these spectra (broken lines), it is not possible to explain the Cu

[361] G. H. Via, J. H. Sinfelt, and F. W. Lytle, *J. Chem. Phys.* **71,** 690 (1979).
[362] R. B. Greegor and F. W. Lytle, *J. Catal.* **63,** 476 (1980).
[363] J. H. Sinfelt, G. H. Via, and F. W. Lytle, *J. Chem. Phys.* **72,** 4832 (1980).

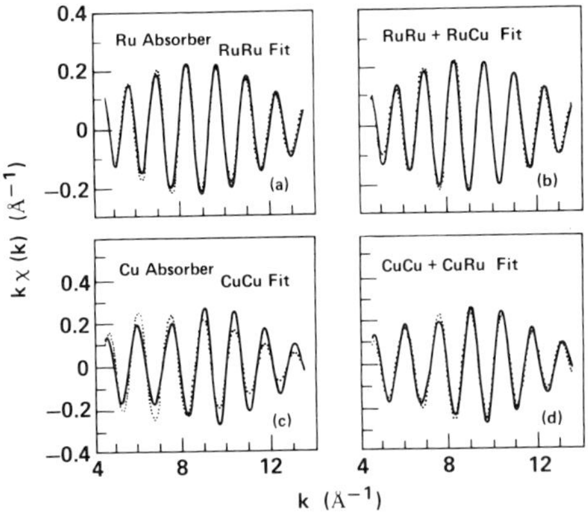

FIG. 44. The nearest-neighbor contributions to the EXAFS $k\chi(k)$ on the K-shell absorption of both Ru (a and b) and Cu (c and d) atoms in a catalyst composed of 1 wt. % Ru and 0.63 wt. % Cu on silica shown as solid lines. The broken lines represent best fits to these spectra, (a and c) assuming total species segregation (i.e., only Ru atoms about Ru and only Cu about Cu) or (b and d) allowing both species about each. The spectrum from the Cu K-shell absorption in particular cannot be explained without a large fraction of Ru neighbors, supporting a model of Cu-clad Ru clusters. [Taken from Fig. 7 in J. H. Sinfelt, G. H. Via, and F. W. Lytle, *J. Chem. Phys.* **72,** 4832 (1980).]

spectrum without Ru nearest neighbors. Specifically, they found that Ru atoms are coordinated with 11 ± 1 atoms which are ≈90% Ru, while Cu atoms are coordinated with 9 ± 2.5 atoms which are only ≈50% Ru. The number and types of these neighbors suggest that Ru atoms form a small core which is coated with a surface layer composed of Cu atoms. This study is a good example of how the special attributes of EXAFS can be used to examine the environment of each atom species individually, and to distinguish the backscattering atom species.

Finally, the effects of cluster size on the metal–metal separation in small clusters of Cu and Ni evaporated on amorphous carbon have been investigated by Apai *et al.*[364] They observed significant changes in two quantities as the metal coverage, and hence the metal cluster size, is decreased from the bulk limit (i.e., diameter > 40 Å): The K-edge threshold energy increases by as much as 1.3 eV, and the nearest-neighbor distance decreases by as much as 10%, a striking contraction. Both of these effects are attributed to the increasing importance of surface energy in determining the structure as the cluster size decreases.

The studies mentioned here are indicative of the sort of information

[364] G. Apai, J. F. Hamilton, J. Stöhr, and A. Thompson, *Phys. Rev. Lett.* **43,** 165 (1979).

about catalysis which one might hope to gain from an EXAFS study, but they hardly represent an exhaustive list. Other interesting studies of catalysts examine fresh and exhaust-cycled Cu–Cr on alumina,[365] Au and Pt on magnesia, alumina, and silica,[366] Pt on Y zeolite,[367] Pt on alumina and silica,[368] Os cluster carbonyl on silica,[369] Co(II) ions in hydrated zeolite-A and -Y,[370] Ni in diamond,[371] and Wilkinson's catalyst, a homogeneous catalyst.[372] For additional discussion of EXAFS studies of catalysts, the reader is referred to a review by Lytle, Via, and Sinfelt.[373]

16. Biological Materials and Chemical Complexes

Biological materials and chemical complexes provide an area where the local and spectroscopic nature of EXAFS can be used to great advantage. Many of these systems consist of large, complex molecules which are difficult to crystallize but which often contain a heavy metal atom at the site of interest. The heavy atom resides in a low-Z matrix, consisting mainly of C, N, O, and possibly some S or P, and is readily accessed in the hard X-ray region for EXAFS studies. A biological example is hemoglobin, a protein which contains one Fe atom for every $\approx$1000 C, N and O atoms. It has four Fe atoms bonded in a planar porphyrin structure. This is the active region, so that EXAFS on the Fe K edge can be used to identify the local structural change when hemoglobin goes from a low to a high oxygen-affinity state.[374] Other biological systems studied using EXAFS include the protein rubredoxin,[171,375–378] the molybdenum–iron protein nitrogenase,[379]

[365] F. W. Lytle, D. E. Sayers, and E. B. Moore, Jr., *Appl. Phys. Lett.* **24,** 45 (1974).

[366] I. W. Bassi, F. W. Lytle, and G. Parravano, *J. Catal.* **42,** 139 (1976).

[367] B. Moraweck, G. Clugnet, and A. J. Renouprez, *Surf. Sci.* **81,** L631 (1979).

[368] R. W. Joyner, *J. Chem. Soc., Faraday Trans. 1* **76,** 357 (1980).

[369] B. Besson, B. Moraweck, A. K. Smith, and J. M. Basset, *J. Chem. Soc., Chem. Commun.* p. 569 (1980).

[370] T. I. Morrison, A. H. Reis, Jr., E. Gebert, L. E. Iton, G. D. Stucky, and S. L. Suib, *J. Chem. Phys.* **72,** 6276 (1980).

[371] J. Wong and F. W. Lytle, *J. Appl. Phys.* **51,** 280 (1979).

[372] J. Reed, P. Eisenberger, B. K. Teo, and B. M. Kincaid, *J. Am. Chem. Soc.* **99,** 5217 (1977); **100,** 2375 (1978).

[373] F. W. Lytle, G. H. Via, and J. H. Sinfelt, *in* "Synchrotron Radiation Research" (H. Winick and S. Doniach, eds.), p. 401. Plenum, New York, 1980.

[374] P. Eisenberger, R. G. Shulman, B. M. Kincaid, G. S. Brown, and S. Ogawa, *Nature (London)* **274,** 30 (1978).

[375] R. G. Shulman, P. Eisenberger, W. E. Blumberg, and N. A. Stombaugh, *Proc. Natl. Acad. Sci. U.S.A.* **72,** 4003 (1975).

[376] D. E. Sayers, E. A. Stern, and J. R. Herriott, *J. Chem. Phys.* **64,** 427 (1976).

[377] B. Bunker and E. A. Stern, *Biophys. J.* **19,** 253 (1977).

[378] B. K. Teo, R. G. Shulman, G. S. Brown, and A. E. Meixner, *J. Am. Chem. Soc.* **101,** 5624 (1979).

[379] S. P. Cramer, W. O. Gillum, K. O. Hodgson, L. E. Mortenson, E. I. Stiefel, J. R. Chisnell, W. J. Brill, and V. K. Shah, *J. Am. Chem. Soc.* **100,** 3814 (1978); S. P. Cramer, K. O. Hodgson, W. O. Gillum, and L. E. Mortenson, *ibid.* p. 3398.

the proteins cytochrone P-450 and chloroperoxidase,[380] calcium-binding proteins,[381] the iron-storing protein ferritin,[382] the molybdenum-containing enzymes xanthine oxidase[383] and sulfite oxidase,[384] the membrane-bound enzyme cytochrome oxidase,[385] and the oxygen-transport protein hemocyanin.[386] An example of a chemical complex is the metal–metal bonded cobalt dimer $[(C_5H_5)CoP(C_6H_5)_2]_2^+$. This material cannot be crystallized, because it is chemically unstable. As a result, EXAFS on the Co K edge can provide structural information not readily available from diffraction techniques. EXAFS has been used to study this metal cluster and the effects of oxidation on the Co–Co distance.[387] Other chemical systems studied using EXAFS range from MnO_4^- ions in aqueous solution[388] to platinum–uridine blues.[389] They include copper sulfate pentahydrate,[390] a variety of transition metal ions in aqueous solution,[391] the magnetic material copper oxalate,[392] bond-isomeric hexakis osmates,[393] hydrolytic polymers of iron,[394] and polymeric conductors.[395] Two other chemical systems, aqueous solutions and catalysts, are discussed in Sections 13 and 15, respectively. In all these cases, EXAFS can spectroscopically access the active region of the molecule and yield information pertinent to the functioning of these systems. In addition, these materials need not be crystalline for the EXAFS studies but, rather, can be in solution in the appropriate active chemical states.

This situation is to be contrasted with that for diffraction techniques, wherein two difficulties arise. Firstly, single crystals are required. These are

[380] S. P. Cramer, J. H. Dawson, K. O. Hodgson, and L. P. Hager, *J. Am. Chem. Soc.* **100,** 7282 (1978).

[381] L. Powers, P. Eisenberger, and J. Stamatoff, *Ann. N.Y. Acad. Sci.* **307,** 113 (1978).

[382] S. M. Heald, E. A. Stern, B. Bunker, E. M. Holt, and S. L. Holt, *J. Am. Chem. Soc.* **101,** 67 (1979); E. C. Theil, D. E. Sayers, and M. A. Brown, *J. Biol. Chem.* **254,** 8132 (1979).

[383] T. D. Tullius, D. M. Kurtz, Jr., S. D. Conradson, and K. O. Hodgson, *J. Am. Chem. Soc.* **101,** 2776 (1979); J. Bordas, R. C. Bray, C. D. Garner, S. Gutteridge, and S. S. Hasnain, *J. Inorg. Biochem.* **11,** 181 (1979).

[384] S. P. Cramer, H. B. Gray, and K. V. Rajagopalan, *J. Am. Chem. Soc.* **101,** 2772 (1979).

[385] L. Powers, W. E. Blumberg, B. Chance, C. Barlow, J. S. Leigh, Jr., J. C. Smith, T. Yonetani, S. Vik, and J. Peisach, *Biochim. Biophys. Acta* **546,** 520 (1979).

[386] J. M. Brown, L. Powers, B. Kincaid, J. A. Larrabee, and T. G. Spiro, *J. Am. Chem. Soc.* **102,** 4210 (1980).

[387] B. K. Teo, P. Eisenberger, and B. M. Kincaid, *J. Am. Chem. Soc.* **100,** 1735 (1978).

[388] P. Rabe, G. Tolkiehn, and A. Werner, *J. Phys. C* **12,** 1173 (1979).

[389] B. K. Teo, K. Kijima, and R. Bau, *J. Am. Chem. Soc.* **100,** 621 (1978).

[390] R. W. Joyner, *Chem. Phys. Lett.* **72,** 162 (1980).

[391] T. K. Sham, J. B. Hastings, and M. L. Perlman, *J. Am. Chem. Soc.* **102,** 5904 (1980).

[392] A Michalowicz, J. J. Girerd, and J. Goulon, *Inorg. Chem.* **18,** 3004 (1979).

[393] P. Rabe, G. Tolkiehn, A. Werner, and R. Haensel, *Z. Naturforsch, A* **34,** 1528 (1979).

[394] T. I. Morrison, A. H. Reis, Jr., G. S. Knapp, F. Y. Fradin, H. Chen, and T. E. Klippert, *J. Am. Chem. Soc.* **100,** 3262 (1978).

[395] H. Morawitz, P. Bagus, T. Clarke, W. Gill, P. Grant, G. B. Street, and D. Sayers, *Synth. Met.* **1,** 267 (1979/1980).

difficult or impossible to obtain for most of the biological materials and chemical complexes. In addition, the crystallization process itself may change the material. Secondly, the molecules and complexes are often large and complicated. For example, the protein nitrogenase has a molecular weight of 220,000. This makes an accurate structural determination difficult and particularly tedious if one desires information about only the active site containing the heavy atom. This point has been discussed in general terms in Section 6,a.

EXAFS is not, however, without its difficulties in studying these materials. Although one is observing the EXAFS on the K or L edge of a heavy atom in a favorably low-Z matrix, it must be borne in mind that the heavy atoms are exceedingly dilute. Examples include hemoglobin containing one Fe in ≈ 1000 light atoms, and nitrogenase containing one Mo for every $\approx 12{,}000$ C, N, and O atoms. As a result, the background absorption is still significant, so that fluorescence detection may be the preferred technique. A second difficulty is the possibility of radiation damage to the delicate biological molecules. These difficulties notwithstanding, its significant advantages have made and will continue to make EXAFS an extremely valuable structural probe of biological materials and chemical systems. Since the emphasis here is on materials problems in condensed matter physics, however, we will not present a complete discussion of the many studies in this area. Instead, we will discuss the results on one of these materials, rubredoxin, as a representative example. For additional studies, one is referred to other EXAFS reviews[396] and to specific reviews of X-ray absorption studies of biological materials.[397] These contain excellent discussions of such studies, as well as lists of references to the original work.

Rubredoxin is a small protein with a molecular weight of ≈ 6000. It contains one Fe atom tetrahedrally coordinated to four sulfur atoms. Prior to the EXAFS studies, X-ray diffraction results[398] indicated that the four Fe–S distances are 2.24, 2.32, 2.34, and 2.05 Å (i.e., three bonds clustered about 2.30 Å, with one short bond at 2.05 Å). It was proposed that this short bond distance stored appreciable strain energy, thereby affecting the redox potential of the molecule. The first EXAFS studies of rubredoxin,[375,376] performed after the diffraction study, used the transmission technique. They showed no anomalously short Fe–S distance. For example, the results of

[396] For example, see Refs. 3–5 and 9 and B. K. Teo, *in* "EXAFS Spectroscopy: Techniques and Applications" (B. K. Teo and D. C. Joy, eds.), Chapter 3. Plenum, New York, 1981.

[397] For example, see R. G. Shulman, P. Eisenberger, and B. M. Kincaid, *Annu. Rev. Biophys. Bioeng.* **7,** 559 (1978); S. I. Chan, V. W. Hu, and R. C. Gamble, *J. Mol. Struct.* **45,** 239 (1978); S. P. Cramer and K. O. Hodgson, *Prog. Inorg. Chem.* **25,** 1 (1979); S. Doniach, P. Eisenberger, and K. O. Hodgson, *in* "Synchrotron Radiation Research" (H. Winick and S. Doniach, eds.), p. 425. Plenum, New York, 1980.

[398] K. D. Watenpaugh, L. C. Sieker, J. R. Herriott, and L. H. Jensen, *Acta Crystallogr., Sect. B* **29,** 943 (1973).

Sayers *et al.*[376] on the same type of rubredoxin molecule as used in the X-ray diffraction studies yielded an average bond length of 2.30 ± 0.04 Å and a root-mean-square deviation about this average of 0.06 ± 0.04 Å. The authors used pyrite (FeS_2) as a known structural standard in deducing these results. The peak width is too small to allow for a short Fe–S distance. A later transmission EXAFS study[377] involved a detailed comparison of oxidized rubredoxin with various standard compounds, including analog compounds with a Fe–S environment approximately that of rubredoxin. The results of this study yielded an average Fe–S distance of 2.267 ± 0.003 Å, with a root-mean-square deviation about this average of 0.00 to 0.04 Å. This also indicates that a short Fe–S distance near 2.05 Å is not possible. Later a fluorescence study[171] was performed on rubredoxin. Due to the dilute concentration of Fe in this molecule, the fluorescence technique gave better S/N than transmission, as discussed in Section 9,a and shown in Fig. 15. This allowed high-quality data to be obtained out to a large value of k (i.e., 13 $Å^{-1}$), giving better resolution in determining a spread in near-neighbor distances. In a least-squares analysis of this data, the spread in distances was determined to be 0.05 ± 0.05 Å. The spread due to thermal vibrations alone was estimated to be about 0.05 Å, so again no anomalously short Fe–S distance was indicated in the EXAFS results. Subsequent refinements of the X-ray diffraction data have reduced the initial spread in Fe–S distances to about 0.1 Å (from 0.25 Å), so that now the results of the two techniques agree within experimental uncertainty. As a result, one can rule out a model for the functioning of rubredoxin which involves a short Fe–S distance.

Acknowledgments

Many people have contributed significantly to the development of the ideas presented in this work. It is a particular pleasure to acknowledge numerous stimulating interactions with J. W. Allen, J. L. Beeby, and P. N. Sen. There have also been helpful discussions with G. S. Brown, P. H. Citrin, E. D. Crozier, P. Eisenberger, S. J. Gurman, J. C. Knights, P. A. Lee, J. C. Mikkelsen, Jr., E. A. Stern, and W. Stutius, among others. We are indebted to C. Herring and M. Rosenblum for a careful reading of the manuscript and insightful comments on various portions of it. One of us (T.M.H.) is grateful for the hospitality of the Department of Physics at the University of Leicester (UK) during the final stages of manuscript preparation. Finally, we thank V. Moffat and I. Lile for their efforts in typing the manuscript.

Some of the materials incorporated in this work were developed at the Stanford Synchrotron Radiation Laboratory with the financial support of the National Science Foundation (under Contract No. DMR 77-27489), in cooperation with the Department of Energy.

Author Index

Numbers in parentheses are reference numbers and indicate that an author's work is referred to although his name is not cited in the text.

C

Cai, J., 50, 51(79), 52(79), 56(79), 88(79)

I

J

K

N

Q

R

S

Tsai, C., 51, 52(85)

Y

Z

Subject Index

F

Q

R

S

Z

Cumulative Author Index, Volumes 1 - 37

D

E

F

G

H

I

J

M

N

O

P

R

S

T

V

W

Y

Z